PALGRAVE STUDIES IN THE HISTORY OF
SCIENCE AND TECHNOLOGY

James Rodger Fleming (Colby College) and Roger D. Launius (National Air and Space Museum), Series Editors

This series presents original, high-quality, and accessible works at the cutting edge of scholarship within the history of science and technology. Books in the series aim to disseminate new knowledge and new perspectives about the history of science and technology, enhance and extend education, foster public understanding, and enrich cultural life. Collectively, these books will break down conventional lines of demarcation by incorporating historical perspectives into issues of current and ongoing concern, offering international and global perspectives on a variety of issues, and bridging the gap between historians and practicing scientists. In this way they advance scholarly conversation within and across traditional disciplines but also to help define new areas of intellectual endeavor.

Published by Palgrave Macmillan:

Continental Defense in the Eisenhower Era: Nuclear Antiaircraft Arms and the Cold War
By Christopher J. Bright

Confronting the Climate: British Airs and the Making of Environmental Medicine
By Vladimir Jankovic´

Globalizing Polar Science: Reconsidering the International Polar and Geophysical Years
Edited by Roger D. Launius, James Rodger Fleming, and David H. DeVorkin

Eugenics and the Nature-Nurture Debate in the Twentieth Century
By Aaron Gillette

John F. Kennedy and the Race to the Moon
By John M. Logsdon

A Vision of Modern Science: John Tyndall and the Role of the Scientist in Victorian Culture
By Ursula DeYoung

Searching for Sasquatch: Crackpots, Eggheads, and Cryptozoology
By Brian Regal

Inventing the American Astronaut
By Matthew H. Hersch

The Nuclear Age in Popular Media: A Transnational History
Edited by Dick van Lente

Exploring the Solar System: The History and Science of Planetary Exploration
Edited by Roger D. Launius

The Sociable Sciences: Darwin and His Contemporaries in Chile
By Patience A. Schell

The First Atomic Age: Scientists, Radiations, and the American Public, 1895–1945
By Matthew Lavine

NASA in the World: Fifty Years of International Collaboration in Space
By John Krige, Angelina Long Callahan, and Ashok Maharaj

Empire and Science in the Making: Dutch Colonial Scholarship in Comparative Global Perspective
Edited by Peter Boomgaard

Anglo-American Connections in Japanese Chemistry: The Lab as Contact Zone
By Yoshiyuki Kikuchi

Eismitte in the Scientific Imagination: Knowledge and Politics at the Center of Greenland
By Janet Martin-Nielsen

Climate, Science, and Colonization: Histories from Australia and New Zealand
Edited by James Beattie, Emily O'Gorman, and Matthew Henry

The Surveillance Imperative: Geosciences during the Cold War and Beyond
Edited by Simone Turchetti and Peder Roberts

Post-Industrial Landscape Scars
By Anna Storm

Voices of the Soviet Space Program: Cosmonauts, Soldiers, and Engineers Who Took the USSR into Space
By Slava Gerovitch

After Apollo? Richard Nixon and the American Space Program
By John M. Logsdon

Frontiers for the American Century: Outer Space, Antarctica, and Cold War Nationalism
By James Spiller

Improvising Planned Development on the Gezira Plain, Sudan, 1900–1980
By Maurits W. Ertsen

Improvising Planned Development on the Gezira Plain, Sudan, 1900–1980

Maurits W. Ertsen

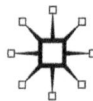

IMPROVISING PLANNED DEVELOPMENT ON THE GEZIRA PLAIN, SUDAN, 1900–1980
Copyright © Maurits W. Ertsen 2016

All rights reserved. No reproduction, copy or transmission of this publication may be made without written permission. No portion of this publication may be reproduced, copied or transmitted save with written permission. In accordance with the provisions of the Copyright, Designs and Patents Act 1988, or under the terms of any licence permitting limited copying issued by the Copyright Licensing Agency, Saffron House, 6-10 Kirby Street, London EC1N 8TS.

Any person who does any unauthorized act in relation to this publication may be liable to criminal prosecution and civil claims for damages.

First published 2016 by
PALGRAVE MACMILLAN

The author has asserted their right to be identified as the author of this work in accordance with the Copyright, Designs and Patents Act 1988.

Palgrave Macmillan in the UK is an imprint of Macmillan Publishers Limited, registered in England, company number 785998, of Houndmills, Basingstoke, Hampshire, RG21 6XS.

Palgrave Macmillan in the US is a division of Nature America, Inc., One New York Plaza, Suite 4500, New York, NY 10004-1562.

Palgrave Macmillan is the global academic imprint of the above companies and has companies and representatives throughout the world.

Hardback ISBN: 978–1–137–56817–5
E-PUB ISBN: 978–1–137–56819–9
E-PDF ISBN: 978–1–137–56818–2
DOI: 10.1057/9781137568182

Distribution in the UK, Europe and the rest of the world is by Palgrave Macmillan®, a division of Macmillan Publishers Limited, registered in England, company number 785998, of Houndmills, Basingstoke, Hampshire RG21 6XS.

Library of Congress Cataloging-in-Publication Data

Ertsen, Maurits W., 1968–
   Improvising planned development on the Gezira Plain, Sudan, 1900–1980 / Maurits W. Ertsen.
      pages cm.—(Palgrave studies in the history of science and technology)
   Includes bibliographical references and index.
   ISBN 978–1–137–56817–5 (hardback : alk. paper)
    1. Irrigation—Sudan—Gezira (Region)—History. 2. Agriculture—Economic aspects—Sudan—Gezira (Region)—History. 3. Agriculture and state—Sudan—Gezira (Region)—History. 4. Regional planning—Sudan—Gezira (Region)—History. 5. Sudan—Economic policy. I. Title.

HD1741.S852G4947 2015
333.76'15096264—dc23
                                                      2015018298

A catalogue record for the book is available from the British Library.

To Tony

Contents

List of Figures   ix

List of Tables   xi

Preface   xiii

Acknowledgments   xv

Introduction
Settling Certain Details Coming to a Deal   1

Chapter 1
Cotton from a Wilderness: The Early Negotiations   15

Chapter 2
A Task of Some Magnitude: Gezira Management Logic   35

Chapter 3
No Man Can Serve Two Masters: Designing Gezira Irrigation   63

Chapter 4
Making the Best of a Rotten Deal: Tenant Realities and Resistance   87

Chapter 5
Another's Week's Toil: British SPS Inspectors and Their Idea(l)s   109

Chapter 6
Move from the Old Grooves: Gezira Continuity and Change after World War II   131

Chapter 7
The Everlasting Rectangles: Gezira and International Development   151

Epilogue
A Typical Battlefield: Understanding Negotiated Development   173

Archival Sources	193
Notes	195
References	267
Index	283

Figures

I.1	Map of the Gezira system (roughly as it stood around 1950)	3
I.2	Cotton harvest in Gezira	4
2.1	Irrigated acreages in Gezira	42
2.2	The four-year rotation in Gezira	46
2.3	Crop areas in Gezira over time	48
2.4	Cotton harvests in Gezira over time	49
2.5	Two different irrigation rhythms	52
2.6	Sowing cotton in the Gezira	54
3.1	Weir on a minor irrigation canal in the Gezira	70
3.2	Syndicate diesel engine	75
4.1	Cotton picking	106
5.1	Sudan Plantations Syndicate cotton inspector standing next to a motorcar	110
5.2	Number of staff employed (vertical) by the SPS (after 1950 SGB) per year (horizontal)	112
5.3	Five-year interval distribution of total years of employment of SPS staff	113
5.4	Sudan Plantations Syndicate cotton inspector's house, probably B-type	115
5.5	Gezira polo	121
5.6	SPS poster with the career of bad tenant Ahmed Abdulla Sakran	124

Tables

2.1	The profit scheme of the 1926 agreement	39
2.2	Overview of the Gezira irrigated areas	41
2.3	Specimen of a tenant's account	57
2.4	Overview of SPS profits in Gezira	61
3.1	The career of R. J. Smith, 1925–1953, Sudan Irrigation Department	64
3.2	List of investments	80

Preface

Although I guess the content of this book is my own responsibility, sharing my thoughts on the Gezira Scheme and its potential meaning for histories and theories of development has only been possible because of the support of many. With the risk of missing a few (apologies for that), I will make an attempt.

First of all, without Sandra, Mafalda, and Merijn, nothing even close to a book would have been possible. Selected quotes from movies and TV series helped me maintain the focus of each chapter and were the result of a proper family effort. The quotes could not be published in the book, but in case you are interested in knowing which ones we considered, just let me know. Furthermore, Mafalda and Merijn helped me in the archives in Durham, each in their own way. Merijn ensured that quite a few of the statistics became available to me in database format; he also drew the map of the Gezira system in chapter 1. Mafalda went through the letters of Ms. Aglen and many of the diaries of Ms. Johnson. I am not sure doing that helped her with her own writing, but if her giggles when reading in the archives sitting next to me are some kind of measure, it must have. Sandra has read all the early drafts and Mafalda all the later ones. In case you see a mistake with an Oxford comma, I can assure you it is not because of any lack of effort on her side!

Jane Hogan and her staff at the Sudan Archive in the Archives and Special Collections from Durham University Library have been of immense help. I will be more than happy to bring copies of the book to Durham, as without their kind and precise assistance, work would have been difficult. The staff at the National Archives in Kew were a great help too, but the smaller scale and the longer time I spent in Durham made the archival work special. Thanks to Veronica Strang, Audrey Bowron, and Simon Litchfield, my stay at the Institute for Advanced Studies (IAS) in Durham was fun and very fruitful as well.

My friends from the field of water and environmental history are too many to list, but discussions with Heather Hoag, Robert Hunt, Thierry Ruf, Vincent Lagendijk, Liesbeth van de Grift, Dolly Jørgensen, Finn Arne Jørgensen, Paul Warde, and Edmund Russell helped me enormously. Once underway in the writing, it were Jim Fleming and Roger Launius who encouraged me to send book proposal and draft texts to Palgrave

Macmillan. Thanks to their encouragement and support, I could prepare the final manuscript with great assistance from Kristin Purdy, my editor. Different people have read parts of the draft texts, including Justin Willis and Glenn Weisz.

Being in the Water Resources Group in Delft has not worked against developing the book either. Not necessarily in terms of time available when in Delft, but especially in Nick van de Giesen allowing me to go elsewhere to study and write, the group's support was clear. My colleagues in Delft have shown great interest in the book project, not in the least because it was a somewhat strange endeavor in the field of engineering and science with its publication culture of short papers with many authors.

The Rachel Carson Center in Munich has been a special place in the whole project. Being a fellow there was simply great. I am sure everyone will understand that my first words of thanks are for my partners in crime, or office mates, Kieko Matteson, Giacomo Parrinello, John Sandlos, Nicole Seymour, Fei Sheng, and Cameron Muir. I could only be in that office because Christof Mauch and Helmuth Trischler allowed me to! The time in Munich was big fun because of the discussions and drinks with Ellen Arnold, Josh Berson, Peter Boomgaard, Emily Brock, Franz-Jozef Brüggemeier, Lawrence Culver, Robert Emmett, Wilko Graf von Hardenberg, Arielle Helmick, Poul Holm, Stephanie Hood, Shen Hou, Eva Jakobsson, Elin Kelsey, Thomas Lekan, Annka Liepold, Uwe Lübken, Francis Ludlow, Jon Mathieu, Kenichi Matsui, Felix Mauch, Anna Mazanik, Xueqin Mei, Jan-Henrik Meyer, John Meyer, Massimo Moraglio, Ruth Oldenziel, Daisy Onyige, Karen Oslund, Chris Pastore, Seth Peabody, Maya Peterson, Katie Ritson, Rachel Shindelar, Vipul Singh, Martin Spenger, Sainath Suryanarayanan, Sabine Wilke, Gordon Winder, Don Worster, Frank Zelko, and Thomas Zeller. It is likely I forgot someone here...

Someone I will not forget is Tony Wilkinson. He has made it possible for me to stay in Durham for a longer period at the IAS, after bringing me to Durham for shorter visits in the years before, allowing me to explore the archives. I will miss his keen interest in anything new, his smartness, his kindness, and his support. Dedicating this book to him is the least I can do to pay my respects to one of the most brilliant scholars and the kindest person I could think of.

Acknowledgments

The research for and writing of this book has been made possible by:

A fellowship at the Institute for Advanced Study of Durham University, United Kingdom, during May–August 2012, when I worked in the Sudan Archive of Durham University.

A writing fellowship at the Rachel Carson Center for Environment and Society in Munich, an international, interdisciplinary center for research and education in the environmental humanities and social sciences, during May–August 2013 and December 2013–January 2014.

Permission for research time and investments by the Water Resources Group at Delft University of Technology, including the fellowships between 2012 and 2014, and several shorter visits to archives in the United Kingdom (Durham University and the National Archives in Kew, London).

Introduction

Settling Certain Details Coming to a Deal

Khartoum, Monday, February 18, 1907, at ten past six in the late afternoon to be exact, it was "all over now"; Charles William Lee Crompton had fallen victim to a fatal illness.[1] Born in 1860, Crompton had been superintendent of the Gezira Surveys between 1902 and 1907. It was his task—with his team of assistants—to survey the Gezira Plain to clarify the boundaries and dimensions of the many properties on it. The country in Gezira was dusty to such an extent that the tablecloth would be "quite brown" after a meal—"when you lift a plate there is a clean or cleaner circle."[2] More than once, Crompton was trapped in a habub (sandstorm). Despite sand, heat, lack of water—"often a bath was a luxury that could not be afforded"—and "the destructive powers of the white ant" destroying the wooden markers, Crompton proceeded with his work.[3]

I could have started this book with someone grander than a dead surveyor. The famous Scottish engineer William Garstin published his plans for irrigation in Sudan in 1904.[4] American entrepreneur Leigh Hunt was one of the first to actually grow cotton as a business in Sudan. Hunt's first cotton farm at Zeidab (north of Khartoum) might have been a proper starting point, as it provided the tenancy model, which shaped Gezira a few decades later. All their stories will be told in this book—some even in this introduction—but I decided to start with Crompton's faith in Sudan for its symbolic value for the story I want to tell. Crompton symbolizes that individual actions are needed to fulfill the central planning of a colonial state (emerging) within a specific (natural and social) environment. Charles William Lee Crompton was one of the first of these individuals—but not the last—to lose his life doing his job of shaping imperial dreams.

Land surveying has been discussed as an early effort of powerful states with ambitions of overarching control.[5] Without a doubt, measuring land in dusty Gezira was an act of symbolic and real power of the developing Sudanese colonial state. Crompton himself simply said he was "carrying out the peaceful occupation of marking out the land in accordance with the most scientific knowledge & practice of this work so that a man may hold his own without fighting for it."[6] Despite this low voice, he was bringing (British) civilization to the Gezira Plain.

Years after Crompton's death, in 1925, the *Daily Mail* celebrated the start of the Gezira Irrigation Scheme. According to the newspaper, the land surveying had been a key element. The land was owned by native proprietors, but they only cultivated some "insufficient grain." When the British came to develop the area, order was needed in this "puzzle-plots territory." After all, the "crazy-pavement plan" had to be rearranged "into regular chequerboard holdings."[7] Proper surveying had ensured that the original proprietors could be compensated for their loss of land to the irrigation scheme, through a 40-year lease of the land. The men who had done the actual surveying—including our tragic hero Crompton—were not mentioned at all, nor was the fact that arriving at the actual boundaries had not been easy.[8]

In *Daily Mail* language, all was planned, executed to plan, and basically inevitable. The main crop in the Gezira model, cotton, was taken for granted as the logical choice. There may have been the problem that cotton was "not the sort of crop these Sudanese are used to—of which you push the seed into the ground and leave the rest to Allah," but that was why a company was brought in: to ensure the cotton was grown—a fine example of oversimplifying—perhaps twisting—history, as we shall see in this book. Obviously, all was in the interest of the Sudanese tenants, those "lucky inhabitants of the Gezira," envied by the "unirrigated" others. Tenants were part of "one of the greatest schemes of Imperial development in modern times" all done by "British energy." The *Mail* correspondent could not resist blaming Egypt, "our supposed sleeping-partner in the Sudan," for only raising "factions and fallacious opposition." [9] Actually, as we will discover, much Egyptian opposition to Gezira came from British engineers (formerly) within the Egyptian Department of Public Works, but let's not move too far ahead now.

Captain Jack Sparrow

This book tells the story of the Gezira Scheme—"one of those outstanding experiments on socio-economic problems"[10]—a large-scale irrigation scheme opened in 1925 under British colonial rule in Sudan (Figure I.1).[11] The typical image of Gezira is of a centrally planned effort by British colonialists, who favored control of tenants and production. That may be the case, and Gezira shows that many planning efforts were involved from many people and institutions, but we would do good in remembering (Captain) Jack Sparrow from the Walt Disney movie series *Pirates of the Caribbean* when trying to understand Gezira's planned development. "Do you think he plans it all out, or does he make it up as he goes along?"[12]

My story of irrigated agriculture and rural society in Sudan in the twentieth century shows how the abstract concept of "development" is shaped by daily actions by people on the ground—in government offices and in muddy fields. I will show how these actions are used in political debates on "development," and how such actions are continuously reinterpreted

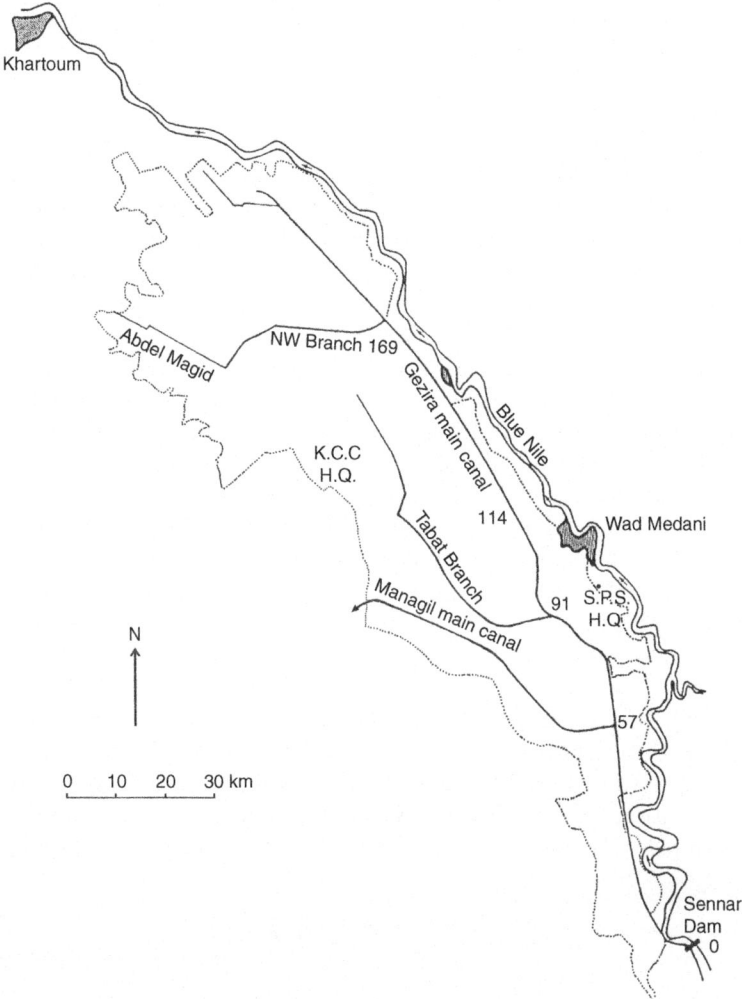

Figure I.1 Map of the Gezira system (roughly as it stood around 1950)

as symbols to support different views on what "development" should be. Projects like Gezira are best understood as continuous negotiations. *Daily Mail* rhetoric proposes smooth planning and inevitable execution of Imperial Development—leaving some Egyptians aside—but inevitability is a human construct. Looking back at Gezira's history, the inevitability of it seems to grow with time.[13]

When Gezira was being negotiated, inevitability was often used as an argument—and certainly attempts were made to create conditions that could not be ignored. As such, later negotiations and decisions were constrained by earlier decisions and actions—in that sense, a certain path dependency (inevitability) is important. Nevertheless, path dependency is a social construction, and we will encounter moments when decisions—and as a consequence possible outcomes[14]—were strongly debated upon by different actors. Gezira's history of development has many moments where decisions could have been different.

The Gezira Irrigation Scheme is located in a very flat plain between the Blue Nile and the White Nile, south of Khartoum. For the British, Gezira's advantage—if not purpose—was its option to develop gravity irrigation from the Blue Nile. Gezira, from the Arabic word for "island" or "peninsula," counts some 5 million feddans (one feddan is roughly equal to one acre), with its most outstanding feature being "its crushing monotony."[15] From 1911 onward, British pumping schemes had irrigated cotton on the plain, but in current irrigation circles worldwide, the term "Gezira" refers to just one thing: the irrigation scheme starting in 1925 with 300,000 feddans currently covering close to 2 million feddans under gravity irrigation from Sennar Dam at the Blue Nile. Gezira's purpose was to grow cotton for the British cotton mills (Figure I.2).

In 1919, after years of negotiations, the Gezira tripartite partnership was agreed upon, stating responsibilities and gains of the three partners involved in the cotton scheme—although the third party had never been present at any formal negotiation table. The Sudan Government was responsible for the main irrigation infrastructure and received 40 percent of net profits from cotton. The concession to manage the scheme and trade the cotton was reserved for the British firm Sudan Plantations Syndicate (SPS); it received 25 percent of the profit in return. The tenants provided labor and received 35 percent of the profit.[16] Perhaps because tenants were not involved in formal negotiations, Gezira had many characteristics of an imposed production regime.[17] The plain was filled with straight canals and square plots;

Figure I.2 Cotton harvest in Gezira. Reproduced by permission of Durham University Library (SAD.1/25/8)

the area was divided into blocks of some 15,000 feddans, each with three British SPS inspectors. A group inspector supervised six to ten blocks. SPS field personnel were "superimposed like the canal system itself on the life of the Gezira"[18] to ensure that the tenants—more specifically male farmers regarded as family heads—grew cotton on ten feddans within their larger tenancy of 30 feddans (which changed to 40 feddans in the 1930s).

A new Gezira policy developed within a context of a changing British colonial policy after World War II, but central planning and control did not disappear. In 1950 the Sudan Gezira Board (SGB), a governmental agency, took over the day-to-day management from the SPS. The Syndicate's share of the profits paid for the SGB and for a bigger emphasis on human development and a new relation with the Sudanese. In the 1950s, the new Managil Extension (to the west of the original Gezira area) doubled the area irrigated from Sennar Dam. Other post–World War II projects, including the Zande Scheme in Southern Sudan, were developed with a close look at Gezira. The first manager of the Niger Agricultural Project in Nigeria—started in 1949—came from Gezira. In Kenya, the British created the Mwea irrigation system in the early 1950s, involving British experts from Gezira. In the 1970s Mwea became the model for developing irrigation in Africa, although its colonial roots were not acknowledged.[19]

Actions Speak Louder than Words

The year that Crompton started his short Sudan career, when Gezira was still void of irrigation canals, one of the key figures on Sudan cotton development—and indeed Gezira, although obviously he did not know that at the time—arrived in Cairo. After a business adventure in Korea, American businessman Leigh Hunt arrived in Cairo in October 1902, on doctor's advice to go to a "kinder climate."[20] Soon he found the Sudan "capable of wonderful development." In 1904, having traveled along the Nile and spoken to Lord Cromer, High Commissioner of Egypt, and Wingate, Governor-General of Sudan, it was clear to Hunt that Sudan could become a large producer of cotton and wheat. He was only eager to ensure that he could be part of that "bright future" to prove that Sudan was "a country where actions speak louder than words."[21]

In a letter to Lord Cromer, Hunt explained that he wanted to establish model farms in different localities, supplying seeds and livestock to farmers "by whose example we should expect to teach the natives" how to develop commercial farming. Hunt was not interested in changing the Sudan into a settler colony (like Kenya), although he mentioned that "settlers must be drawn from the outside."[22] Hunt was not impressed enough by the Sudanese to depend upon their labor and was eager to "introduce the American negro planter." This would also serve to solve a "most embarrassing domestic problem" in the United States, the "Negro problem." Hunt was convinced that "the American negro if carefully selected" would serve as an example for Sudanese farmers. Hunt's reasoning was perfectly simple. He considered the

Sudan climate suitable for American blacks, who were to him experts in cotton growing—and tobacco, sugarcane, other potential crops. They would not "make serfs of the natives," something other nationalities ("Turks, Spaniards, Italians") apparently would.[23] Actually, some American blacks indeed came to Sudan, but as far as the records go, however, malaria quickly killed them.[24]

More important for our story is Hunt starting the Sudan Experimental Plantations Syndicate (SEPS) in 1904, to develop cotton planting on a commercial basis. Originally Hunt was thinking about big money, 600,000 to a million pounds. He approached the firm Wernher, Beit & Co. for financial support. Being closely related to the South African mining industry, the firm was used to large investments, but even Wernher, Beit & Co. considered the investments Hunt asked for too high. The fact that the whole plan was "entirely in the experimental stage" would not have helped either. The firm was prepared to invest on a smaller scale, with the promise that after five years a larger investment would be considered. If the experiment was a success, a new firm would take over whatever was there from the SEPS.[25] At the first meeting of the SEPS on September 2, 1904, Frederick Eckstein became one of its directors. Eckstein was the brother of the representative of Wernher, Beit & Co. in South Africa, chairman of the Central Mines Investment Corporation and soon to become a major player in the Gezira story. Hunt himself was appointed as Managing Director in the Sudan.[26]

The SEPS did not really have a flying start, even though Hunt was as energetic as ever. Already in October 1904, he had arranged a pumping plant—in cooperation with a "Mr. Smart, the Engineer."[27] In July 1905, a land purchase from the government was arranged.[28] The Syndicate was not allowed more than 10,000 feddans along the Nile at Zeidab, and we learn that exact ownership of the lands was not clear yet. In case "any native rights" were found on the lands, Hunt had to pay compensation. We also get a glimpse of what the colonial public-private partnership in Sudan might mean. It was up to the SEPS to build the whole scheme, but the government had to agree with all arrangements. In case of issues with land ownership, it was the government's responsibility to "conduct the negotiation with the native." The license holder, the SEPS, would supply water to tenants "subject to Govt control." The concession was to end in 1915.[29]

In October 1905, the SEPS Board discussed its financial situation in view of the possibilities for a cotton scheme. It was decided to focus on the existing property before extending activities. Among the problems encountered at Zeidab were the alkali soils and the fact that many "white men in our employ" died of malaria.[30] MacIntyre, who would become Managing Director of the SPS later, came to Zeidab in September 1906 to replace an engineer who had died "before a pump ever ran."[31] Eager as always, Hunt suggested to request an extension of the concession,[32] but with these continuing problems, Wernher, Beit & Co. sold its stocks back to Hunt.[33] On his turn, he managed to convince D. P. MacGillivray, a "very able

Scotchman" who was a manager at a Cairo Banking House—and whose first names remain a secret—that cotton could be a commercial success.[34] MacGillivray bought Hunt's interest, and the London Well Group represented by Eckstein "bought back into the Company" as well.[35] On June 17, 1907, the Sudan Plantations Syndicate Limited (SPS) was formed.[36] Hunt resigned and MacGillivray was appointed as Managing Director,[37] with Eckstein on the Board.[38]

Understanding Gezira

Colonialism is typically reserved for states,[39] but it was the SPS that played a major role in Gezira and made quite a bit of money from it.[40] In 1928, the SPS managed seven ginning factories with 560 gins in Gezira, "the largest number of Gins directly under one management in the world to-day."[41] Along kilometers of canals, thousands of tenants farmed the fields. The setup of Gezira, with a private firm managing the scheme on behalf of a government, may not be found in many colonial settings, but it does mean that agreements and fights between different interests are explicitly spelled out in the sources. Within a spirit of cooperation, the government and SPS had different ideas and never fully trusted each other. Disagreements about how to proceed, what to focus on in policy, and who should pay are obviously a normal part of government, with different governmental bodies negotiating all the time. Indeed, the separation found in Gezira—blurred as it sometimes is because of internal disagreements—offers an excellent representation of colonial development contexts. It serves a historian looking for a good story, too!

If any image emerges from the scholarship on Gezira over the years, it is of a centrally planned, British colonial effort, with some changes over time, but continuously based on strong control over tenants. Depending on the author's political and/or academic ideas, some flavor and judgment is added. For Arthur Gaitskell, the first SGB manager after a long career with the SPS, Gezira was the model to go for when planning development in African nations—the fact that he did not have any experience elsewhere did not really hinder him. For Tony Barnett, the sociologist who studied Gezira in the 1970s, Gezira represented failed development, as it had not supported a transition to capitalism and it kept tenants in vulnerable and dependent positions. In the 1990s, anthropologist Victoria Bernal discussed Gezira as instrument and symbol of colonial oppression of Sudanese farmers.[42]

To be fair, all three authors have a point. Gezira was centrally planned by British colonial power assuming control of tenants and production.[43] The Gezira scheme may have brought benefits to many, not in the least government and SPS but also to groups of tenants—but not all tenants have done well economically. Furthermore, Gezira's strong focus on cotton caused problems when world market prices for cotton went down and emphasized Sudan's dependent position within the global economic system. However, reading all three accounts together⁴⁴ does not yield a fuller understanding of Gezira. How can Gaitskell's success be defended when many tenants remained in debt, with problems sustaining themselves? On the other hand, how can Gezira remain a symbol of strong oppression—or failure for that matter—when a large group of tenants has succeeded in making significant economic progress? How can something be a failure anyway when it still exists and many people live in it?

I argue that we have to look anew at Gezira to see what it may stand for, in terms of colonial control, farmers' actions and resistance, and the broader development debate. Taking a new look at Gezira forces us to reunderstand British colonialism itself. The colonial project cannot be understood as something that rolled over colonized landscapes and societies. Idea(l)s of planned irrigation and profit in Gezira had to be realized by African farmers and European officials.⁴⁵ "Empire worked on the ground because it relied on vast support staff of clerks, technicians, teachers, and medics who handled the day-to-day tasks of colonialism."⁴⁶ Tenant farmers responded to SPS control in various ways, but basically tried to minimize efforts in cotton. Tenants were not loyal, silent receivers of colonial wisdom; farmers had their agendas and agency. In turn, SPS inspectors were closely controlled by SPS management. After all, there were production goals to be met. Working schedules of inspectors were strict, which frustrated many of them. British inspectors entertained themselves in clubs and at home—including drinking heavily when visiting each other.

Studying Gezira also allows gaining new insights on the concept of societal development. The changes and continuities in Gezira-related policies and actions—and how these were seen by contemporary and future scholars and stakeholders within the twentieth century—shed light on (irrigation) development within the Gezira Scheme itself but also in the general debate on development in the Sudan and Africa. The Gezira project is a prime example of "contested development"—perhaps not in exact alignment as proposed, but certainly in spirit. Development activities "are not born successful (or unsuccessful); they become so over time and for many different reasons in different places."⁴⁷ A colonial project like Gezira is best understood in terms of continuous negotiations of different kinds. We will encounter official negotiations between the SPS and different governmental agencies, but we will also see how tenants and field staff negotiated new realities on the irrigated fields every day. Gezira's continuities and changes are "characterized by punctuated equilibriums—long periods of stasis that give way relatively suddenly [...] to a new alignment of prevailing interests and ideas."⁴⁸

A White Men's Enterprise

Gezira's negotiated history was a centralized plan seen by a state, provoking responses by several actors—British and Sudanese alike—who tried not to be seen, with tenant responses primarily based on their available weapons of the weak.⁴⁹ Negotiations in expensive hotels and Gezira mud were directly linked, both shaping Gezira reality and setting the scene for new negotiations.⁵⁰ I may not go as far suggesting that Sudanese tenants controlled the Empire's management of irrigation as Egyptian farmers might have done in the Ottoman Empire. Having said that, though, the close relations between Egypt producing food for the empire and the Ottoman center distributing the food does clearly show that imperial relations were not abstract forces, but shaped by daily actions of humans—farmers, administrators, and rulers.⁵¹

Although colonial powers transformed agrarian realities in British India and the Netherlands' East Indies too,⁵² Africa was generally treated as a blank slate in terms of development ready for large scale targeted development efforts—although not necessarily in terms of social realities, as we will discuss in this book. Quite a few African colonial schemes were modeled as tenancy settlement projects; many postcolonial African irrigation policies adapted the settlement model—especially its strong production control—to a certain extent. After decolonization, the modernization mission did everything but disappear. Options for controlling farmers were less, however, as the old power relations were no longer accepted. Postcolonial schemes transformed "coercion and force" into "extension and training."⁵³

Specific for our Gezira story is how modernization was shaped. The Gezira commercial model is pretty unique. The economic crisis and cotton diseases of the 1930s were key events, resulting in a different Gezira—for example in irrigation management. The impact of World War II on daily management in Gezira was huge, as many British inspectors served in the army and Sudanese daily managers—without the power of the inspectors—were appointed to replace them. Each time, positions had to be formed again, common answers to the question how to produce the cotton needed to be found, as the mutual interests of (at least) the Sudan Government and SPS ensured continuing on the road of modernization.

Gezira was a "high-energy environment" in terms of institutional changes, "the conflicts to which they inherently give rise, and the role that development projects play in helping or hindering them."⁵⁴ A suitable—although perhaps less original—main title for the book could have been *Controlling like a Company* as the SPS played such a key role. Tempting as the *Controlling* title is, I did not go for it. Another title emphasizing the "white enterprise" in Gezira was suggested to me by a former SPS inspector, who wrote in one his letters in the early 1980s that he thought it "incredible" that Gezira was not generally recognized as "a model for future generations on how to develop a desert country without destroying the fabric of local society & handing over the peasantry to be exploited by unscrupulous or incompetent foreigners in the interests of their self servicing fellow country men."⁵⁵ After presenting this view on the history of Gezira at one of the biannual Sudan reunions for former SPS personnel, several people asked him to write more about Gezira, but he felt "too old to start collecting the necessary material" and not "in touch" with the debate anymore. He doubted if anyone would print something with "the temerity to ascribe virtue to a white enterprise."[56]

The white enterprise would actually work nicely in combination with "The White Man's Burden," the poem by the English poet Rudyard Kipling.[57] The poem was referred to in a 1946 letter of the Governor of the Blue Nile Province. Writing to British government employees—including SPS employees—on the required changes in attitude toward the Sudanese after World War II, and realizing it was "out-of-date to quote Kipling," the Governor had nevertheless been "unable to resist adding a few verses" as they were "strangely applicable to our needs today."[58] The poem—though open to a variety of perspectives and interpretations clearly linked to colonialism—was published "prophetically in the year in which the Sudan Civil Administration began."[59] Gezira as a "white men's enterprise" links the development mission of British colonialism and the business mission of British cotton. Gaitskell would not have objected to these worlds being combined, as he praised the thoroughly planned approach of Gezira uniting social development and sound business. Barnett would have agreed, as for him the cotton interests that dominated in developing the Gezira hampered modern development.[60] Bernal would not mind either and emphasize that it was definitely white men constructing colonial control over the Sudanese.

An Area of Sudden Charm

However, the title should show my emphasis on development as negotiation. As others have used exactly that same word for their own books,[61] I decided to link planning and improvization in the title. Improvised and/or planned, Gezira was clearly a male world, with most early British employees being single men. British women working for the SPS were rare—almost nonexistent in the sources. When they entered the world and the archives of Gezira, it was after World War II and typically as spouses.[62] The first British woman in Gezira doing actual work for the SPS—as a government employee—was Geraldine Mary Culwick in 1949. Her job was to study nutrition in Gezira; she was probably the first British person to visit more than a few Sudanese tenants at home and definitely the first Brit to live in a Gezira village. There were many SPS inspectors who considered this weird—if not subversive. Culwick was an excellent observer of Gezira social relations, and not only the Sudanese ones she was supposed to study. When her transport from the station in April 1949 went wrong due to the "usual Sudan muddle," she dryly observed "how seldom any arrangements ever work out according to plan here!" After reading some older documents, she considered "the sense of uncertainty and impermanence which pervades" in these papers "a shock." This may sound a little naïve, but just a moment later she pointed out that it would be "ridiculous to suppose that an organization so complex as the Scheme could take shape without acute internal struggles between the various interests involved, with their differing points of view." I cannot agree more with her and share her observation that "the record is seldom dull."[63]

The period covered by the sources centers on the period 1905 to 1960.[64] Although, obviously, partially a function of my own search abilities, I am pretty confident that the sources allow reconstructing a "three waves model" of Gezira. Between 1900 and 1915, we encounter early negotiations between government and Syndicate, smaller pump schemes and a first agreement just before World War I. Between 1916 and 1930, the Gezira Scheme was actually constructed in earth, concrete, and further detailed agreements between government and SPS. First experiences with the gravity canal system and early extensions were realized as well. The period between 1931 and 1955 is one of constant flux, with major changes after the crisis in the early 1930s, after World War II, in 1950 with the termination of the SPS concession, and in 1956 with Sudanese independence. As I will show, however, there was enough continuity too in this period.

Despite the rich material, there are many issues I will not include. I keep away from more general policy-related debates in Sudan, the United Kingdom, and the world. Obviously, I include UK colonial policy and Sudan's economic policy in the period I study, but Sudan's planning procedures, its road to independence, and its new position in the world only find a place in the book when linked to Gezira.[65] Furthermore, there is a chapter specifically about the Sudanese tenants, their strategies, and responses, but on the scale of the whole book, the Sudanese are relatively absent. This is definitely because of sources available, but at least partially a deliberate choice as well. I would like to correct the image of British colonialism as one monolithic power with replaceable colonizers just filling in the blanks. To show the importance of the colonizers on the ground, actual people representing British colonial rule in Gezira are introduced.[66] Next to the SPS inspectors, I include the irrigation engineers. Being one myself, originally trained to work overseas in development projects, I do feel related to them.

As a result of such considerations, I may be just writing a story of one large irrigation project in Sudan. This may even be a good idea, as the—unsatisfactory—publications we have on Gezira are all written with a larger story in mind, in terms of desired development or lack of it. Obviously, my focus on day-to-day actions of different actors may be mundane, but not necessarily "uneventful, uninteresting, or uninstructive."[67] I do have a bigger story to tell, by emphasizing the importance of daily human agency on the ground when constructing development.[68] In a rare moment of agreement between myself and the SGB, Gezira "appears from the air, to be a monotonous checker-board pattern, only relieved by subtle variations of colour," but "proves on the ground to be an area of sudden charm, of unexpected felicity."[69]

Agency and Structure

There is so much mundane to engage with that this story of unexpected felicity counts seven chapters and an epilogue. The first two chapters explain how Gezira came about and its particular setup as a tenancy settlement scheme. In chapter 1, I discuss the negotiations between 1900 and 1925, emphasizing the differences and similarities between the goals of the different groups. This chapter is basically a chronological account of negotiations and ends in 1925 when Sennar Dam was used for the first time. Chapter 2 starts with the extensions of Gezira, to break away from the standard image that opening Sennar meant immediate stability. The extensions are a major example of parallel planning and improvization. In the same chapter, I discuss the management arrangements in Gezira in general terms. The agricultural calendar, the Gezira production rhythm, and the profits are included, as well as changes in the calendar because of lower profits and changing ideas about profits.

The next three chapters discuss the daily realities in the scheme, from the perspectives of engineers, tenants, and inspectors. The irrigation engineers appear in chapter 3—the SPS will appear as well, as it had its own ideas about the canal system. The chapter uses engineer Smith's career in the Sudan Irrigation Service and the major issues he had to deal with in his long career, including changing the system during construction, realizing the extensions, and maintaining the canals. The Gezira tenants are the center of debate in chapter 4, although SPS and government will appear again. The SPS inspectors appear in chapter five. Chapters 4 and 5 discuss what was supposed to happen with tenants and inspectors, respectively, how they should behave within their version of the Gezira theatre, and what role they carved out for themselves. The final two chapters discuss Gezira after 1945. Chapter 6 shows how after World War II Gezira was renegotiated with the context of a changing British colonial policy. Chapter 7 discusses how after World War II the international development cooperation scene was built on colonial foundations, with Gezira and its (former) experts playing major roles.

Together, these chapters should have enough to offer for scholars working on colonial histories, obviously first and foremost of Sudan and Gezira, but for example also of colonial cotton projects and policies,[70] and certainly not restricted to British colonization in Africa.[71] Scholars working on state planning and its responses, including issues like development cooperation, planned development, or free markets—either with a focus on Africa or not—would find the book relevant too.[72] After reading my Gezira story, it should never again be a surprise that development plans are not realized, and outcomes are blurred and continuously negotiated upon.[73] In explaining how Gezira reality was contested, interactions between the environmental context and human agency are relevant as well—for example when discussing the Gezira crop rotations or the sediments in the canals. As such, the book could be interesting for environmental historians.[74]

Finally, I have tried to write a book for those thinking about agency and structure. In the Epilogue, I reframe Gezira as a discussion on development, with changes and continuations in its meaning. When taking up the concept of development itself, I want to link it to notions of actor-network theory. Gezira development was constructed within networks of negotiation that were continuously (re)created by human actions engaging with other acting humans, often through nonhuman intermediaries.[75] In making these notions operational, this book is an example—if not the first book actually doing so—of integrating actor-network theory in a historical study. Surveying land in Gezira was definitely an early act of state control, but as with all outings of state control, the actions of real humans had to carry through that control. In development, Captain Jack Sparrow meets Bruno Latour.

Chapter 1

Cotton from a Wilderness: The Early Negotiations

The year 1904 was an excellent year for new initiatives in Sudan. Hunt established his Sudan Experimental Plantations Syndicate, Garstin published about irrigating the Gezira, and the Egyptian Public Works Ministry created its Sudan Irrigation Branch. Charles Edward Dupuis became the first Inspector General of Irrigation in Sudan.[1] In 1908, he presented plans for irrigating the Gezira Plain from the Blue Nile, building on Garstin's ideas.[2] Dupuis showed that a canal reaching the Gezira Plain close to Wad Medani could command some three million feddans. Much was to be done, however, before such an area, "vastly in excess of possible requirements of the Sudan for many years to come," could be irrigated. The population on the plain would require "a good deal of education" too.[3]

At first, there was more in Sudan than Gezira to be developed. Between 1900 and 1905, the Sudan Government issued 15 mining concessions, but as most of the mining companies were not economically viable, after 1905 the government concentrated on agricultural activities.[4] The Kerma plain in Dongola Province included 20,000 irrigated feddans as early as 1909, with possible extensions of 40,000 feddans. Its farmers might be "naturally backward where anything new is taken up," but Kerma would yield 30,000 tons of millet, maize, lubia, barley, and wheat.[5] From 1908 onward, Gezira became a key attraction for Sudan investments in irrigation and railways.[6] The railway line from the While Nile to El Obeid should pass the Blue Nile at Sennar, even though large parts would go through "more or less uninhabited country." With a dam at Sennar and the Gezira canalized, however, the situation would "altogether alter."[7]

The Gezira could produce the revenue needed for Sudan "to become satisfactorily self-supporting,"[8] but Wingate did not expect public money either from Egypt or London becoming available soon. The Sudan had to "induce British Capital and thus the British Capitalists to have a vested interest in the country"; powerful men like Lord Milner, Lord Erroll, and Sir William Mather had to be convinced to invest in Sudan.[9] Cotton was a major item on the development agenda. Once the Sudan Plantations Syndicate (SPS)—possibly because of its activities at Zeidab—suggested that long staple cotton could grow as a winter crop between July and April, a business model based on large-scale cotton cultivation became feasible without jeopardizing Egyptian water demands too much.[10] The Sudan Government could ask Egypt to be "given a fair chance" to stimulate "sound British Capital to invest in the Sudan."[11]

This chapter will show—in a rather chronological way—how Gezira plans were shaped through British, Egyptian, and Sudan interests, how investment options for Gezira were created, and how happy everyone was when in 1925 Sennar Dam brought water to the irrigated area. Our first focus will be on the experimental cotton station at Tayiba, which started in 1911, was taken over by the Sudan Government in 1913, but returned to SPS management that same year. Once SPS and Sudan Government had defined their respective positions in their mutual desire for Gezira in 1913, negotiations shifted to the financial aspects of realizing the large project. Finding the money was hampered by the outbreak of World War I; in the meantime, the details of the agreement between SPS and the government had to be refined as well.

Together, SPS and the government realized "a very much more elaborate scheme of development" than Dupuis had outlined in his 1908 report. Plans were gradually developed in strict confidence, but were ready when the government was "called upon" to define the Gezira development.[12] The government and Syndicate operated within a relationship that was "relatively cosy" though tense at times. Mutual self-interest was the guiding principle.[13]

Possible Diminutions of Control

Within the Condominium, Sudan depended both politically and financially on Egypt; any plan involving Nile water should guarantee Egypt's claims to that water. This explains the early focus on winter crops, like wheat,[14] as Sudanese summer cotton would require the same summer water that Egypt's thirsty crops needed. At the same time, Wingate expected that increased British investment in Sudan—especially for cotton—would stimulate the Egyptian government to somewhat lower restrictions on Nile water.[15] He was careful, however, to maintain the leading role of the British Consul General in Egypt, if only because the Consul would not appreciate "a possible diminution of his own control."[16]

Around the time that Wingate and his staff were thinking about Sudan economic policies involving Gezira, the idea to grow cotton as a winter crop was brought forward, most likely by MacGillivray.[17] In May 1910, MacGillivray and Alexander MacIntyre (then the SPS manager at Zeidab) went to the Gezira area.[18] Later that year, a delegation including Mather, MacGillivray, and J. Arthur Hutton (Chairman of the British Cotton Growing Association [BCGA]) visited the Foreign Office in London to discuss options for the

SPS conducting an experiment whether winter cotton could actually grow in Gezira. The Foreign Office (FO) preferred the Sudan Government "to go a-head without delay with large irrigation works and experiments of its own," which would be "an answer to all enquiries." With this clearly unfeasible, a Syndicate project offered "the best prospect for making a start."[19] Garstin advised to experiment for at least two seasons.[20]

In November 1910, the Syndicate firmly declared its "ultimate policy" to "undertake and finance" large-scale irrigation in Gezira, with the goal "to supply water to the natives on their own lands" and "to develop Government lands" on terms to be discussed with the government.[21] Delegations from SPS and the Sudan Government met nine times in December 1910 and January 1911 to discuss how to proceed. The total experimental area was to be 3,000 feddans, but a start would be made with 1,000. The Syndicate preferred an experiment of three years; the government suggested four. The representatives could not agree on financial arrangements—specifically on the SPS request to receive more than 5 percent remuneration. Furthermore, the parties disagreed what would define the success of the experiment[22] and whether success meant that the SPS would have a "definite preferential claim" to develop the larger Gezira scheme.[23]

On January 12, 1911, the negotiations were broken off. Already on January 9, Macdougall Ralston Kennedy (Director of Public Works in Sudan) had ordered a pumping station in England and arranged 100 agricultural laborers from Upper Egypt for Gezira.[24] This suggests that—whatever happened with the negotiations—the Sudan Government was determined to go ahead themselves with the cotton experiment in Gezira anyway.[25] It is also clear that the Syndicate was eager to participate in the Gezira project. Soon after January 12, the tone of the SPS changed.

From a syndicate doing everything itself, the Syndicate changed into one of three partners, next to government and local landowners. The partners would share the profits of converting Gezira from "an ill-cultivated and comparatively unproductive area into a well-cultivated, highly productive area." The Syndicate offered to provide the necessary irrigation works "on fair terms." Assuming the experiment was successful, the SPS even promised to provide the funds to build Gezira. The government would just have to ensure that land became available.[26]

The Foreign Office in London doubted the financial capacity of the "Zeidab people" to fulfil their promises and refused to entrust "the larger scheme" to the SPS.[27] Nevertheless, the SPS running the experimental farm on behalf of the Sudan Government was acceptable.[28] Given the small size and budget of the government, it was better that the Gezira test was executed "by Capitalists with experience of farming." Success would allow the Syndicate purchasing 10,000 acres at a set price within the—future—irrigated area.[29] Any SPS management of the larger Gezira scheme was not part of the deal yet.

The details of the experiment were discussed to avoid "divergence of opinion" on the scientific soundness of the experiment. The option that the government would do specific tests on soil properties and crop growth was discussed, but not considered essential.[30] In March 1911, a "satisfactory agreement" was reached to start experimental cotton growing at Tayiba in Gezira.[31] The government would build the irrigation infrastructure and provide the pumps; the Syndicate would manage cultivation "on a rental basis to Tenants."[32] Irrigation would start on July 15 and end in February. One-third of the fields would have cotton, another third winter cereals, and one-third durra (a local grain crop), lubia (a local fodder crop), or lie fallow. After four years, or earlier if possible, a separate committee would decide on the success of the experiment.[33]

A Plan So Cunning...

The Tayiba Experimental Station opened on April 7, 1911.[34] Irrigation started on July 15.[35] With "the energetic action of the troops" sent by the government and technical support from Kennedy, 1,000 feddans were developed in the first season.[36] Mather expressed "feelings of satisfaction," praised the Sudan Government and Eckstein, and was happy that "the Natives [were] anxious to come in rather than to oppose the undertaking."[37] Another happy actor was the British Cotton Growing Association, who joined the Syndicate in 1911.[38] Wingate welcomed a positive public statement on Tayiba, though it was "undesirable to emphasize" that Khartoum and London liked one private firm over others.[39] The final statement conveniently excluded any detail on the agreement.[40]

With Tayiba farming underway, the future prospects of a larger Gezira Scheme became more important than ever. As a 1911 Christmas present, a study on the Blue Nile suggested that Gezira could include five million acres of irrigated cotton between July 15 and October 15. Even when irrigation was applied until the beginning of April—thus needing some of Egypt's water—the minimal potential area would still be one million feddans.[41] The location of the future dam was studied in more detail as well. When visiting the Blue Nile in 1911, irrigation engineer Percy Marmaduke Tottenham studied the granite bar in the Blue Nile at Sennar and decided it was the perfect place for a dam controlling irrigation water for Gezira.[42]

Obviously, grand plans need big money, which was a slight problem. The year 1912 might see a million pounds becoming available, and possibly in 1913 a similar sum, but nothing was clear.[43] A response to the financial constraints was increasing the number of pumping areas similar to Tayiba before starting with a dam and canals.[44] At Tayiba, Gezira was "made to 'blossom like a rose' so successfully that it 'would be positively wicked to delay one moment longer than is necessary' to develop 'this wonderful plain.'" Reaching the promised land had to solve the problem that most of the land was owned by Sudanese.[45] The Sudan Government had to step in and ensure Sudanese land ownership would not block the colonial cotton dreams. How they did so is discussed in chapter 4.

In the meantime, the question regarding which of the two "partners" in Gezira would be the one to develop cotton as a commercial enterprise remained squarely on the table. The Sudan Government might not have had many options to develop Gezira itself but was not prepared to leave all action to the SPS. However, too much pressure could result in the SPS leaving Sudan. This would be a disaster, although the threat was probably low. Once cotton would prove to be "a success," it was "improbable" the Syndicate would leave Sudan.[46]

On its turn, the Syndicate did everything it could to show its good intentions. When Wingate complained that "the attitude" of the SPS on water rights "made any sort of negotiation out of the question,"[47] Lovat and Eckstein were quick to assure both Lord Kitchener (High Commissioner of Egypt) and Reginald Wingate (Governor-General of Sudan) that water control should remain a government responsibility.[48] The SPS was more than eager to finance the larger Gezira scheme; the Directors planned "to husband" SPS resources to be ready for future Gezira action.[49] Like two hedgehogs making love, both sides showed strong desire and extreme caution.

...You Can Put a Tail on It...

Future lovemaking depended on whether Tayiba was a success or not. In 1913, Edgar Bonham Carter (Legal Secretary in the Sudan Government and as such heavily involved in the negotiations) feared that the SPS would claim the experiment a huge success after only two seasons with relatively good conditions—no huge rains or pests.[50] The Syndicate was clearly carrying out the terms of the agreement, but their interests "must necessarily influence their presentation of facts."[51] What to do? Ending the agreement after only two years was impossible without consent of the SPS.[52] Another concern was how to find the large numbers of tenants needed to grow the future cotton. Many Tayiba tenants came from Zeidab, but more were needed. One could "hardly expect from a commercial company" that it would provide expensive training for newcomers. Sudan officials were, however, not in the position to start a training program at Tayiba either.[53]

Wingate was clearly not happy with negative opinions about the SPS—he labeled them as "incorrect and misleading"—and emphasized the need to show that cotton could be grown as a winter crop before any "definite educational scheme could be established."[54] He did not completely trust that the government could access all the Tayiba data needed to draw clear conclusions either. Therefore, he did recommend taking over Tayiba before the four years had passed. Wingate was not alone in his ideas on Tayiba, but the ideas of future SPS land ownership in Gezira had changed—if not in Sudan, then clearly in Egypt and London. In 1913, the miracle deal of 1911 seemed "a most unfortunate one," as the Sudan Government had offered 10,000 good feddans with "nothing of great value in return." [55]

The SPS response to any government take over was unknown, but it was unlikely that the Syndicate would accept 30,000 feddans of rain land.

However, the 10,000 irrigated feddans did not exist yet;[56] the 27,952 feddans of government land in Gezira were "probably unsuited for irrigation."[57] Legal options to change the Tayiba arrangement were unclear, however, and none of the governments had any desire to chase away vital private capital. Yet, the last thing the British officials wanted was Sudan becoming a second Rhodesia with a chartered company running the colony.[58]

Wingate and his officers did what they could to find a "more or less ingenious way" to get "out of the impasse" on who would be involved in a future Gezira scheme.[59] Early 1913 was the moment, however, when Cairo—Lord Kitchener—decided to take the driver's seat.[60] Kitchener's letter to Wingate was clear: Sudan officials were "not to give any personal support to commercial enterprises" as the officials were "hardly equipped to deal with experienced City financiers." Permanent land concessions were not allowed, and the Sudan Government was to take over Tayiba immediately.[61] Wingate welcomed "the soundness of Lord Kitchener's view," but did not see how the government could avoid giving the SPS "their pound of flesh"—the 10,000 or 30,000 feddans. He asked Cairo to "show me the way."[62]

Wingate could be assured, Kitchener was engaging himself fully with Gezira. In his letters, Wingate was clearly unhappy with Cairo's interference, but knew that the so-called autonomy of Sudan was practically nonexistent.[63] Kitchener did not always share plans with Wingate, and when Wingate realized that Kitchener planned to realize dams on the Blue and White Nile "more or less at the same time," he panicked for a brief moment as he was afraid to be pushed "by the Cotton Growing Association and others" to continue with Gezira. However, the dams would only start in two or three years—after 1915—allowing Wingate "much more breathing space" to start "proper agricultural development."[64]

On Tayiba, Wingate realized that the SPS wanted clarity on its promised land, but any clarification had to come from Cairo. Actually, Kitchener hunted Lee Stack, the Sudan Agent in Cairo, for details on the deal with the Syndicate. Kitchener agreed with Wingate and Stack that the agreement had to be honored, but he wanted the Sudan Government to "get out of it as cheaply as possible."[65] Kitchener hoped to "induce" the Syndicate to accept the 30,000 feddans of rain land, although he did not expect it to happen. The alternative was offering 10,000 feddans in the Bageir area, for which Tottenham was designing an irrigation plan. Stack warned that irrigation had to be possible within ten years after the end of the agreement, but Kitchener argued that "a lot might happen in 10 years."[66] Actually, the Syndicate was probably not eager at all to call Tayiba a success after two years and take the land. The 10,000 feddans on offer were "useless." Under the Cairo pressure, the Syndicate felt "forced to agree that Tayiba was a success" and was "in the soup."[67]

In spring 1913, the government planned to take over Tayiba management. It would be the place to train the Sudan officials needed to realize more extensive developments elsewhere in Sudan.[68] The Agricultural Department was still "excessively weak" and the Irrigation Department still "a somewhat weak adjunct of the Egyptian Irrigation Department."[69] A certain Mr. Turner, who had worked for the SPS at Zeidab, applied for a government position. Just a few years before, he had been rejected by the Department of Agriculture for lack of a diploma. Given the need to have government people in Tayiba as soon as possible, Turner suddenly became "a qualified man."[70] Taking over all SPS staff directly would have been great for the government, but obviously the Syndicate did not agree.

...And Call It a Weasel

Mr. Turner was not the only factor required to produce Gezira development. Huge amounts of money were needed too. A £3 million London-backed loan for Sudan should bring the cash to build railways and irrigation schemes.[71] With money on the agenda, it was obviously uncomfortable that Hutton wanted Wingate "to know exactly" that "any alterations in the existing management" of Tayiba would give the BCGA "a very awkward position." When recommending the loan to the prime minister, the BCGA had "naturally assumed" that SPS management at Tayiba would continue for at least two or three years longer.[72]

These pressures, the continuing issue of how to compensate the Syndicate, and the wish to push Gezira further made Kitchener approach the SPS with a new offer for both the larger scheme and Tayiba.[73] The new proposal was discussed on July 21 between Kitchener, Wingate, Lovat, Eckstein, and MacGillivray.[74] Developing Gezira irrigation infrastructure—including leasing land from owners—would be the government's responsibility. The Syndicate would act as a government agent in arranging tenancies and directing cultivation "as done at Tayiba." In addition, Kitchener proposed a distribution of the gross profits: 35 percent for the government, 25 percent for the Syndicate, and 40 percent for the tenants.[75] The SPS accepted the new proposal, although Eckstein wanted clarification on some aspects "to prevent any misunderstanding in the future."[76]

The proposal made the government and Syndicate partners—despite the status of the Syndicate as government agent. The government would build the canals leading to 1,000 feddan blocks; the SPS would be responsible for subsidiary canalization in these blocks. Government and SPS agreed that the subsidiary canalization should be done by the tenants, "of course under the supervision of the Syndicate." MacGillivray mentioned that rapid development of 100,000 feddans would require steam ploughs, trashing machines, and other heavy machinery in the cultivation process. This was accepted and financial arrangements were adapted. As with the buildings, the machinery would be taken over by the government after the concession period.[77]

This concession period was discussed at some length. Kitchener made it perfectly clear that without a considerable SPS contribution to capital needed, there would be no Gezira. He demanded £1,000,000; the Syndicate offered £500,000, which was accepted by Kitchener (and thus by Wingate).[78] Eckstein asked ten years' use for each unit of 10,000 feddans. Kitchener offered a fixed ten-year period, which could be renewed and would start once the financial contribution from the Syndicate was available. The government would ensure that water reached the fields three years after that moment, leaving seven years for the SPS to make a profit. Eckstein protested that this arrangement would not offer enough financial compensation for the SPS, as at least four more years would be needed to make a profit. In this light, the conditions for possible renewal of the agreement became even more important.

The immediate result of the meeting on July 21 was the Syndicate retaking Tayiba management on August 1, 1913, under the "new basis," including the division of percentages.[79] Wingate sent a telegram to Khartoum to confirm this, but details would "be settled later."[80] Lovat was a little disappointed that Wingate was just sending the telegram. Lovat wanted "to commit to paper what was settled yesterday,"[81] after all, even "a phrase written in the best faith might easily lend itself to misinterpretation, if no opportunity for discussion and correction were provided."[82] Wingate clearly did not want any new discussions. He assured Lovat that the new agreements would take effect once the loan had passed Parliament. Extending the ten years' concession with another five years was possible when the Sudan Government was "entirely satisfied" and "no political or other considerations" had arisen. Wingate also declared the new start at Tayiba in August as temporary; in case the agreement would "not mature on the lines indicated," the government would step back in.[83] The SPS should make its financial contribution available soon.

In the meantime, MacGillivray was only too happy to confirm that he was sending SPS people back to Tayiba.[84] Obviously, time was short; the SPS people could perhaps reach Khartoum on August 3. They had to inform the tenants of the new working arrangements, at least about the tenants' 40 percent of gross receipts, but tenants "should not be told how the balance is divided between Government and ourselves."[85] Wingate suggested similar secrecy, as settling "certain details" was still needed, making it "very desirable to say as little as possible about the project."[86]

Once Bitten, Twice Shy

Step by step, a new reality was created in a joint government and Syndicate effort. Despite disagreements and different visions, the Tayiba crisis resulted in a new agreement and a new roadmap toward realizing the larger Gezira scheme. The new goal was to find the money to realize the shared cotton vision. As with other agreements between Sudan civil servants and Syndicate business men, Sudanese farmers were not represented in the formal negotiations. This is a major thread through Gezira history: two partners with continuous disagreements stick together for financial reasons, with the third party not present at the negotiations.

Sticking together worked fine, but finding the money took time and effort. Just days after the agreement in July, Wingate optimistically referred to a Treasury insurance the final passing of the bill through parliament would not be difficult "in any way".[87] The Loan Act was to be passed on July 24, 1913, through the House of Commons, but had to wait until August 15, 1913.[88] The Syndicate was less optimistic, as finding their half million was "no easy matter," especially since about £35,000 extra in cash or interest was required "to gild Kitchener's pill."

The terms of renewal were key for finding Syndicate investments: how certain would those investments be? Lovat thought a solution was "not beyond the wit of man," but given uncertain future political realities, a clear renewal clause was needed. The SPS "City friends" would dislike any terms "too wide a term to make the renewal clause of any real value."[89] Wingate expressed his surprise to Kitchener about the SPS problems finding the money, but was not impressed that Lovat "should jib" at the renewal clause.[90] Kitchener made it very clear that the SPS had to "produce the cash to go on with" Gezira.[91]

The discussions about renewal clause, loan, and SPS contribution were a new round of the general Gezira game. The key question was how to ensure that the joint effort would go through, while realizing the most positive conditions for one's own interest. It was a case of "once bitten, twice shy," especially after the troubles on Tayiba.[92] Obviously, our two hedgehogs in love were not alone. Rather influential bystanders—the Cairo and London governments—were even more cautious than the Sudan Government for "the scrutiny of hostile critics." Tenants' welfare and their "thoroughly fair, not to say, philanthropic" treatment had to be ensured.[93]

Wingate and Kitchener pressed for an agreement before Kitchener's departure half September 1913 from London to Cairo.[94] Wingate assured Lovat that details, including the renewal clause, would be arranged by a Gezira Commission.[95] Needless to say that Wingate asked the Sudan Government representative to safeguard Sudan Government interests "as much as possible."[96] Just a few days later, government optimism grew when Kitchener told Wingate that the Treasury in London was able to "let him have half a million," which should yield "stirring up Eckstein & Co. to produce the other half million."[97]

Kitchener was "anxious" to ensure that the Syndicate's financial contribution would not be part of the actual £3 million loan.[98] Furthermore, he wanted an agreement "with little or no detail" and many points "settled verbally." Including specific targets for the irrigated area was not his idea, but guaranteeing exclusive SPS use of that area was in order. However, he did press for details about the Syndicate's duties. Only government-approved tenants should be selected. The Syndicate should not make more than 5 percent profit on activities like selling seeds to tenants. Wingate was a little more hesitant to "interfere too closely in such questions," but agreed tenants needed protection from "undue extortion."[99]

In an attempt to counter at least partially the strong influence of Kitchener, Lovat pressed for the presence of Wingate at the negotiations on Gezira in London. Wingate's "gentle and tactful handling"[100] was needed. Again,

Wingate did not have any intention to join and asked MacGillivray to meet him in Scotland before negotiations started.[101] MacGillivray did not oppose to an agreement on broad lines with details in a separate note later, but he insisted on including two issues in the agreement: the loan and the irrigated area.[102]

For the Syndicate, a guarantee for 100,000 feddans with cotton was key to ensure their "friends" remained interested in providing the money. Words—"tiny aspects"[103]—were important. It was the government's "intention" to develop irrigation; differences between "about" or "not less than" 100,000 feddans were discussed. Government representatives realized that "some area must be specified," but rather liked phrases like "it is our intention" as it did not commit "too much." It was "a pity" the Syndicate insisted on a specific number.[104] Obviously, the representatives did understand the SPS desire to "stick to what was practically agreed upon" at the July meeting, namely 100,000 feddans and £500,000.[105] The representatives realized that Wingate and Kitchener wanted "no mention of the area," but defended their compromise as having as "little liability as possible."[106] Kitchener even considered blaming the SPS for realizing insufficient irrigated area once water was available.[107] SPS insistence on clarifying the irrigated area was rather smart.

The draft agreement stipulated that the government intended "to bring under irrigation a portion of the Gezira" as "a joint undertaking in partnership" with the SPS. The government provided the main works; the SPS was the single partner throughout the concession period. It would pay for management, supervision, collecting, storing, and marketing of the crops. Furthermore, the aim was to "direct and assist the cultivators." The SPS should consult the government "on all matters of importance affecting their joint undertaking"; a text in the margin of the official document reads, "This is to be amplified by letter."[108]

With Treasury permission for the loan,[109] there was every reason to be satisfied with Gezira progress, but MacGillivray—backed up by the Board of Directors of the SPS—kept insisting on clarifying the area to be irrigated.[110] In November 1913, Kitchener accepted the draft agreement with some further suggestions—for example on tenant recruitment (Europeans could not be tenants) and profits. He wanted every case of unsatisfactory Syndicate management to be recorded in a Secret document, to be used as "Defaulter Sheet" when negotiating renewal of the agreement.[111]

On November 21–22, 1913, MacGillivray met Kitchener several times. Kitchener told MacGillivray that all issues had to be settled with the Sudan Government, as any other decision process "would be to put the cart before the horse." Kitchener "merely expressed his views," although he did not think that his ideas were "open to serious objection." The 100,000 feddans question was still a key issue, with MacGillivray insisting on including it and Kitchener refusing to do so. MacGillivray proposed to have two Syndicate directors present at the meeting planned for January 1914 to visit the dam site. Once the Syndicate would see "the magnitude of the work" definite mention of the minimum area would become "less vital." Wingate was relieved that the Syndicate was "coming down a little,"[112] even though he considered "the divergence of views" still "fairly wide" requiring "a certain amount of give and take on both sides."[113]

Apart from insisting on defining the irrigated area, the Syndicate was satisfied with the draft agreement, as it allowed a Gezira plain shaped as an irrigated plantation under SPS management. Financial security to build the cotton dream had not been achieved yet—there was still no London-backed loan. The Syndicate stressed the importance of visiting Gezira in January 1914 before committing to the plans.

A Most Desirable Undertaking

There was clearly one person who seemed to have no doubt at all about the status of the Gezira plans. Kitchener was pushing ahead. In November 1913, Wingate was "patiently awaiting" how building Gezira infrastructure would proceed. There was nothing clear about the £500,000 from the Treasury, but Kitchener seemed "pretty easy in his mind about it."[114] In terms of power relations within the Condominium, Cairo was in the driver seat, although Khartoum had (at least partial) access to the breaks. In February 1914, Wingate was "glad" that Gilbert Falkington Clayton (his private secretary) had "secured a copy" of Kitchener's letter to the Foreign Office of January 31. In the letter, Kitchener explained his ideas on Gezira after his visit to Sudan in January 1914 with Murdoch MacDonald of the Egyptian Department of Public Works (the delegation was completed by engineers Arthur Lewis Webb and H. H. Macclure, plus Wingate, MacGillivray, and Lovat).

Kitchener was convinced that the Gezira scheme was financially sound. Cotton revenues would be huge and indirect revenue from railways and customs would be considerable too. Sennar Dam would have the potential to irrigate a much larger area than the original cotton scheme. The experts supported these claims, as irrigating a larger area was simply a case of widening the main canal and leveling and preparing land for irrigation. The experts did not doubt that Gezira was a "most desirable undertaking, and certain to prove a profitable one."[115]

Kitchener estimated that the total Gezira costs would not exceed £2,000,000—with the dam taking three-quarters of that sum. Adding £100,000 each for irrigating the Tokar and Kassala areas respectively, and £800,000 for extending the railway, Kitchener could finally show why the £3,000,000 loan for Sudan was needed.[116] Kitchener was also happy to announce that discussions with the Syndicate had "reached a satisfactory solution."[117] Within the shared spirit of success, government and SPS had agreed to open another 2,500 feddans area in the south of Gezira at Barakat, on the same profit-sharing arrangement as Tayiba, until the larger Gezira scheme would be ready[118] and water delivery from Sennar Dam would start in 1917.[119]

The new agreement with the Syndicate was inspired by experience at Zeidab. Wingate was "more than pleased" with Zeidab. Using the Zeidab way of providing water—"a most instructive one"—in Gezira, there could be "little doubt" that Sudanese farmers would not be "far behind" their Egyptian neighbors.[120] Perhaps impressed with the huge possibilities of Sennar Dam, Wingate suddenly seemed to have realized what it meant to grant the Syndicate exclusive rights to manage all the land that could be irrigated from Sennar. Did the Syndicate have a monopoly over an unlimited area at the costs of the Sudan Government?[121] Kitchener agreed with Wingate to redraft the specific clause on this point,[122] but Kitchener did still not want any specific figures on irrigated areas in the agreement.[123] A new text was sent to the SPS representatives.[124]

With future arrangements clarified for the moment, it was time to arrange the construction work itself. In January 1914, the National Debt Commissioners in London granted the Sudan Government £500,000, repayable on January 3, 1919. Constructing Gezira could finally start.[125] Such a large project was new for the Sudan Government, which had to develop its own internal regulations. This topic becomes even more interesting with official Sudan dependence upon Egypt and personal frictions within the small British community.[126] Obviously, Cairo engineers would supervise Gezira construction, but Wingate wanted Cairo to accept "whatever system the Sudan Government may devise."[127] MacDonald was key, as Sudan Irrigation Adviser and major engineer of Egypt's Public Works. He should not have too much power, for example, to arrange lower rates on Sudanese railways and steamers—both of which would go against Sudan interests.[128] Luckily, Clayton was soon able to confirm that MacDonald was "most anxious to meet the wishes of the Sudan Government in every way" as long as his responsibility for technical aspects was clear.[129] Especially when Wingate himself would take action, he would "get a good deal of irrigation."[130]

The intriguing relations between Khartoum, Cairo, and MacDonald became even more important when a committee under the guidance of Major Pearson studied the effects of a projected dam on the White Nile.[131] Its findings were "of a somewhat alarming nature." The dam would flood part of Sudan and required considerable (financial) compensation for Sudanese inhabitants, but the benefits of the dam were for Egypt only. Wingate wanted "to get MacDonald to see it through our spectacles." MacDonald was to be treated "with entire confidence" and Wingate hoped MacDonald would "not hesitate to do the same in respect" to himself. In this case of Sudanese and Egyptian interests "distinctly conflicting," Wingate wanted to avoid "the somewhat tortuous official channels of correspondence" and discuss the matter directly with MacDonald.[132] Just a few days later, Wingate asked Clayton not to discuss the Pearson committee with MacDonald, as he might go to Kitchener.[133] Apparently, involving Kitchener had to be prevented at all cost.

The Merest Flea-bite

In the meantime, in spring 1914, the Syndicate had discussed the draft Gezira agreement[134] and sent their proposed amendments to their legal advisers and the Sudan Government.[135] On July 27, 2014, Wingate received an encouraging letter from Chancellor of the Exchequer David Lloyd George to develop Sudan as a cotton-producing country.[136] A day later, Austria declared war on Serbia, and what we now call World War I started. The war pushed any business to the background, but exchanges on the Gezira project continued. In October, Wingate expressed his appreciation with Tayiba and Barakat. Tenants appeared "satisfied and content" and the cotton promised good financial returns.[137] On November 13, MacGillivray was expected to have the final SPS decision "in his pocket."[138] However, on November 24, 1914, it was made clear that reaching a Gezira agreement was impossible[139] until "the return of normal conditions to taking up negotiations again at the point where they were broken off."[140]

On January 14, 1915, Kitchener and Wingate visited Sennar; they liked what they saw, even though much work was suspended due to the war.[141] Of the £500,000 made available by the British government in 1914, some £135,000 had already been spent. After the outbreak of the war, work on the main canal was continued, as the contractor had already ordered excavating machinery and manual labor (from Egyptian convicts) was available.[142] There was, however, "reason to anticipate" that the Syndicate would not be able to start their construction work soon.

In February 1915, MacDonald published a note on Sudan irrigation development, with specific emphasis on Sennar and Gezira, for which he estimated some 4,000,000 feddans available for irrigation.[143] Tayiba results suggested that half of this area would be available for cotton. The remaining area could be used for basin cultivation of grains. This idea was still lost, as just a few months earlier, the Syndicate decided to stop growing wheat at Zeidab, sealing the fate of that crop for many years to come.[144]

There was some debate on the cotton irrigation schedules in Tayiba, when on March 13, 1915, Kennedy announced suspending of pumping after March 15.[145] MacGillivray was quick to ask for additional pumping, as he feared considerable loss of harvest and "much disturbance amongst tenants"; the Sudan Agricultural Department supported his request.[146] MacDonald instructed Kennedy to continue pumping "until further notice;"[147] MacGillivray expected irrigation to be needed until March 31.[148] In July, MacDonald found the time to visit Sennar Dam and go further up the Blue Nile to Roseires, accompanied by MacGillivray.[149]

Both the Syndicate and the Sudan Government remained optimistic about British financial support for Gezira. Lovat imagined work on the dam would start in 1915, as such "a large revenue producing operation should not be held up longer than it can possibly be avoided." He was sure Kitchener would do his best in the Cabinet.[150] Wingate was also mildly optimistic, as

Gezira expenses represented only "the merest flea-bite" compared to the war expenditures.[151] Kitchener was less optimistic, but urged the SPS to "carry on in the Gezira" and "especially look after our black partners."[152] Talks within the Condominium on Gezira and the loan continued, but without clear results.[153] In August 1915, Wingate complained that the upcoming winter months were "again wasted," but he could not do anything about it.[154] He occasionally visited Tayiba and Barakat to support these "training grounds" for the larger Gezira.[155] Wingate realized that the war limited financial options, but that directly after the end of war—whenever that would be—official loans would still be difficult. With a claim that "cotton possibilities in the Sudan are very great" and that investment in Gezira—"in what is practically a British possession"—would help solving cotton shortages in the Empire, he urged the London government for another advancement of £500,000.[156]

The changing financial conditions of the war made the Syndicate encounter even more problems with finding their share of the money. The Syndicate concluded that "the necessary capital should be raised as part of a general Government loan."[157] This obviously required a change in the draft agreement, as the Syndicate obligation of finding half a million pounds might have to be waived. Perhaps the concession period could be changed from eight into "a full twelve year's run."[158] In October, the SPS decided to offer the Sudan Government a loan of £500,000 after all, assuming that a Gezira agreement would be reached "to the mutual satisfaction of the Government and the Syndicate." The outstanding points "left for settlement" included the concession period, the cotton area, irrigation rules, taxes, and the financial input from the SPS.[159]

In December, Wingate sincerely thanked the Syndicate for "the goodwill shown by the Syndicate in this attempt to assist the Government," but refused the offer.[160] London had decided that nothing could be financed for Gezira until after the war. Negotiations in London between Khartoum, London, and the Syndicate were needed.[161] Bonham Carter, Bernard, and MacDonald would arrive in London in March 1917 to meet the Treasury, the Public Debt Commissioners, and the Foreign Office. The SPS appointed a committee including the Chairman and "any Directors who may be available in town" to "prepare the ground for the arrival" of the Sudan delegates. That year, 1917, was to become a year of harvesting.[162]

Not Letting Them Forget Our Business

One key player would miss out on the harvest, if it would ever come. Kitchener had drowned on June 5, 1916, when his ship the *HMS Hampshire* sank close to the Orkney Islands. In the next General SPS Meeting, Eckstein praised Kitchener for all his support in bringing Gezira from dream to reality. His use of words like "patience and care" seemed a little out of place, but Eckstein was certainly right that Kitchener "induced" the British government to arrange financial support.[163] Eckstein referred to the ongoing negotiations as well. Just a few weeks earlier, Hutton had complained that despite "a great deal of talk about developing the resources of the Empire," he would like to see "whether these people who talk so much really mean business."[164] The circumstances for great campaigns to induce Gezira action did not seem to be right, though. Spending money on faraway Sudan in times of war was not something to be too open about. Any Gezira spending had to be prepared behind the scenes.[165]

MacDonald, Bernard, and Bonham Carter arrived in London a few months later than planned—on June 13, 1917 to be exact.[166] MacDonald would discuss his project idea—published in Nile Control, the grand version of his 1915 note—with a Committee of Experts consisting of Garstin, Webb, and McClure.[167] MacDonald was very positive; the delegation "could not have arrived at a better time." Cotton was "most prominently in the public eye," as the war had shown that cotton for the Lancashire Mills was not guaranteed. MacDonald and colleagues were only too prepared to show that Sudan "deserved special treatment."[168] Special treatment was not given, however. Nile Control might have received positive responses, but "spending money while the world is in such a state of disorganisation" was not what the Treasury had in mind.[169]

"[N]ot letting them forget our business,"[170] the Sudan representatives mobilized all the support they could find. Detailed budget estimates were prepared with the help of the Syndicate.[171] Asking for money for after the war had ended helped, as the Treasury was "relieved" that money was not needed "immediately."[172] However, to the "great surprise"[173] of the delegates, money from Britain still did not become available. Raising capital in Egypt returned on the agenda,[174] but with Egypt as potential donor, it was imperative to show that a Gezira Scheme would not hurt Egyptian interests.

The Committee of Experts supported the concept of Nile Control. The "one continuous chain" of Nile projects in Egypt (delta) and Sudan (White and Blue Niles) were "so interwoven" and "one whole problem" that building one would "inevitably be followed by that of the others."[175] Similar to the Aswan Dam, the Sennar Dam was "of prime importance and magnitude." Sennar was capable of irrigating at least 600,000 feddans with 200,000 feddans under cotton. MacDonald even suggested that there was sufficient water to irrigate 900,000 feddans,[176] based on a higher Sennar Dam providing storage for Sudan's water requirements after January 18.[177] Obviously, this made the scheme even more expensive, but Gezira was still the only scheme "within present practical consideration." With Gezira using summer flow, Egyptian interests demanded that White Nile storage would be constructed simultaneously.[178] In September, Garstin confirmed his support, but he doubted whether Egypt would be prepared to finance Gezira.[179]

In summer 1917, as a formal response to the cotton crisis, the Empire Cotton Committee (ECC) started to study how to stimulate cotton production. With Hutton in the committee, the Sudan representatives received ample attention. The hearings on Sudan read like one big party. "Even to the unprofessional eye" Gezira was a wonder of irrigation. Labor would be easily available, as people "migrate easily."[180] As Gezira plans had already been accepted by Khartoum, Cairo, and London in 1913, and had gone through both Houses of Parliament, it was just a matter of ensuring the plan was executed. Gezira was not one of the "new schemes." MacGillivray explained the partnership, including the need for "the other two partners [to] keep the black man very happy and contended."[181] The ECC was impressed and urged that Gezira should proceed "with as little delay as possible."[182]

They Want Cotton

While mobilizing all this support, the Sudanese and Syndicate representatives did not forget to agree on Gezira themselves. In August 1917, the delegates discussed the outstanding points in the drafts. The ideas from fall 1916 still guided the discussion—nine months had not changed the basics much. Kitchener was still inspiring discussions; especially the renewal clause "had been insisted on" by him. Reasons for nonrenewal included "bad farming" and "political reasons," concepts the Syndicate delegates did consider rather vague. On changing them, however, the Governor-General and High Commissioner had the final word.[183]

Another topic important to the Syndicate was the exact moment water would become available from Sennar. MacGillivray insisted upon being told 18 months in advance that water would start flowing to Gezira. MacDonald argued that this could only be done with sufficient certainty 12 months in advance, at the end of the construction season in June. The compromise was setting 24 months for an announcement with "reasonable probability" and 12 months for the "definite and certain notice." On another aspect, MacDonald considered that irrigating cotton between July 15 and April 15 was feasible, but the government had to sanction this. The period of agreement remained undecided. One option was 12 years without renewal, another ten fixed plus four years renewal—the SPS preferred this potentially longer period.[184]

In September, the parties met again. This time the Syndicate insisted that "it was useless to discuss questions of finance" without an agreement. The Sudan delegates insisted on finding money first but were prepared to negotiate further on the agreement "subject to confirmation by the Governor-General."[185] Just two days after the meeting, the SPS offered the Sudan Government a 500,000 pounds loan—the second time.[186] And again, the government thanked but did not accept, even though the National Bank of Egypt was not prepared to lend more money to Sudan than it had already made available.[187]

A final delicate issue was raised by the government in October. The original 1914 agreement blocked the government from developing cotton on its own outside the Gezira canal area. With the "great and increasing demand for cotton in the United Kingdom" this clause was reconsidered.[188] After a "long discussion," the Syndicate agreed on the "unwisdom" of the original arrangement and accepted cotton outside Gezira as long as the Syndicate was "safeguarded against the withdrawal of their tenants or staff by any such basin scheme."[189]

In October 1917, MacDonald had regained—or managed to keep—his optimism, as finances were "likely to be forthcoming."[190] Indeed, in November Wingate received news hot from the press that "half a million for Sudan" would be found. The news had not even yet reached the Foreign Office;[191] FO confirmation came three days later.[192] The Treasury did not offer any money, but the Sudan Government was allowed to spend the remaining part of the original loan. The Chancellor was willing to arrange an additional 250,000 pounds from the National Debt Commission, but only if he "would still be in the same position after the upcoming elections." The Sudan Government happily accepted this offer.[193]

Many were very pleased, but some were "considerably astonished" about the Treasury showing "how they want cotton."[194] Hutton was excited to inform the whole world, but the Chancellor asked him to make "as little public as possible," as the decision "might be a rather awkward precedent" creating a run on scarce London money.[195] Thus, the BCGA could not express its gratitude to the Sudan delegates "in a public manner."[196] The Sudan delegates returned in Cairo on October 30, and were congratulated by Wingate with the "eminently successful termination of what must have proved very delicate negotiations." The report of the delegates only proves too well that lobbying Gezira—the project that had been decided upon, the project that logic dictated, and the only one under practical consideration—was hard work![197]

Hard work it was, but the apparent success was clear enough, or so it seemed. In May 1918, the Sudan Financial Secretary confirmed that the Sudan Government accepted the London drafts, with some small revisions "in red ink."[198] Obviously, everyone was aware that this success was again only a next step in the right direction. The budget for Gezira had increased to £2,550,000 for the major infrastructure, plus £840,000 for the SPS for smaller canals, buildings, agricultural machinery, and agricultural loans.[199] Finding this money was still a big issue in 1918.[200] Again, the Syndicate offered a loan to the Sudan Government, this time for £5 million, but Wingate wanted loans with a British Treasury Guarantee.[201]

When MacGillivray passed away in February 1919, the SPS Manager in Sudan, MacIntyre, took over the negotiations.[202] The delegates reached an agreement in October 1919. Next to the arrangements already known—like partnership, supervision, and water delivery—how to ensure that the "considerable number of officials in direct contact with the natives" by "British and others of a lower grade of education" would work out was arranged. The government was to be informed about all appointments. In case tenants complained, any official could be removed "at least to a post where he would not be in contact with the natives."[203]

The Syndicate tried to create "a compensatory advantage" in terms of preferential treatment at the end of the concession period. The Sudan delegates "absolutely refused" to accept such terms and even threatened to reduce the Syndicate's 25 percent share. The ten-year concession period would start on the day that the Gezira main infrastructure could irrigate 50,000 feddans of cotton.[204] Anticipating the start of the larger Gezira system, the government and the Syndicate agreed to develop additional pump irrigation areas, which would become part of the larger Gezira system. Hag Abdulla (6,000 feddans) started in 1920;[205] it doubled the SPS cotton-growing area and offered "valuable training ground" for tenants and inspectors. Wad el Nau Pumping Scheme (10,000 feddans) was opened in 1922.[206] The Syndicate stakeholders granted their Directors a bonus for their "gigantic work" and their "heavy responsibility." After all, they had to guard their stakeholders' interests and satisfy their two partners-in-crime.[207]

We Can't Call Him as a Witness

The agreement between the government and SPS was "an accomplished fact," but passing a multi-million pound loan through Parliament, especially one with money reserved for activities of a private company, was not trivial. Six million pounds were divided in 4,500,000 pounds for Gezira main infrastructure, 400,000 for the SPS, 400,000 for Tokar, and 700,000 for the National Bank of Egypt.[208] Convincing London and Cairo that investing in Sudanese irrigation was good did not become easier when public protests grew. Kennedy wrote a series of letters protesting against Gezira plans.[209] Although being referred to as "the evil genius of engineering matters in the Sudan," Kennedy was also regarded as not important—"men of his weight should not be too seriously considered."[210] The political situation changed, however, when Mr. Aswan, William Willcocks, joined the opposition. Willcocks would visit Sudan[211] to "surprise the world."[212] He did not suffer from a lack of self-confidence. Wingate became even more careful—or hesitant to decide—than ever. Any suggestion that the government "deliberately refused to take advantage of opinions volunteered" by this distinguished engineer had to be avoided.[213]

Wingate and the Egyptian government established the Nile Project Commission to deal with the Kennedy-Willcocks tandem.[214] The reports of the deliberations in 1920 read like a comedy—especially given the seriousness of the topic. Answering Kennedy why his suggestions were not used, Burnett (his successor as Director of Public Works in Sudan) answered that he "found it so difficult to take [Kennedy] seriously." Maxwell, the defendant for the Sudan Government, asked Willcocks about documents being genuine. On Willcocks's answer that "[t]he Lord knows, I don't," Maxwell dryly observed: "Yes but we cant call him as a witness."[215]

The Commission was clear: the accusations of Kennedy and Willcocks that the Nile projects were wrong or—worse—that numbers had been falsified were not founded on evidence.[216] Despite this unconditional support, MacDonald complained that the Commission "was not as successful as it ought to have been," as Willcocks had "managed to persuade some of the

Public out here that it was biased against him." With the High Commissioner "upset about the idea of a newspaper attack" and continuing discussions, the Nile Commission report did not end the debate.[217] Its report, however, was used by those in favor of Gezira to "hocus" and "hoodwink"—Kennedy's words—the Sudan Loan Bill through Parliament.[218]

For the SPS, all Gezira money was money well spent. In 1921, the Syndicate had developed some 60,000 feddans in Gezira, with 12,000 "fully canalized."[219] With work on the dam and main canal in full swing again, water from Sennar would perhaps become available in the 1924/25 season. Success came at increased costs, as the original estimate of £5 million was far below the £10 million actually spent, due to increased labor and material costs and some technical problems with Sennar.[220] Already in 1921 the option to ensure a positive return on this huge investment by increasing the area under cotton in Gezira was considered. However, enlarging the area to 450,000 feddans required an extra £2 million, which was impossible at that time—but not later, as discussed in chapter 2.[221]

Problems with the contractor postponed the earliest date of completion of the dam from 1924 to 1925.[222] The Sudan Government reached an agreement with a new contractor, and Eckstein's "unshaken confidence in the soundness and ultimate success of our undertaking" was rewarded in 1923, when the Sudan Government gave the desired note on expecting to finish the dam and canals in July 1925.[223] In 1924, final confirmation was given. There was "no reasonable doubt" anymore; the Gezira gravity scheme would start in July 1925.[224]

A Camel in the Morning

The first water from Sennar Dam reached 70,000 irrigated feddans in July 1925; in August of that same year, 65,000 feddans were sown with cotton.[225] All the necessary tenants to grow the cotton had been found, "which is a healthy sign."[226] Starting in 1904 with £80,000, the Syndicate was now a firm with £600,000 in capital, a Stock Exchange value of £5 million pounds, 2,500 shareholders, and 10,000 tenants. Lovat was almost sentimental, when remembering that just a few years earlier "we used to ride out on a camel in the morning and see the whole of our cultivation. To-day you go out in a motor car, and it takes you the whole of a day to inspect a single one of the 20 divisions into which the area is divided." He concluded that the "magnitude of this enterprise is really almost impossible to conceive unless you have actually visited it."[227] In the next chapter, we will visit the Gezira area to discover its routines.

Chapter 2

A Task of Some Magnitude: Gezira Management Logic

Well before the first 300,000 feddans were irrigated from Sennar in 1925, enlarging the irrigated area in Gezira became a shared desire of the Sudan Government and the Sudan Plantations Syndicate (SPS). Extending the irrigated area would allow more extensive use of Sennar Dam and its canal system—the two most expensive elements—and increase revenue. Sudan officials were convinced that additional revenue was needed to pay back the huge loans. The SPS obviously did not object to more cotton, as long as the SPS would manage it. Each extension was the result of new negotiations between the Sudan Government and SPS, with two issues coming to the front time and again: whereas the SPS would offer the best management and where to locate the extension areas?

As soon as the pumping schemes in Gezira had shown that water requirements were less than projected and Gezira could potentially be extended up to some 450,000 feddans within the original water allocation,[1] the government and Syndicate alike started developing their own ideas about extending the original plans—that were still under construction. Even though soils outside the original 300,000 feddan area were of lower quality, extensions would still provide financial revenue. The area west of the first planned irrigation area ("Old Gezira") would require a feeder canal from the main canal right through the original irrigated area. A northern extension only required a longer main canal and would profit from "more intelligent natives" compared to the western area.[2]

However, any suggestion that the Sudanese irrigated area could be larger than the highly contested original value was reason for caution within the British colonial administration. In a letter of May 22, 1924, the High Commissioner of Egypt stressed "the desirability of avoiding any public reference" to any possibility for a Gezira area above 300,000 feddans "in advance of negotiations" between London and Cairo.[3] Details of negotiations with Egypt on sharing Nile water have been discussed by others;[4] to cut a long story very short, the 1929 Nile Waters Agreement reserved the natural flow between January and July for Egypt. In the other months, water could be stored for Gezira; its Main Canal could draw 126 cubic meter per second (m^3/s). In 1936, Sudan and Egypt agreed that Gezira could start taking 168 m^3/s in 1940; widening the Main Canal for this discharge was completed in 1956.[5] With these discharges, the Gezira command area increased to some 1,000,000 feddans in the early 1950s.

In addition to enlarging Gezira, agreements with Egypt resulted in detailed operational rules for Sennar Dam and thus Gezira irrigation rhythms. As such, the larger political issue of relations with Egypt was closely related to daily routines of cotton growing in Gezira. The tenancy model is one of the distinguishing features of Gezira, but, as will become clear below, its birth was in Zeidab after a failing business model of hired labor. The idea of partnership linked closely to scheme management—who was supposed to do what at which moment—including the hierarchical relations that were either assumed or created. Agricultural procedures in Gezira were set and controlled by the Syndicate, with tenants working within a strict control system built on the cotton production cycle.

Given the theme of this book, it should not come as a surprise that scheme management did not necessarily proceed as planned. Each season anew, rain was a major disturbing factor in planning agricultural activities. The different rhythms and tasks concerning irrigation, agriculture, and (field) maintenance will be discussed, including how they changed over time. Arrangements on paper will appear as less strict in Gezira's daily reality. When an economic crisis and cotton diseases struck at the same moment in the 1930s, many agricultural operations were changed. Changes in crop rotation actually opened up new possibilities for extending the irrigated area. After the crisis, the tenants were more than ever treated as a collective of laborers under an tenancy contract—not as the independent smallholder partners under a sharecropping arrangement as claimed by both government and Syndicate.

A Good Day's Work for Lancashire

As the latest agreement with the SPS allowed irrigated cotton outside the Syndicate concession area in Gezira under government management, the Sudan Government was tempted to manage the extensions itself. Government cotton systems would be a prompt response to criticism in Sudan and the United Kingdom on the major role the SPS seemed to play in developing Sudan. This criticism was partially fed by fear that the Gezira was mainly ensuring that the SPS earned "large dividends," instead of serving "the real interests of the native cultivators and the general population."[6] Distrust between Civil Service and SPS—between civil servants and inspectors—was real in those days: 'Have you done a good day's work for Lancashire?' was a question that did not get a very polite answer in 1923 from an officer working on the Gezira to whom it was put."[7] The Sudan Government considered extending the existing irrigated Gezira area as a "sound course" to "counteract any possibility of the Syndicate acquiring too dominating a position."[8]

When explaining this strategy, Governor-General Stack emphasized "how limited" the Syndicate's "powers and rights" were in the larger plan for Gezira. First of all, the potential irrigable area from Sennar Dam was about 3 million acres, but the SPS concession was only for 300,000 acres. In addition, the concession was limited to ten years with a possible extension of four years. The SPS would go to much trouble to earn that extension for financial reasons, "the strongest possible incentive," giving the government "a strong controlling position."[9] Sudan's Financial Secretary George Ernest Schuster agreed with the need for extending the cotton area, obviously for the increased financial gains to be made, but also simply because it was possible to irrigate an area of 450,000 feddans with available Nile water. Whether the government should run the extensions or the SPS, however, was not immediately clear to him.

Schuster had no doubt that the SPS would be ready to participate in a larger area, but he assumed that in return the SPS would claim a longer concession period to earn back the larger investments. This might be politically less acceptable, due to the SPS image of "profiteering."[10] One clear and desirable advantage of government-led irrigation would be that a model different from "the Syndicate's system of spoon-feeding, drilling and closely supervising the native cultivators" could be developed. Such a model would focus on establishing local Sudanese organizations responsible for irrigation. However, although the contract with the SPS did not ban government-managed irrigation outside the concession area, that same contract specified that such efforts should not compete with the Syndicate. Schuster concluded that government-managed irrigation would have to be quite similar to "Syndicates lines,"[11] offering only very limited alternatives for Sudanese social development.

A leading role for the Sudan Government had its "obvious disadvantages" anyway. For a start, the government did not have the people to run large irrigation systems, "a task which would overstrain the Government machine." Schuster even doubted whether the government had the expertise; he "certainly would not care to entrust a scheme like this to our own Agriculture Department." With time, the government would learn how to run irrigation on its own, but it would be wiser to see how the SPS cotton area would work out. The government could learn from SPS experience, and buy time to prepare for future extensions. This strategy was further enhanced by the Treasury in London being "dead against the Government undertaking any extension on its own account." Further cooperation with the Syndicate was required.[12] Therefore, it was already clear before the Gezira system started irrigating from Sennar in 1925 that any extension would be a shared activity of SPS and government.

A complicating factor in determining extension options was the relation between water demand, water availability, and irrigated area in Gezira. The original numbers, based on MacDonald's Nile Control, were not valid anymore.[13] In the pumping schemes, water needs seemed to be only 30 percent to 50 percent of the original numbers. Schuster estimated that the new figures would allow irrigation of up to 1 million acres in Gezira, but as exact water needs in the larger Gezira area were not known yet, that remained to be discovered. Determining the outcomes from this "vast experiment" would take time. For sure, it would provide many lessons affecting the design of future extensions, but the matter was complicated.[14]

Schuster wanted the Sudan Government to appear not too eager to extend Gezira. Not "working against time" would "greatly facilitate" the government's position in negotiating with the Syndicate.[15] The fact of the matter was that the Sudan Government had signed a concession with the SPS granting it a certain water supply per feddan; as such, any change in the water allowance per feddan, whether based on new data or not, required new negotiations with the SPS. Schuster had clear ideas about the conditions he would like to offer to the SPS for extensions, but he preferred to "play a waiting game," if immediate agreement on new terms would be difficult.[16]

In February 1926, despite or because of the waiting game, basic agreement between the Sudan Government and the SPS was reached on extending the irrigated area.[17] In general, the new concession for the Syndicate was similar to the original Gezira arrangements, with government providing canals and SPS providing management—as long as only British citizens approved of by the Sudan Government became director of the SPS.[18] The distribution of gross profits could be changed if needed. The SPS was to "instruct the cultivators and supervise the cultivation." The tenants were—as always—supposed to provide the labor for the cotton and were not involved in any official negotiations.

It took until the end of 1926 to formally announce what government and Syndicate had actually agreed upon. A total extra area of 150,000 to 200,000 feddans was foreseen, depending on "the degree to which economy in the consumption of water" could be realized.[19] In light of the report by the 1925 Nile Commission just one year before, which changed the restriction on irrigation in Gezira from acreage to water use "in terms of volumes and seasons only,"[20] acceptance by the High Commissioner in Egypt and the London government was expected. However, before everything being absolutely certain, given the sensitivity of any agreement on public opinion in Egypt, the Sudan Government asked the Syndicate to refrain from announcing the agreement in public.[21] After postponing the scheduled General Meeting in 1926 a few weeks, however, SPS chairman Eckstein could explain on December 8, 1926, that despite "the sacrifices" the SPS had made, the agreement was beneficial to the Syndicate as "we have now a fairly long life in front of us, and a larger area."[22]

The sacrifices Eckstein mentioned were a lower SPS share of the profits. In the old concession, the SPS share was 25 percent; the government share was 35 percent. The first proposal of the Sudan Government for the new agreement was to change the SPS percentage to 20 percent, allowing 5 percent to the government to pay back the loans. In the first instance, the Syndicate had not been prepared to accept anything lower than 22½

Table 2.1 The profit scheme of the 1926 agreement

	Added feddans	Government	SPS
1926–1928	NA	37.5	22.5
Per 7/15/1928	45,000	38.5	21.5
Per 7/15/1929	75,000	39.5	20.5
Per 7/15/1930	30,000	40	20

percent as its share.[23] In the end, the SPS did agree with the 20 percent but managed to build in some pressure on the government. SPS profits would only be brought back to 20 percent in relation to the growth of the irrigated area (table 2.1).

In return, the Syndicate obtained a longer concession period, with the new concession ending on June 30, 1950. The agreement did not include specific options for SPS activities beyond 1950, but it did not exclude those either. Two moments were defined when the government could decide to take over Gezira management before the end of the concession period—June 30, 1939, and 1944, respectively. The Syndicate would manage any extension that became available before the end of the concession period; the government would guarantee a minimum irrigated area of 450,000 feddans.

North by Northwest

In September 1925, before the agreement between government and Syndicate, irrigation engineer Harmood Victor Carruthers Johnstone could already show a rough plan for a Gezira extension to the north.[24] SPS Manager MacIntyre approved the canalization plan for the first 45,000 feddans on November 18, 1926, with the usual reservation that he "had not yet passed it agriculturally." Until then such reservation meant that MacIntyre would propose excluding some small areas of unsuited land—even a plain as flat as Gezira included its minor depressions.[25] In December 1926, however, much to the surprise of the Governor-General—and we can only assume of the engineers as well—MacIntyre "felt doubts as to the quality of the soil over the whole Northern area."[26] He proposed going west,[27] or developing 80,000 feddans in the south.[28]

Government experts actually agreed with MacIntyre that northern soils were not as good as in the concession area, but not that these lands should be excluded from an extension. When the government checked with its legal experts the room for maneuver it had, these came back with the (balanced) opinion that the SPS could not block a northern extension of at least 45,000 feddans, but that it had to agree with the exact shape. The Governor-General did not really try to hide his anger about MacIntyre's move. The SPS had "had ample opportunity" to check the area—"in fact the senior member of their staff had traversed it many times."[29] However, he realized that caution was needed, as realizing the extensions required close cooperation with the SPS. Not extending Gezira was simply not an option.

It was not really surprising then that on March 9, 1927, the Sudan Government announced that the Gezira would be extended both to the north and to the west. The "Western compromise"[30] included 55,700 feddans in the north, a new branch canal bringing water to 80,000 feddans in the west, an additional 14,300 feddans through rearranging existing areas, plus an additional 30,000 feddans for the Kassala Cotton Company.[31] The results of the disagreement between government and Syndicate suggests that MacIntyre did what the government wanted to do: play a waiting game. The SPS Board was very pleased with the result and gave MacIntyre a bonus.[32]

However, the compromise on concession period and location of extensions did not end negotiations. Early in 1928, MacIntyre wanted to ensure that it was clear who would be responsible for what. The note covering MacIntyre's proposals counts 19 pages; his ideas were all considered, although "practically all the points had been raised at a previous discussion."[33] He went as far as claiming that the "[g]overnment sluices and pipes should end outside the toe of the canal banks, provision being made for heavier banks where required for future clearances."[34] MacIntyre would like the agreement "to retain as much as possible" the idea of partnership from the original agreement. The Financial Secretary responded that he did not think the SPS "could claim a general protection against all Sudan Government Legislation."[35] Negotiations must have been slow and complicated at times!

The SPS used lower soil quality of extensions as an argument to open "their mouths rather wide" and demand the original 25 percent share in all extensions.[36] Such a general demand was not acceptable for the government, as extensions were not similar.[37] Nevertheless, the urgency of the government in developing new extensions—for financial reasons—and its willingness to accept better terms for the SPS if needed was shown again in April 1929, when the Governor-General urged his civil servants to reach an agreement with the SPS on the exact extensions in the north and west.[38] Giving away too much would weaken the negotiation position of the government. Keeping details of plans "deliberately vague" would avoid being associated closely to any particular form of agreement with the SPS, especially concerning shares of the profit.[39]

In May 1929, reaching a balance between urgency and caution became the task for Sudan negotiator and Financial Secretary Arthur Huddleston. He was given a free hand to reach a new agreement with the SPS, with the explicit request to ensure the SPS profit percentage should be as low as possible. Huddleston should not to be "too influenced" by fear of failing negotiations either and should not accept any "undue concession."[40] Apparently, Huddleston did his best to avoid being influenced: in June, MacIntyre complained to him that an agreement had not been reached yet, despite the great work of the SPS. The cotton areas were "one large agricultural school," with Sudanese "who never did practically any work in their lives" being "induced to take up work under modern methods to better themselves and the whole Sudan."[41] In addition to the (public) view he had on the Gezira project,[42] the complaint also showed "how intensely" MacIntyre disliked the government "running the north."[43]

Predictable as ever, the two parties agreed and split the difference: the SPS share was set at 22.5 percent, plus a guarantee for a minimum share when cotton yields were low.[44] In the end, the government was as pleased as the SPS. A good working relation with the Syndicate was maintained, Gezira would be extended with areas bringing in government revenue, and on top of that the government had increased its own freedom for future action in adjoining areas. Although still not completely free in setting terms "which might attract away the Syndicate's tenants," the government had successfully claimed that it was inevitable that governmental terms would be different, but that it was the government itself to be the judge on this issue.[45] The Western compromise confirmed the relation between SPS and the Sudan Government in Gezira—happy partners, but always with their own agenda.

At the end of 1929, when Gezira was being extended to 450,000 feddans, the government set the maximum irrigated area in Gezira within the existing agreement on Nile water at 600,000 feddans.[46] Halfway through the 1930s, Gezira was actually extended to about 800,000 feddans, using the "freed" water resulting from a different crop rotation, as will be explained below (figure 2.1). In 1935, Sudan's Irrigation Director A.N.M. Robertson suggested that readily available and cheap extensions had gone.[47] Gezira's potential area of 1.7 million irrigated feddans was only "third class land" and therefore less suitable for extensions.[48] In 1938, this general strategy was confirmed: irrigating anything other than good land in the Sudan was to be avoided. However, in a conceivable future "pressure of population, coupled with a higher standard of cultivation" could make it necessary and possible to bring third class land under irrigation.[49] Table 2.2 shows that—after several extensions of Gezira after World War II—this future had (to) become reality. The areas in table 2.2 are not including the large Managil

Table 2.2 Overview of the Gezira irrigated areas

Area	Start	Maximum feddans	Comments
Southern	1925	259,973	Larger before 1931 rearrangement
Northern	1928	191,385	
Western	1929	215,854	
Central	1930	190,180	
Abdel Magid	1937	39,098	Receiving water from Sennar, outside SPS area
North Western	1951	99,407	First new area under Sudan Gezira Board
Total area	1925	985,232	Not the sum of individual areas as areas have changed

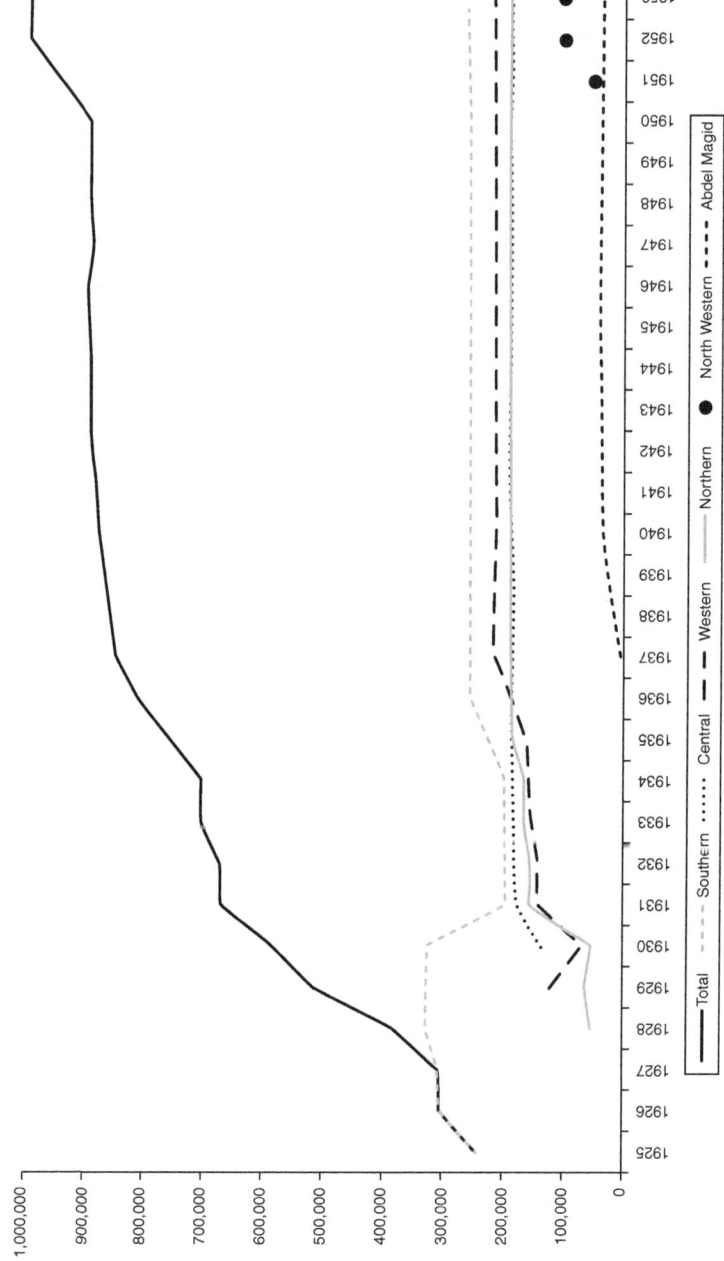

Figure 2.1 Irrigated acreages in Gezira—Please note definitions of zones change in 1931

extension after World War II, which almost doubled the irrigated area from Sennar—discussed in chapter 6. Now, we turn our attention to what was achieved within the Gezira irrigated area in the 1920s and 1930s.

The Foretelling

In 1925, the Syndicate was very happy that Sennar was completed and able to deliver water under gravity. Obviously, not all was perfect, but "every big undertaking" had some "little mishaps" in the beginning.[50] The main goal, extending the cotton growing area from 20,000 feddans to 80,000 feddans—with 100,000 feddans in the next season and 131,000 feddans under cotton in 1928[51]—had been achieved. All feddans were cultivated, as the SPS had "no difficulty in finding Tenants" for the cotton fields, a "healthy sign."[52] The successes of the first Gezira years in the first place allowed the Syndicate to trade lower profits from extensions for a longer concession time of 25 years.

In 1926, during the adjourned[53] Ordinary General Meeting in December 1926, the sky was the limit—and a painting of Eckstein in his honor a symbol of success. The Syndicate and its shareholders had "indeed every reason to be pleased." Sennar Dam and the "somewhat complicated canal system" functioned well. Crop damage from diseases and insects was very low and the first crop "exceptionally good." It was "a much better crop in fact that we can expect to have every year," a warning that became reality only a few years later.[54]

SPS Director Arthur Asquith praised Gezira management, its cropping procedures, and its ways with tenants. A major operation was ploughing the feddans to be planted with cotton; given the jump in acreage, the Syndicate—actually its Sudanese plough operators—had to plough some 1,000 feddans each day. The subsequent hoeing was done with "many hundreds of bulls."[55] As water came from far away Sennar instead of a Syndicate pumping station close to the irrigated fields, irrigation water had to be arranged beforehand through a system of so-called indenting—details of which follow in chapter 3. Cotton picked by tenants and laborers had to be transported to ginning factories along a newly constructed light railway. The factories produced 750 bales every day, which had to be transported to Sudanese ports and finally Britain. Luckily, all this work was possible because the Sudanese tenants "had absolute confidence" in "being fairly dealt with" by the Syndicate.[56]

The first tenancy arrangements of the SPS had been drafted at Zeidab. Originally, the Syndicate employed laborers to grow cotton and other crops. Soon the SPS discovered, however, that this model "could not possibly compete" with Sudanese farmers producing on land they had leased from the SPS.[57] Being a production company with hired labor was simply not profitable: "the show did not pay."[58] In 1908, the SPS asked governmental permission to remodel its policy: it would organize its own land into tenancies and supply water to local land owners as well.[59] Holding a tenancy came with an enforced rotation of cotton followed by wheat and fallow or legumes (usually lubia) on a 30 feddan holding. An agricultural manager and inspectors (three in the 1910/11 season) supervised cultivation. Tenants were "required to follow the technical advice offered." Land preparation, threshing, and ginning was done for the tenant.[60]

The government received tax revenue and the Syndicate a water rate. The SPS became landlord-manager, because tenancies gave "a more certain return."[61] The tenancy model was "less subject to risk than actual development of land." In 1910, the Syndicate had 254 tenants—23 Greeks, 80 Egyptians, 18 from Dongola, 109 local, 2 Syrians, and 1 British on 724 acres. The "not over-industrious" Sudanese farmers "worked very hard," although Greeks showed "more intelligence." The SPS Board was remunerated for these successes: in 1910 the directors could divide 600 pounds between them.[62] Government officials came "to look at the place" and the number of applications for tenancies was greater than the SPS could cope with; "fresh natives were constantly clamouring for water on the same terms."[63]

In addition to Zeidab showing to SPS and the Sudan Government that cotton of the desired quality could be grown in Sudan, the Zeidab experience also structured the relations between government, private capital, and "the local cultivator."[64] I do not argue this because the terms of tenancy in Zeidab were similar to those applied in Gezira. Zeidab tenants received water and in return handed over 50 percent of the gross yield in cotton and 50 percent of the monetary worth of each feddan cultivated in wheat, dura, and green crops. Fifty percent of a tenancy had to be planted with cotton. The Syndicate did not guarantee successful water delivery at all costs.[65] The original Zeidab model was applied in Tayiba as well, but as we have discussed in chapter 1, the division of profits was changed in 1912.[66] It was this new financial distribution that brought in the real money for Syndicate and government.

The new Tayiba profit scheme was (obviously) designed without consulting the tenants, but they were definitely giving their opinion in the Tayiba fields. MacGillivray had expected "some slight difficulty with the Tenants" once the new profit arrangements would be announced end of July 1913,[67] but he was surprised that many tenants left Tayiba—although new tenants took their place. A tenants' delegation approached the provincial government requesting to be allowed to complete cultivation on the original terms. The tenants were clearly not accepting the change of terms during the season, which had started on July 15. MacGillivray considered it "imperative to refuse" the tenants' protest as it would be too expensive, which does suggest that the new arrangement was indeed less profitable for tenants. A firm policy toward tenants was "the only line to take" to avoid that a similar protest would arise "when it might not be so easily dealt with."[68] Wingate first wanted to check whether others within government circles agreed with MacGillivray;[69] nothing should be done until the governor's opinion was clear.[70]

The governor praised the tenants for being "reasonable throughout" and "sufficiently alive to their own interests to grasp clearly" the reduction in profits under the new arrangements.[71] The governor quoted a tenant saying that he did not know "what may be done next. We may be handcuffed and marched off to prison for all we know."[72] In case finishing the year under original conditions was impossible, the tenants demanded financial compensation when abandoning their fields. The governor thought this "a perfectly reasonable request," but did not want to criticize "a scheme planned by much abler heads than mine" with financial arrangements he did not know about either. He was clear to point out, however, that the process had "somewhat seriously shaken the confidence of the native in our fair dealing."[73]

In August, after a discussion with MacGillivray in person—both were in Scotland that month—Wingate announced that "some compensation" for tenants was required "to safeguard the good word of the Government," which was everything,[74] especially to "materially assist towards the further rapid development."[75] The costs were to be divided between government and the Syndicate, "in a manner to be settled hereafter"; within the larger plans, costs would be "too insignificant to bother about" anyway. Tenants staying or taking over vacant tenancies had to work under the new arrangement. In September, MacGillivray could report a total of 71 tenants, of which 69 were new.[76] Wingate was relieved with the "satisfactory" news[77] but had emphasized earlier that this "compassionate treatment" was an exception.[78]

So, to sum up, Zeidab was the source for the Gezira tenancy model and was drafted when the original Syndicate design proved to be unprofitable.[79] As much as Zeidab paved the way for Gezira as "joint undertaking,"[80] however, having a satisfactory cooperation model was not the key condition to develop the larger Gezira cotton area. The chronology of negotiations from the first chapter suggests that that train had already left as early as 1908, three years before the new Zeidab model showed its profitability. Gezira was the goal, Tayiba simply had to show that cotton could grow in the Gezira plain. Developing Tayiba whatever happened was decided upon by the Sudan Government as early as 1911—Kennedy had already ordered the pumps.

Drowning by Numbers

In Gezira, a tenancy consisted of a number of fields (howashas) of ten feddans. The original tenancy included one field of cotton, one field with "leguminous crops," and one fallow field.[81] In the early crop arrangements, the "leguminous" field would be half in durra and half in lubia (the actual leguminous crop). Crops rotated over the three fields, with one crop or fallow each year on another field to ensure that fertility would be maintained. Fields with the same crop would be in the same Number, allowing easy crop management—for the SPS. Numbers would receive irrigation water from the smaller canals. The system of crop rotation was "deliberately bald and impersonal" to show the " fundamental change" for the people of the Gezira. Each tenant had fields in different Numbers. As crops would shift each year, so could Numbers and fields. The way a tenant managed a field was "now less a matter of his own whim and judgment,"[82] as all tenants had to do specific actions on the same days. The resulting rotation of crops, fields, and tenants resulted in a somewhat complex system. "To a newcomer the miles of canals and cotton fields (Numbers) seem very confusing."[83]

In the 1930s, for reasons we will discuss below, a four-year rotation was introduced, with two fallow years before a cotton crop. A ten-feddan howasha for cotton remained the basis for a tenancy; without cotton, one could not be a tenant. A four-year rotation meant that a normal tenancy would become 40 feddans. Eight Numbers of nine plots each formed one "Rotation Unit," with a Syndicate inspector responsible for a certain number of such Units—his maroor. Within a Unit, two Numbers would have cotton and one durra—lubia had disappeared (figure 2.2). Durra was cultivated on five feddan fields; in order to keep all durra in one Number, two groups of nine tenants shared fields. This standard setup was occasionally changed because of "odd shaped pieces of land." Some Units were spread along more than one canal.[84]

The changes in rotations in the 1930s when Gezira was well underway resulted in the absence of one standard rotation. The Gezira archive is well-stocked with reports on the different rotations in tenancies in Gezira. Once the four-year rotation had been introduced, some areas continued to use a three-year rotation—like Abdel Magid, Hosh, and Wad el Nau Blocks[85] for reasons explained in chapter 4. In the Abdel Hakam Block, both rotations were applied to test whether two fallow years were indeed needed for cotton.[86] Even in the standard areas with four-year rotations, the positions of cotton and durra were not similar in every Unit. In those areas designed

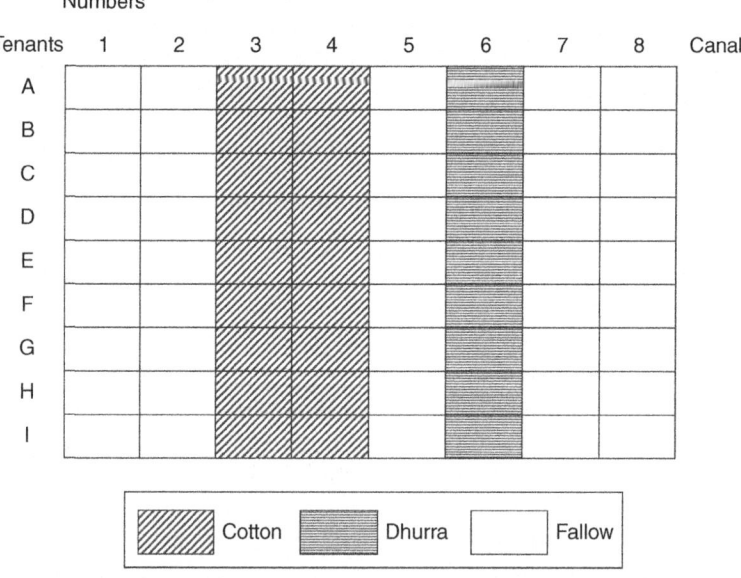

Figure 2.2 The four-year rotation in Gezira[87]

for a four-year rotation, cotton was never sown on a Number alongside last season's cotton Number, but older Units with canals feeding Numbers in multiples of three could not have such a regular cropping pattern.[88]

World War II saw the introduction of wheat—to produce food for the empire—and so-called fringe cropping—planting crops in empty areas within the scheme to maximize the cultivated surface. Discussions on crop rotations continued well into the 1950s, after the SPS concession had ended. A three-year rotation was changed to an experimental six-year one in the Seleima Block—including some new crops introduced after 1950. In the huge Managil Extension—discussed in Chapter 7—a three-year rotation was applied on tenancies half the size of the original Gezira ones.[89] All these changes resulted in different cropped areas in Gezira over the years (figure 2.3).

Opening Sennar Dam with its large cotton area increased cotton availability, which became the main export product of Sudan. In 1919, total exports had a value of £2,740,759; in 1929, total exports were worth £6,526,000 of which £4,981,732 came from cotton. However, cotton exports fell from £3,252,076 to £641,718 between 1930 and 1931. In government circles, "deep pessimism" was expressed—the "golden goose had more or less overnight become an albatross."[90] To his "deep regret," MacIntyre could not pay dividend to his shareholders in 1932, "for the first time in the last nineteen years."[91] One year later, a small profit was made, but no dividend paid yet,[92] which had to wait until 1933.[93]

One major reason was the economic crisis of the 1930s. The 1929 cotton harvests on new fields (roughly equivalent with Kassala Cotton Company fields) were a good 4.5 kantars[94] per feddan, but the crop was marketed right in the years "of the catastrophic fall in cotton prices."[95] In older Blocks, cotton harvests had gone down from between 3.3 and 4.9 kantars per feddan between 1926 and 1929, to 2.1 and 1.3 in 1930 and 1931, respectively—it only returned to 4 in 1935. Figure 2.4 shows the dramatic drop in cotton yields in the early 1930s—even though the 1931 crop was quite good. In the Abdel Magid Block, which started after the crisis (see chapter 4), cotton yields dropped as well after the first few years.

The low cotton yields of the 1930s were caused by diseases like "leaf curl" or "leaf crinkle" (which were associated with the White Fly) and Black Arm hitting hard. The abundance of rain in the early 1930s was probably one cause; especially Black Arm seemed to like wet conditions. The lubia crop was accused of being "a source for the prolific propagation of White Fly." Lubia was sown in August on fields where cotton was cut out in May; irrigation on lubia revived the old, infected cotton roots. The flies became infected and migrated or were blown into new cotton fields spreading the disease.[96] In response, lubia was almost banned. It did return some years later, but never in its former amount (figure 2.3).

The major response was introducing a four-year rotation with cotton followed by two fallow years.[97] The downside of a four-year rotation—which actually was the Zeidab rotation—was a decrease in the cotton area; luckily

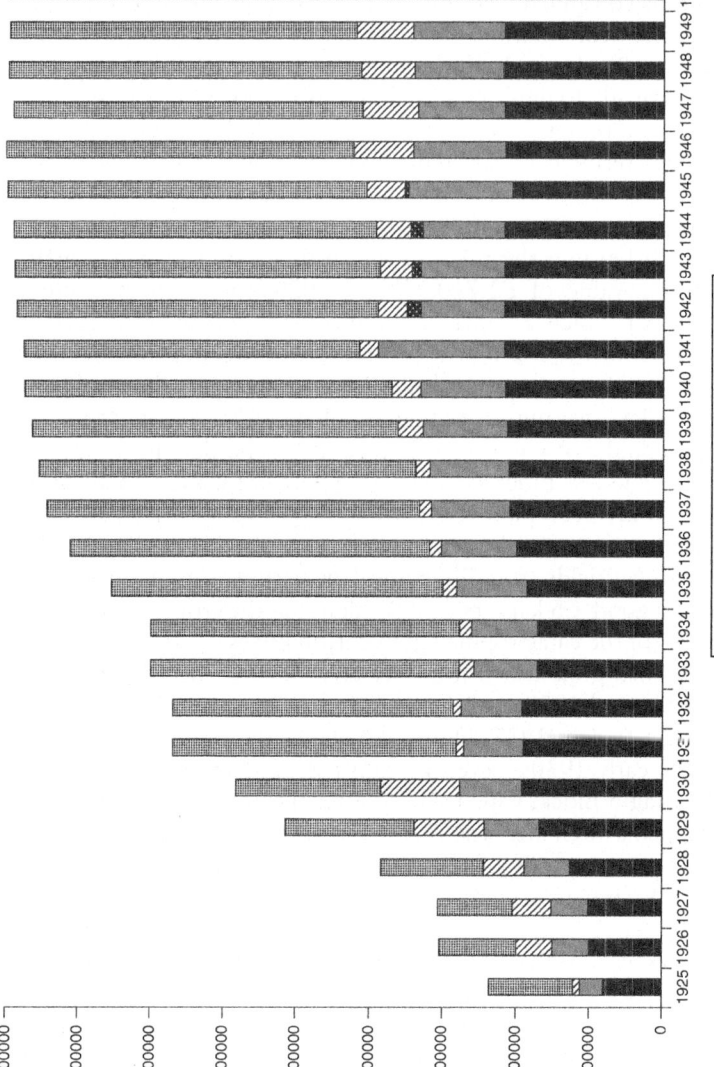

Figure 2.3 Crop areas in Gezira over time

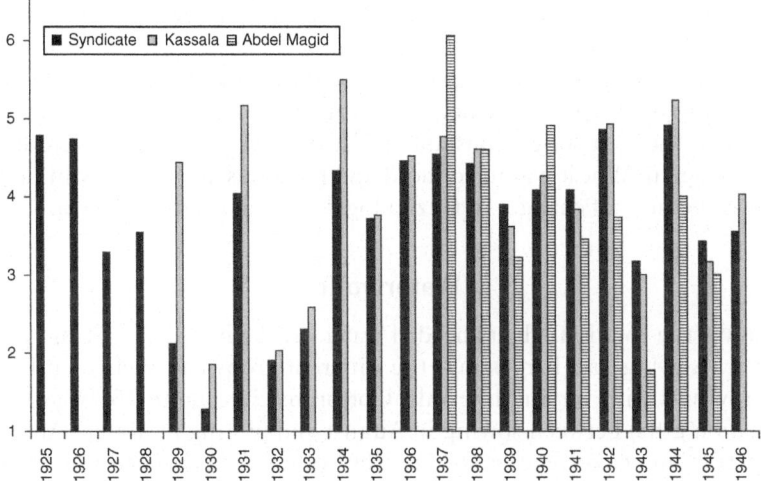

Figure 2.4 Cotton harvests in Gezira over time

the irrigated area was still being extended. In the new rotation, the cotton area would still be 156,000 feddans.[98] Crops "required by the cultivators"—food and fodder—were to be grown on lands set apart, so outside the rotation schedule.[99] This emergency cropping pattern could not be sustained, however, as it put pressure on the fallow turns. For the 1933–1934 season, a new version of the rotation was announced, which would not only ensure the necessary two fallows but also bring crops like durra back into the rotation.

A four-year rotation could reduce pests, as longer non-irrigated fallow after a cotton crop would prevent regrowth from remaining cotton roots—as long as it did not rain. Especially early rain brought up much old cotton, which had to be cut away, requiring much labor and "constant supervision."[100] Fully eliminating a pest like leaf curl, however, required removing the old cotton "by the roots." The challenge was doing so "without incurring prohibitive expense." Roots had to be removed, not just cut. Given the enormous amount of plants to be removed—each feddan included 13,000 "sturdy plants"—the task was "of some magnitude."[101]

The new rotation was sanctioned in an agreement between the government and Syndicate in February 1933. That same year, MacIntyre was confident that leaf curl could do no "appreciable harm" anymore, but the same could not be said for Black Arm.[102] Durra stalks and forage crops for livestock had to be used on the holding.[103] The new rotation should give higher average cotton yields and lower costs for ploughing and weeding. Furthermore, "larger and better food and fodder crops" should compensate tenants for the lower cotton prices. After all, keeping "tenants satisfied and solvent" was "of paramount importance for the success of the scheme."[104]

The overall business trade-off in the rotation change was going from a higher number of cotton fields in a three-year rotation toward higher cotton yields per field in a four-year one.[105] In every rotation, however, removing the old cotton was a major extra demand on tenants' labor. The new rotation changed the cultivated areas in Gezira (figure 2.3). In 1931, the cotton area dropped slightly, but the lubia area dropped dramatically. In 1933, the cotton area dropped again as the new four-year rotation came into use, but after that drop, the area stayed constant. In later years, fallow areas were being filled up again. Wheat was introduced and lubia was allowed back in, not in the least because of changes in Gezira logic—discussed in chapter 6.

Waterworld

Whatever the rotation, plants needed water on the hot, dusty Gezira plain. Between sowing and harvesting, the different crops had their own irrigation rhythms with varying intervals. Cotton received up to 15 "waterings" (irrigation gifts) between sowing in August and picking in March–April. In the period immediately after sowing, the young plants needed water about every 12 days, "particularly when the weather is hot." Toward December, with cooler weather and higher plants, 14-day intervals were used; the final interval during the picking months was 16 days. Durra was sown during the end of July or early August and would typically receive water three to five times, with an interval of some 20 days, before being harvested in October or November. Lubia was grown in less fixed periods and received on average four to eight irrigation gifts[106] with intervals of 21 to 25 days.[107] As any other irrigation system, Gezira had to match all these demands and rhythms for different areas from one source, Sennar, through one canal system.

The agreements on Nile water with Egypt gave Gezira entitlement to a certain amount of water in a clearly defined time frame and included specifications for the regulation of Sennar Dam. Each year before January 1, details had to be sent to Egypt; obviously, that was standardized "fairly well." During the flood season—July 15 to October—Sennar water levels could be raised up to 417.20 meters above mean sea level (msl), although quick fluctuations in Nile flow meant that regulation could "only be carried out by rule of thumb." In October and November, the reservoir could be filled, resulting in a target level for December and January of 420.70 msl. Fluctuations of five to ten centimeters were accepted, as "too meticulous" regulation would "defeat its object" and result in "undesirable fluctuations" in Nile flow downstream. Between February and May the reservoir would be gradually emptied. Each year, this resulted in specific amounts of water available for Gezira crops. For example, in 1933, water available from Sennar was divided into a flow of maximally 111.37 m^3/s in the flood season and a volume of 588.5 million m^3 in the storage season.[108]

The management strategy for the system of fields, canals, and Sennar Dam was to allow as few changes as possible—in other words, being as predictable as possible. Given the size of the Gezira gravity canal system, regulating structures could not "be handled like bath-room taps, and shut and opened at a moment's notice without upsetting something somewhere."[109] Gravity made water flow from Sennar through main, major, and minor canals to fields. At various locations along the way, control structures were managed by staff from the Sudan Irrigation Department—or SID, which we will meet in chapter 3. The object was "a steady uniform flow in the whole canal system, throughout the week."[110] Too frequent changes in major canals could easily result in unbalanced water supplies to minor canals.[111]

The irrigation engineers supplied water to 36 Syndicate Blocks. The aim was "to supply the correct amount of water" assuming "reasonable demands of the Syndicate or their representation." Minor canals brought water to field canals. These "Abu Ishreens" irrigated fields in Numbers. In the original arrangement, the SID was to operate canals "down to and including the Abu Ishreens heads" and Syndicate inspectors would distribute water within Abu Ishreens to fields. However, already before Sennar came into operation, "efficient working required" SPS control of minor canals. Water demands in Abu Ishreens could be different each day, which could require quick changes in minor canal settings.[112] Syndicate inspectors were assisted by ghaffirs—canal operators opening and closing pipes, regulating weirs, who were supposed to be "au fait" with their canals and irrigated areas.[113]

The inspectors decided how to distribute water to different Numbers according to crop needs. The water they distributed had to be asked for by themselves one week ahead every Tuesday. Inspectors could not ask for the maximum of all Abu Ishreens together, as minor canals were not designed to meet all peak demands simultaneously. Each Block was allotted a maximum quote. In times of high demands, any "tendency on the part of the cultivators to flood their crops unnecessarily" had to be controlled.[114] An inspector would send requests in cubic meters of water per feddan per day for his Block to the relevant subdivisional irrigation engineer. Inspectors could also indicate "Nil" or "No change." The engineer translated all requests—"indents"—into cubic meters of water per day for a subdivision and added his results to the requirements of subdivisions downstream of his own. The resulting indent was sent to the subdivisional engineer directly upstream. Finally, the main subdivisional SID engineer at Wad el Nau reported the total indent to the Resident Engineer at Sennar; the Divisional Engineer at Wad Medani was informed too.[115] The final amount of water released from Sennar included water losses encountered along the 57 kilometers of Main Canal between Sennar and the main distribution point.

Indenting typically tried to follow average water demands and irrigation rhythms, but frequent adaptations were needed. A major disturbing factor was rainfall. The rainy season corresponded with the start of the cotton season, when cotton was most vulnerable to water stress and excess. Rain required short-term action. A rain shower in a Block basically meant that water asked for should not be used on the field. Rain also made Gezira roads change into unpassable muddy swamps (as we will see in chapter 5). Consequently, an average irrigation rhythm did not exist (see figure 2.5). Rainfall could make indents asked a week before useless. On Saturdays, adjustments to Tuesday requests could be sent in for changes between minor canals. Water flows in the Major Canal system should not be changed.[116] Thus, all the water asked

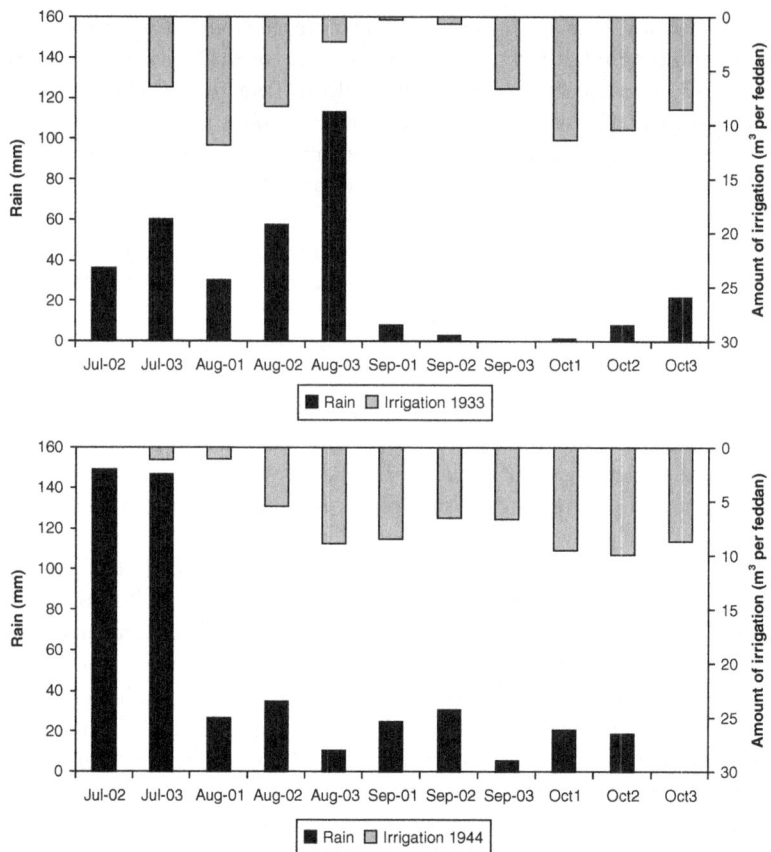

Figure 2.5 Two different irrigation rhythms (per ten days, e.g. Jul-02 is July 10–20)

for on Tuesday would still flow to the inspector's area the whole week. This water had to be "uniformly and completely absorbed on the land" until the irrigation engineers would adjust canal settings upstream.

In special cases, inspectors could indent for more or less water on days other than the official ones,[117] but there were restrictions. Indents received after 2 p.m. were included in water delivery only on the next day. After informing the subdivisional engineer of the water reductions needed, including suggestions regarding which minors could handle the extra water, an inspector had to find an immediate solution for excess water. The drains that were constructed in the Gezira in the late 1920s removed rain water from fields[118] and were not supposed to be escapes for canal water.[119] With the total amount at his disposal being the same, all an inspector and his ghaffir could do was balancing water flows between Numbers, by opening some pipes earlier than planned, others somewhat later. This might not be ideal for the crop, but still better than doing nothing. In case of acute danger of drowning a cotton plot, a fallow Number could be used as an escape.

Rain was not the only disturbing factor. Inspectors could indent too early or too late; they could also open pipes to the Abu Ishreens too late or close too early. Sometimes mistakes were made in adjusting the weirs in the minor canals. Another reason why theory and practice of indenting did "not always agree" was canals silting up or silt blocking pipes. Numbers might be "high"—in literal sense, when a pipe was higher than required for optimal flow. Higher water levels in the minor canal would push more water through such a pipe, but a high water level might increase the risk of a breach. Such a breach—even the threat of it—should be met by opening extra pipes in order to reduce the water level quickly. Damage to cotton had to be avoided, but a durra field could be used as escape. As we will discuss in chapter 3, irrigation was only done during the day, which allowed for some flexibility in timing when it was "convenient to water at night."[120]

Inspectors had to learn "which pipes are normal, which are slow and which are fast," and how to deal with them.[121] During the rainy season and the young cotton crop, such knowledge was vital. Later in the season, when rain had ceased and temperatures were lower, irrigation rhythms stabilized. Obviously, the logics of indenting and the behavior of the canal system became clearer over time, but even with experience, not all indents worked well. In 1942, two cases of "doubtful indent" occurred, with one of them resulting in tennis courts of the K-88 Club running dry. The original indent of 4,000 m^3 per week was raised to 5,000 m^3 twice a week;[122] we will discover in chapter 5 why tennis courts were important in Gezira. All actions—planned or not—"should of course be reported" and included in the Maroor handbook.[123]

Chains

Stable irrigation rhythms did not necessarily mean that the work load was lower after the busy sowing and rainy season.[124] The Gezira cropping calendar was a labor calendar as well, in terms of what action had to be performed when. By whom was not the question: the tenants were to perform most of the activities prescribed. Additional laborers were hired during the two labor-intensive periods in the cotton season—sowing and picking.[125] Before a new cropping season, the fields to be planted with cotton were prepared. Steam engines (later diesel ones) managed by Syndicate personnel ploughed the fields, after which the fields were ridged (figure 3.2). Sometimes bulls were used for ridging fields, which was cheaper but considered "inferior to engine ridging." Low lands required larger ridges as "safeguard for the young cotton plant" against rain storms. After ridging, each field was given a regular "messa" (grid) of "gedwels" (plot channels) and "tagnets" (minor field ridges) by tenants. During land preparation, the field channels had to be cleared out by the tenant.

Early rain was welcomed, as it softened the soil, making field preparation easier and allowing early weed growth before cotton was sown. Early weeds meant that "establishing a clean crop" was much easier. A season with late rain was "usually a difficult one," as "thick growth of weeds" occurred when the young cotton started to grow as well. Typically, cotton should be sown between August 10 and the end of the month, with some extension possible into September in case of rains. Sowing cotton earlier would enhance the risk of Black Arm, but late sowing diminished crop quality. Sowing dates varied per field. Numbers not adjacent to last year's cotton fields were sown first. Low-lying fields were sown later, when rain was less likely to occur. Supervision of tenants, control of water, and labor supply were all factors that lengthened the sowing period beyond the optimal duration. Stricter—in SPS terms "better"—supervision did require a sowing period moving outside the two boundaries mentioned. The sowing period compromised between disease in and quality of the cotton.[126]

Young cotton seed had to germinate, which meant securing wet soil conditions at or very closely after sowing. If done properly, prewater (dry) sowing resulted in better germination than postwater (irrigation-based) sowing. Dry sowing had the additional advantage of less "tendency to hurry on the work" to avoid the soil drying out too quickly. However, dry sowing should not be done "too far ahead of water" either. Postwater irrigation was "the sounder agriculturally," assuming sowing was done properly. With dry sowing, one person could do the work. Postwater sowing required two persons; any tenant sowing on his own had to "be stopped at once" to avoid bad germination because of the hole drying too quickly.

Usually, postwater sowing was applied on fields sown early and prewater sowing on fields sown toward the end of the sowing period. The basic method of sowing was the same for post- or prewater operations. With a sowing stick (seluka) holes were made, after which the seed was thrown in the hole. Typically, a male worker made the holes and a woman or a child did the actual sowing (figure 2.6). Many tenants contracted laborers to do

Figure 2.6 Sowing cotton in the Gezira. Reproduced by permission of Durham University Library (F.B. Hunt; SAD.692/19/3)

the sowing. Depending on the contract, laborers sometimes hurried to complete one field to be able to move to another one. Controlled sowing, with contracts per day and people working as a group, checked "the tendency to hurry on with the work." It was slightly more expensive, gave better cotton results, but was "not very popular" with tenants and laborers, making supervision "absolutely essential."

Rainfall during sowing upset the rhythm as much as it did with irrigation. One can imagine that rain disturbed the planning of dry and wet sowing, including messing up fields that were just (partially) sown. Any "spasmodic rainfall" required modification of sowing plans. A badly germinated crop or one damaged by heavy rain required resowing, but tenants did "not take much interest in re-sowing" as inspectors should have. As tenants would only "philosophically sow again a howasha which has been completely destroyed by rain," inspectors were encouraged to ensure that tenants received "a general ruling" or to "issue separate instructions" when needed.

Once the cotton was in the soil, "every effort" was needed to check weed growth. Mechanical weeding was very effective, but might be too effective as it ruined field layouts and damaged the young cotton. Manual weeding had to be "hurried along" as rain could break up the work and stimulate additional weed growth. After sowing, re-ridging was typically needed but not always possible. Thinning of upcoming cotton was also a task required from a tenant during the growing season. Timing and quality of these operations had to keep pace with the irrigation rhythm. Again, this required "continual supervision" ensuring "that the work keeps moving."

At the end of the growing season, cotton had to be picked, which required massive labor. As always, preparations had to be "made in good time" before the first cotton was ready. Cotton receiving stations had to be prepared, cotton sacks marked and issued to tenants, as well as baskets and balances for weighing the cotton. Tenants were supposed—and "urged"—to arrange pickers, both from inside and outside the scheme. Typically, picking would start on half a maroor, allowing an inspector to "see at a glance" whether picking was "proceeding according to plan." With picking cotton taking a few months and cotton still maturing not allowed to dry out, irrigation continued during the picking season. Careful planning was necessary, as the picking itself required dry soil. Cotton picking was "tedious work," and pickers had to be checked continuously for "careless work." Needless to say, all picking activities and results had to be kept in the "Cotton Book."

With picking over and irrigation ended, time came to remove all cotton remains. After the cotton crisis, old stalks and roots had to be pulled out with a "pincer-like tool (kamasha or hadeed)." All cotton plant remains were collected and burnt. This "formidable task" required "much supervision"—again. Tenants were often instructed to bring the plant material to a central collecting point to allow closer control. Even less popular was baling, the removal of the standing water from field canals. Baling as a control measure against mosquitoes was introduced as a tenant's duty once health issues became important—relatively late in the history of Gezira, as we will discover in chapter 6. After September 15, baling had to be done after each irrigation turn, together with clearing field channels and roads from weeds. Sometimes a tenant had the "misfortune" to flood a road during irrigation, which made weeding unnecessary.

Prescriptions for the durra and lubia crops were less strict. Tenants were considered "more knowledgeable" about its cultivation. Tenants preferred a crop like durra—"the staple food for man and beast in the Sudan"—above cotton. Nevertheless, the Syndicate still saw much "carelessness in cultivation and selection of seed" and many tenants growing durra "without much urging." Durra irrigation started before cotton and its cultivation overlapped with the cotton-sowing period. The second irrigation gift for durra had to be planned after all the cotton had been watered for the first time. Harvesting durra was often done under contract, although labor shortage was an issue. Lubia was principally a fodder crop. Therefore, tenants that worked with the SPS bulls were allowed to grow it, as were other "good tenants." Lubia was "delicate" and vulnerable to water shortage and excess, but the "average tenant" would take "less trouble" in cultivating it. Lubia was the crop least popular with the Syndicate.

A Bird in a Gilded Cage

A good tenant was a tenant not needing much supervision when producing the cotton. If a tenant failed to do the work required by the inspector, or did not hire enough external labor to perform a task, it was "the duty" of the inspector to arrange the work being done. The tenant would be charged for work by extra laborers—"tulba." The literal meaning of tulba was "tax." In case a tenant continued to disobey the orders of the inspector, an interim tenant—"wakeel"—could be brought in with all costs, including the wakeel's wages, charged to the original tenant. All charges to tenants, including seeds, fertilizers, cotton sacks, baskets, balances, rakes, and root pullers, were administered on the tenant's balance sheet. Throughout the season, monetary advances were made and each tenant was debited with his share of the cost of ploughing and other expenses "apportioned between the three partners."[127] Costs were balanced against the profit share for each tenant. The SPS managed all accounts; table 2.3 gives an example.

Sharing costs and benefits between the partners was defended as it provided the incentive for "careful supervision" by the government and Syndicate, who were directly interested in actual cotton results.[128] As explained earlier in the chapter, however, the profit sharing system—managed by the SPS in close cooperation with the government—might have been "the only possible method" to make a profit on the Gezira scheme. Fixed charges would "have appeared very frightening" for tenants, as they might be too high when yields were low.

Managing the accounts of tenants instead of allowing them to run their own business was a guiding principle in Gezira. The government and Syndicate arranged matters between themselves, with the third party—the

Table 2.3 Specimen of a tenant's account[129]

Date		Nos	Days	£E.m/ms	Date	Bags	1	2	3	4	5	6	Total		£E.m/ms
1925					1925										
July 27	On A/C of Profits	477		7.000											
Aug. 13	Balance of Profits	72		4.300	June 30									Appreciation on 13.85 K.	16.941
Oct. 1	On A/C of Profits	30		6.900	July 1									Blce brought forward from O.B.F. 34	11.324
Oct. 17	Tarrad	228		1.000	Nov. 30									Final Apprec	11.319
Nov. 12	On A/C of Appreciation	484		10.000	1926										
Nov. 25	Cost of Ploughing	7.90/67		9.064	Feb. 13	2	2.72						2.72		
Nov. 27	Cost of Cotton seed	7.93/67	26	4.000	Feb. 28	7	7.12	1.10					8.22		
Dec. 7	Tarrad (twice)	155	131	2.000	Mar. 13	4		4.38					4.38		
1926					Mar. 13	2	1.21	1.08					2.29		
Jan. 4	Final appreciation	57	961	11.350	Mar. 28	18		3.07	14.95				18.-		
Mar. 5	8.22 K. Cotton	48	281	4.100	Apr. 12	16			11.02	3.78			14.80		
Mar. 19	5.57 K. Cotton	542	164	3.300	Apr. 30	12			5.45	5.57			11.02		
Mar. 26	Cost of Abu / 6 dug	7.162/78	265	0.685	May 17	2				0.96	1.08		2.04		
Apr. 06	18 K. Cotton	41	587	12.600	June 1	1					0.89		0.89		
Apr. 22	14.80 K. Cotton	522	705	10.300	June 30	64	11.1	9.63	31.40	10.3	1.97		64.36		
May 07	11.02 K. Cotton	48	1044	11.000											
May 25	2.04 K. Cotton	508	780	2.000										Cotton prices 40%	104.209
June 7	89 K. Cotton, on A/c of Profits	970	1629	10.850											
June 30	Cost of Abu / 20 dug	J203/84	6573	1.913											
June 30	Interest	212/85		1.095											
	Balance transferred			30.336											
				143.793											143.793

tenants—in the background. That is not to say that tenants were not seen as important. Far from that: Gezira arrangements had to be "fair and really acceptable to the cultivating tenants." This was not just an issue of "moral principles," but also simply good business, as "developing the concession area would be impossible" otherwise.[130] In the debates on tenants, we encounter ideas of arrangements being good for tenants but also of tenants in constant need of being reminded of what was good for them.

When asked about the Gezira partnership and percentages in 1917, MacGillivray explained to the Cotton Commission that Gezira would not work unless government and Syndicate would "keep the black man very happy and contended." He was also very clear that a tenant was "really" the laborer, working "under the supervision and instruction, if I may call it so, of the management."[131] As late as the 1950s, when the cotton drilling times of the SPS were supposed to be over, inspectors were informed that although a tenant was not a laborer, close supervision "by the field staff" was "still very necessary." Advice like "Patience pays. Don't lose your temper. Don't swear at Tenants or labourers, never strike anyone" was given.[132] In itself these hints might not have been superfluous, as there is evidence that SPS inspectors set fire to durra or tenant's houses for neglecting cotton.[133]

The Tenancy Contract stipulated that tenants had to cultivate their fields "in a proper manner" and "to the satisfaction of the Syndicate." Tenants had to "obey the reasonable orders" of SPS inspectors, on the risk of missing the share of the cotton proceeds. A tenant could grow durra "sufficient for his own requirements," but not sell the grain nor make it more important than the cotton. In cases of "crime or dishonesty towards the Government or Syndicate," a tenant could be evicted "without any compensation." New tenants could then enter, although that was subject to the debts from that tenancy to the Syndicate.[134] In the original Gezira contract, a tenant could be evicted in Clause 12. In 1941, Clause 13 was the one on eviction.[135] The reason for this change had to do with debts and evictions as well, only in a slightly different way.

When the economic crisis hit Gezira in the 1930s, many tenants decided that they did not want to be a tenant anymore and left their tenancies. Many of those remaining could not pay back the loans the SPS had provided them with. Suddenly SPS and government were confronted with a huge group of unprofitable tenancies. As an immediate measure during the depression, the government and SPS lent £60,000 to the tenants—as a group.[136] Over the years, a total of £546,000 was transferred to the tenants' shared account to cover the so-called "Bad Debts."[137] Obviously, this was not something that could be kept up for long. Furthermore, repayment of the debts had to be ensured and something had to be done to avoid them from occurring again.

In February 1931, MacIntyre proposed to increase the tenants' share temporarily and secretly, as it "might easily be misunderstood" by tenants.[138] The government did not agree with such an secret accounting approach;[139] any measure required should be told to tenants. A temporary measure was not what the the government had in mind anyway. In April 1933, it proposed to treat tenants "as a corporate body and not as individuals."[140] Creating a special collective fund would allow SPS and government to cover debts of individual tenants.[141] As the SPS did not want to change the Tenancy Agreement too much, as "it might upset the tenants,"[142] the next few years saw little progress. The government even threatened to terminate the concession, even though the disagreement was more on details than on principles. The Gezira Scheme was simply too vital to both Sudan and SPS to spoil the profits.

In 1936, the Supplemental Agreement "to provide a fund for the assistance of Tenants in bad years" was almost ready.[143] In 1937, the Tenants Equalization Fund was formally created.[144] The fund cancelled individual liability for the bad debts, which removed "a millstone" from all tenants and even rewarded some tenants "for having 'stuck to it'."[145] The principle was simple enough: in years with cotton profits, a lump sum was transferred from the total tenants share "at the joint discretion of the Government and the Syndicate" to the fund. In years without profit, sums were taken from the fund to cover the tenants' 40 percent share. Both the government and Syndicate claimed that criticism that the fund was "advantageous to the Government and the Syndicate at the expense of the tenants" was "quite unjustified." Obviously, the SPS and the government did profit. They could "appropriate" their own share without problems for the tenants, whose main benefit was stability and predictability. The fund would prevent the "disastrous conditions" after the crisis. Perhaps "exceptionally large profits" were not possible anymore, but suffering "such hardships" would not happen again either. Less tenants would "give up in disgust."[146]

Treating Gezira tenants as a collective may have been brought to new heights with the Tenants Equalization Fund, but it was certainly not a new concept. Already from the start of Gezira—and Tayiba—tenants were treated as a collective for certain purposes. Ploughing was charged as a share in a collective expense. The problem of securing profits of tenants—including which of the other partners had to pay—was not a new topic either, as in 1926 the issue was already raised by the Syndicate. Who would have to pay when cost of living would rise, or low cotton yields and/or prices would occur? How to deal with tenants being discontent, especially because of "political or religious propagandists"? What to do if the government developed a "desire to promote political well-being"?[147]

In 1926, Eckstein had known the answer. Low-profit margins for tenants meant low-profit margins for the SPS, so the SPS "would not be in a position" to offer any compensation to tenants. On political matters, the government's and the Syndicate's ideas "might well be different." Political disagreement meant that the government would have to pay compensation.[148] Despite these first ideas, when the real crisis came, negotiating the positions of the government and SPS took many years. The years after 1935, ironically not in the least because of higher durra yields, had brought enough financial resources to reach the agreement in 1937. In 1938, however, the Syndicate was only too happy with the new arrangement, which still allowed "as much reward as possible to individual enterprise subject to certain co-operative features which are regarded as being essential to its success."[149]

New agreements between the government and Syndicate on collective funds were drafted in the 1940s. All in all, a series of funds with different names and specifications were created, but all were based on government and Syndicate agreeing that this was the way to go for the benefit of the Gezira Scheme as a whole.[150] In discussions on Gezira after World War II, both in and outside Sudan, the benefits of collective arrangements were brought forward—not in the least by Gaitskell himself—but those enthusiastic about the collectivistic regulations somehow continuously failed to mention that tenants were never asked about them.

In the summer of 1946, a six-week strike by tenants marked the end of a period of relative stability. The strike was a response to a rumour that one of the collective funds (the Reserve Fund) was going to be used for paying governmental policies. The government informed the tenants' representatives that money was actually taken from the Tenants' Welfare Fund, a fund that only covered the 1930–1931 debts.[151] A Gezira Special Committee was established to advise on the matter. After investigating, the committee recommended distributing £400,000 of the Reserve Fund to the tenants before the end of the year. Cotton sowing resumed. The high profits of the late 1940s would have at least helped to avoid new strikes.

Money

The collective arrangements of the late 1930s were a direct response to the problems Gezira suddenly encountered in making a profit. The Tenants Equalization Fund secured the government and SPS against individual tenant debts, as these were collectively insured against tenants' future profits. Without discussing this aspect too much in detail,[152] it is clear that money was made indeed. For the Sudan Government, Gezira was the economic cork keeping the country afloat. Gezira profits allowed Sudan to become more independent from Egypt. When the post–World War II larger-scale development plans we will discover in chapter 6 were contemplated, Gezira was to produce the money—the other plans were made possible by Gezira cotton revenue anyway. Gezira was "the backbone of the colonial economy."[153] The second partner, the Syndicate, made no profits in 1930, 1931, and 1932. The years 1931 and 1932 missed out on dividends, but in 1933 profits recovered somewhat, and dividends were paid, though still less than in the 1920s. The 1930s were still somewhat "shaky and uncertain,"[154] but at the end of the concession period, in the cotton price boom after World War II, the profits skyrocketed, as table 2.4 shows.

Whether Gezira was such a success for the tenants is less clear. The diversity within the tenant community was considerable, but the debate of tenants' success is also pretty ideological—as shown in chapter 4. However, given the many tenants and laborers coming to the Gezira area, one would assume that money was made.[155] Obviously, the crisis years were not the

Table 2.4 Overview of SPS profits in Gezira[156]

Year	Cotton area feddans	Yield kantars per feddan	Price pence per lb	Net return pounds	Direct expenditure pounds	Profits pounds	SPS Share pounds	SPS expenses Pounds	Dividend pounds
1926	80,031	4.80	18.00	2,340,616	718,925	536201	582,281	215,763	375,000
1927	100,058	4.70	18.00	3,356,629	721,412	840,504	759,319	278,888	450,000
1928	100,768	3.30	19.70	256,3402	814,009	584466	573,954	223,557	562,500
1929	131,292	3.60	18.40	3,269,162	970,504	725080	699,630	237,226	562,500
1930	174,164	2.30	7.90	885,905	1,027,245	0	192,702	188,060	225,000
1931	196,799	1.40	6.40	393,940	1,023,103	0	85,552	226,237	0
1932	194,935	4.10	8.50	2,270,988	992,539	243,366	495,807	190,348	0
1933	195,941	1.90	8.10	875,347	896,219	0	193,299	214,288	90,000
1934	175,834	2.30	8.60	1,025,324	892,907	84,425	226,841	220,407	135,000
1935	176,150	4.50	8.20	2,187,920	952,483	27,1914	470,425	250,355	198,000
1936	185,758	3.70	7.90	2,077,858	930,526	272764	451,386	280,524	247,000
1937	199,770	4.50	8.60	2,908,401	956,217	500,101	618,832	311,647	309,375
1938	207,242	4.60	5.90	2,091,913	1,027,844	239,044	451,104	290,667	247,500
1939	206,274	4.50	6.20	2,252,945	1,032,313	228,486	485,285	274,878	198,000
1940	206,880	3.80	9.60	2,722,407	714,654	358,585	581,958	240,528	198,000
1941	207,594	4.00	8.90	2,952,244	696,740	440,388	621,993	217,809	198,000
1942	207,121	4.00	9.10	2,922,591	698,365	488707	616,261	214,870	247,500
1943	206,486	4.80	9.30	3,697,480	716,792	693,224	776,878	221,545	297,000
1944	206,571	3.10	10.60	2,614,936	737,323	578,700	533,584	237,131	495,000
1945	206,578	4.90	10.60	4,280,156	752,519	1111991	837,611	260,047	544,500
1946	196,541	3.40	10.30	2,605,760	773,543	963,682	516,938	270,189	618,750
1947	206,176	4.00	19.20	6,789,924	855,843	198,4013	1,372,042	301,716	618,750
1948	206,346	3.40	38.50	11,753,038	918,832	421,3545	2,339,120	336,423	618,750
1949	206,778	4.30	38.50	13,819,832	966,628	4,576,516	2,741,206	401,243	618,750
1950	206,737	4.60	41.30	1,611,8155	1,134,693	58,20630	3,189,374	541,108	618,750

time for profits; many tenants resigned and those who stayed complained about the "alien agencies"—the Syndicate and the government. In the 1934 season, Gezira was 2,000 tenants short. In 1937, when the Gezira Scheme was back to normal and extensions ready, demand for tenancies exceeded supply.[157]

I am not sure, however, that success is an interesting concept anyway; it is certainly not an issue one could easily agree on. Success and failure depend on perspective. Many were deeply happy and many were deeply unhappy about Gezira. The following three chapters discuss the (un)happiness of three groups: irrigation engineers, tenants, and Syndicate inspectors. These chapters show how success was negotiated, how some happiness was traded for other happiness, and how perceptions of success resulted in new negotiations. Agents striving for their happiness made clearly defined plans appear less agreed and less clear in practice—as we have already seen a few times. Success proves to be negotiated, man-made, and not stable.

Chapter 3

No Man Can Serve Two Masters: Designing Gezira Irrigation

When he could find a moment in his hectic 1946 working schedule, the director of Irrigation and Irrigation Advisor to the Sudan Government, R. J. Smith, looked back "with nostalgia to the carefree days of pre-war extensions to the Gezira, with the confident planning and certain execution of works."[1] He was in a good position to compare the postwar years with those before World War II, as Smith had started his career in Sudan in 1925, just before the start of Gezira irrigation from Sennar. With the exception of two years in the 1930s, when he was in the Gash Delta in the east of Sudan, Smith made his career through Gezira. As much as the Sudan Irrigation Department (SID) itself, his career had been shaped by and for Gezira. Between 1945 and 1953, he was Assistant Director for Development (1945–1946) Director of Irrigation between (1946–1953). As such, Smith was a major player in Gezira irrigation development after World War II—table 3.1 shows his career.

As Director of Irrigation in Sudan—and as such head of the SID—Smith was the head of more than 80 people in 9 divisions plus headquarters, all employed to manage, design, construct, and think about irrigation systems in the Sudan. Compared to 1925, when Smith had entered the service of the Sudan Government, the department had grown enormously, as had his own importance within the SID. The 80 people—not all of them engineers, some positions were not filled either—did not include the numerous lower Sudanese field staff (ghaffirs) that made the many changes to the gates and arranged daily maintenance and repairs. To give an idea about their numbers, Tabat subdivision alone had close to 60 ghaffirs employed in 1933.

This chapter focuses on irrigation infrastructure and the engineers as the main group responsible for it. To enable large-scale cotton production, a largescale irrigation system was constructed by the SID. I will go into the procedures and issues of designing, constructing, and managing Gezira irrigation infrastructure before and after World War II—in which Smith played such a decisive role. At the same time, the chapter is one huge example of planning as continuous improvisation resulting from continuous negotiations.

Table 3.1 The career of R. J. Smith, 1925–1953, Sudan Irrigation Department

Year	Position	Location
1925	Appointed as engineer—officially not yet in SID	–
1927	Assistant Sub-Divisional Engineer	Wad Medani
1928		Abu Usher, Wad Medani Division
1929		Tabat, Wad Medani Division
1930	Sub-Divisional Engineer	Merebeia, Wad Medani Division
1931		Tabat, Abu Usher Division
1933		Sennar, Sennar Dam Division
1935		Aroma, Gash Board
1936	Chief Engineer	
1941	Divisional Engineer	Wad Medani, Projects Division
1945	Assistant Director	Development, Wad Medani
1946	Director of Irrigation	Wad Medani
1951	Irrigation Adviser and Director of Irrigation	Khartoum
1953	Retirement	

We will discover why one of the first tasks for Smith in 1925 was changing the irrigation infrastructure in Gezira—not on paper, but in the field where canals had already been constructed.

The change was the result of huge differences of opinion between the Syndicate and Sudan irrigation engineers on what a suitable irrigation system for Gezira exactly entailed. The debate centered on whether Gezira should apply continuous irrigation or only daytime irrigation. The SID engineers had their own ideas about a proper irrigation system, which was to be based on continuous flows throughout the system, "similar to the system of perennial irrigation in Egypt."[2] The emerging problem was that the engineers appeared to prefer something different than the Syndicate had in mind. The SPS wanted to copy the rhythm of the smaller pumping systems, with irrigation during the day and no watering during the night, to the large gravity system in Gezira. Although the engineers did not agree with the idea, they did modify the canal system to the desires of the SPS. The differences of opinion between SID and SPS on irrigation matters reemerged continuously in Gezira in discussions on extensions. The resulting system of night storage is indeed a key feature of Gezira; it is the material manifestation of the differences of opinion between irrigation engineers and Syndicate management.

This chapter will show—against popular images of engineering endeavors being totally planned and then executed as planned—how creating the Gezira irrigation system amounted to not only improvised planning but also planned improvisation. Smith referred to the prewar extensions as "carefree," but that does not mean they were developed just like that—as I have already shown in chapter 2. Furthermore, the extensions of the late 1920s increased the canal system considerably, whereas the changes in crop rotation in the 1930s changed water demands and as such flows in the canals. The night-storage system now seen as so typical for Gezira was created only when the first canals were already dug and the main distribution works in place.

Concentrating Forces

Keeping the irrigation system working, with all its changes, sediments, and weeds, was the task of the Sudan Irrigation Department. For SID, Gezira was not only a major task but also its reason for existence: SID was formerly established in 1926 because of Gezira. Although the engineers did not always get their way in the negotiations with the Syndicate, for them Gezira was symbol for and proof of SID existence and importance. In 1904, after Garstin had visited Sudan, the Egyptian government established a Sudan branch of the Irrigation Department of the Public Works Ministry. Dupuis was the first Inspector General of Irrigation in the Sudan. When he became Under-Secretary of State for Public Works in 1909, Tottenham succeeded him as Inspector General of Irrigation in the Sudan.[3] Therefore, the first engineers supporting the Sudan Government in their irrigation plans and projects were British irrigation engineers in the service of the Egyptian Ministry of Public Works, delegated from the Egyptian Ministry to work for the Sudan Government.

The staff list of the Sudan Government from 1914 shows under "Egyptian Department in the Sudan" ten names of engineers working on the Gezira canalization, with another four working in the White Nile area and three more in the Dongola region. In those early years, especially because of his close involvement with Gezira, engineer Murdoch MacDonald, whom we already encountered in chapter 1, was the main person linking Egypt and the Sudan. His position—and some of the discourse it created—made clear to many, however, that "[n]o man can serve two masters, least of all Egypt and the Sudan in matters affecting the Nile."[4] Whereas personal relations between Wingate and MacDonald seem to have been very good, in general, Wingate did not appreciate the status of the irrigation engineers in Sudan. He complained that he had "little or no control" over them.[5] In 1922, the Egyptian engineering staff in Sudan was given a new "but not very clearly defined" status, but things stayed much the same as before. Egypt placed its staff at the disposal of the Sudan Government, but more as a loan between friends than by any formal arrangement.[6]

In May 1925, just weeks before Sennar Dam would deliver its first water to Gezira, W. D. Roberts—Inspector General of Irrigation in the Sudan and supervisor of the Gezira works—asked the Egyptian Ministry of Public Works to send four experienced engineers to manage Gezira after July 1 that same year. On July 20, three of them were in Gezira. In August, F. Newhouse—then the Acting Inspector General —enquired where the fourth one was. After receiving the news that this appointment was still pending, he declared that after October 20 the Egyptian engineers were not needed anymore. On October 1, a telegram from the High Commissioner's Office in Egypt to the Sudan Government asked to keep the Egyptian engineers in the Sudan a little longer, or even better, include them in the new Sudan service.[7]

A few years earlier, in 1923, Dupuis had recommended that Sudan should set up an Irrigation Department of its own, as "details of operation" in relation to the Gezira concession of the Syndicate had to be settled clearly and preferably in advance of the start of the big gravity scheme. This became particularly acute once, as we will discuss in much more detail below, "difficulties had arisen over the control of the minor distributaries and in the matter of watering by night."[8] In May 1923, Governor-General Stack confirmed in a letter to High Commissioner Edmund Allenby that Sudan preferred its own Department of Irrigation. Obviously, Egypt would be allowed to keep a say in matters concerning the Nile.[9] Anticipating a Sudan department, the Sudan Government made engineer R. M. MacGregor from the Punjab Irrigation Service the Irrigation Adviser of the Sudan Government in 1924.

In March 1924, that is after the appointment of MacGregor, Financial Secretary Schuster made no attempt to hide his disappointment about the appointment, which he considered "inopportune" and "avoidable." Egypt would consider the appointment as "an attempt to split from the Egyptian Government Service"—most likely a correct interpretation—even though Egypt had placed its staff "unreservedly at the disposal of the Sudan Government." Obviously, Schuster did realize "the extreme importance" of a competent Irrigation Service in Sudan, but he would like to see "some title of less sinister import" than "Adviser" for MacGregor.[10] Perhaps keeping the appointment a secret was a good strategy.[11]

The Sudan Irrigation Department was formally established on January 1, 1926. MacGregor, who had remained Irrigation Adviser despite the discussions, became the first head. Harmood Victor Carruthers Johnstone, in charge of design and construction of Gezira became the administrative head of the department. A deputy director for future extensions—in the person of Major Croker—was immediately appointed as well. Just before the formal establishment of the SID, the Egyptian department in Sudan included 46 engineers, with 8 at Headquarters, 11 in the White Nile area, 12 in the Blue Nile Division, and 15 in the Gezira Division—Smith was one of them. The 1927 SID Staff List showed a lower number, but could still show a sizeable labor force with its five engineers in the Sennar Division, seven engineers plus two clerks at the Wad Medani Division—Smith was one of them—and 11 engineers plus a surveyor in the Projects Division.[12]

More importantly than the slight decrease in number of employees, SID had become a department of its own, "uninfluenced by the views of Egypt."[13] The working methods and lines of responsibilities within the SID were put down in the 1927 Manual of Orders of the Sudan Government Irrigation Department.[14] In 1934, the position of administrative head was changed into the formal head of the SID. Furthermore, the position of Irrigation Adviser was made part time with a focus "on larger irrigation questions, particularly involving Egyptian or other outside authorities."[15] In 1937, the position of Irrigation Adviser became full time once again, but its responsibilities did not change.

In the 1934 Manual of Orders, we find that the SID budget was divided into two posts: Gezira Irrigation System and Minor Works. This may sound like good news for irrigation works other than Gezira (the Minor Works), but up to 1934, attention for these minor works had been "on a very small scale" only.[16] A 1927 note had already explained that the SID was pushed to such limits that any other area besides Gezira had to wait. Despite the "utmost sympathy for local interests elsewhere," the SID could not "do otherwise than concentrate its forces in the Gezira."[17] The birth of Gezira was its own birth. "The Department developed with and for Gezira."[18]

SID engineers had to deal with many issues simultaneously in the early days in Gezira. They had to bring the new scheme into operation, but also had to start remodeling some 1,000 kilometers of canals to deal with the night irrigation issue, "a factor unforeseen in the original design." If that was not enough, the SID started extending the area "without any breathing space" by itself instead of using external contractors to do the construction work.

Negotiating Night Storage

At the start of his long career within the SID, in 1925, one of the first tasks engineer Smith must have been assigned was changing a large part of the canal infrastructure in Gezira. The original design of the engineers anticipated on the Gezira canals flowing continuously up to field level. All canals above field level would carry water for 24 hours per day. This meant that fields would have to be irrigated day and night as well. The engineers had planned this continuous flow following their well-known practices from India and Egypt. The 1919 Agreement between the government and the Syndicate had even stipulated that the Gezira area would be irrigated "similar to the system of perennial irrigation in Egypt."[19]

For the engineers, this system was logical: a gravity irrigation system should mimic natural flows as much as possible, to avoid issues like erosion and sedimentation. Furthermore, in an area as large as Gezira, cutting off the water during the night at Sennar was not an option, as filling the larger canals every day was impossible. Only during one month, May, canals should be empty for maintenance. In addition, cutting water every night in smaller canals would result in huge unproductive drainage flows. It was clear: water had to flow 24 hours each day. As their Indian and Egyptian colleagues, Gezira tenants were expected to irrigate their field whenever the water came, even at night. Obviously, some kind of schedule would inform them when it was their turn.

Along these lines, construction of the Gezira canal system started in 1910, with construction of Sennar and parts of the Main Canal. In 1922, a new contractor was granted the work, but the basic design did not change. Parts of the infrastructure were already in use for the pumping schemes built in Gezira to train new SPS inspectors—and make a profit—but these canal systems were quite small compared to the full 300,000 feddan area. With construction and first use already well under way, it is not hard to imagine that the irrigation engineers were rather unpleasantly surprised when the SPS refused to accept continuous irrigation. When the SPS exactly objected formally to the system of continuous irrigation for the first time is unclear, but years later, when the Sudan Government and SPS were negotiating about possible measures in the crisis years of the early 1930s—discussed in chapter 2—Huddleston could not resist complaining about the timing of the SPS.

"[W]hen the works for the irrigation of 300,000 feddans were well advanced towards completion, it transpired that the watering of the land could not be carried out at night, whilst it was obviously impossible to close the canal every night, as was the custom with the pumping schemes then operating in the area. In the Government's view this situation had arisen on the agricultural side of the undertaking, i.e. in the sphere assigned to the Syndicate."[20]

At the time he wrote this, Huddleston used this argument to point to additional budgets the government had to grant to change the canal system in the 1920s, which would be reason to ask for fewer financial efforts from the government for crisis solutions discussed in chapter 2. In his answer just two days later, MacIntyre stated that the reference to the early 1920s was "irrelevant." He claimed that the SPS was not to blame for the changes at late notice, as it had informed the SID about the need to avoid night irrigation "from the commencement, again and again."[21] The Syndicate had imagined Gezira as a larger version of the pumping schemes, with irrigation during the day and pumps switched off at night. In the small pumping systems, filling canals in the morning was not an issue. In the large Gezira system, however, it was.

Roberts and MacGregor made their position clear in January 1924—just 18 months before Sennar should start delivering water.[22] They acknowledged possible reasons for the SPS to maintain the pumping rhythm in the large Gezira system, ranging from unwillingness of tenants, possible discomfort at night, physical difficulty to control water on fields in the dark, to health hazards for tenants. They even understood the SPS concern that the Gezira cotton-growing model required "constant supervision of the field waterings," which were "impossible to give at night."[23] Supervising tenants only during the day would require less British staff too, as double shifts were avoided.

Acknowledging reasons, however, did not mean that they accepted them. Roberts and MacGregor were "firmly convinced" that the Gezira should be based on a system of "watering to be carried on at night." Reconstructing the ideal irrigation system would create problems with silt from the Blue Nile. A major advantage of continuous flow was that most of that silt would remain in suspension in canals and only settle on fields. With changing flows and eventually night storage in smaller canals—with almost stagnant water for many hours—silt would settle in the canals. The main point for the engineers was, however, that they simply did not see why Sudanese farmers could not do what their colleagues over the world apparently could: irrigate at night. They went as far as claiming that they did not know any system "of comparable size where consumption is closed down at night and the entire population retires to rest."[24] They could not believe Gezira tenants, who "had already accepted immense changes in daylight farming,"[25] would be unable to irrigate during the night.

At first, Roberts and MacGregor did not consider it their problem to solve either. The irrigation engineers had delivered irrigation infrastructure that was "quite workable." The problems now created were problems "on the side of the actual field irrigation," which was the responsibility of the Syndicate.[26] Obviously, the presence of a managing agency like the SPS, with a production strategy based on close supervision, was the main reason why night irrigation was avoided in Gezira, and the engineers knew that. Keeping good relations with the SPS was key as it had been before. Therefore, "the only positive remedy lay in the construction or remodeling of works forming part of the Government's side of the undertaking."[27]

In their 1924 note, Roberts and MacGregor stated somewhat less polite that the problem of night irrigation was "so strongly urged upon us"[28] that they could not refuse to reconstruct their original design. Accepting the need to remodel their ideal system did not silence the engineers on the issue, though. Every time the issue could be brought up—for example, when extensions to Gezira were discussed—the engineers did bring it up. Whenever they could, the irrigation engineers claimed that night storage was "a radical departure from recognised practice."[29]

Realizing a Radical Departure

How to change the original design from continuous flow to night storage was not clear immediately. The 1924 note elaborates a little on possible escapes of night water from the main system to areas north of Gezira, but ideas did not go much further than that at the time. The Sudan Government invited an outsider to give his opinion on reconstructing Gezira irrigation and thinking through the entire idea of controlling water flows. On June 4, 1924, Arthur Douglas Deane Butcher—Director of the Delta Barrage in Egypt—presented his ideas on how to avoid night irrigation and the consequences this would have on water control in the larger canal system.

Butcher's answer to the night irrigation problem was a system of water control and distribution "far more rigid" compared to Egypt with its "less exacting conditions." Gezira had special features "both agricultural and engineering."[30] In engineering terms, blocking flows anywhere in the system during the night would mean that canals in certain parts should be enlarged, as these should be able to pass the extra water. Which Gezira canal could carry the night water that would not go to fields? The larger canal branches were designed for a maximum flow of 20 m^3 per feddan, but

Figure 3.1 Weir on a minor irrigation canal in the Gezira. Reproduced by permission of Durham University Library (F.B. Hunt; SAD.692/19/8)

Butcher considered enlarging these canals to the required 28 m³ per feddan too expensive. He turned his attention to the minor canals (figure 3.1).

In Gezira, minor canals brought water to the Abu Ishreens, which on their turn brought water to the fields—as explained in chapter 2. Butcher proposed that the pipes to Abu Ishreens could be closed during the night and that the minor canals could serve as reservoirs. If their dimensions were carefully designed and constructed, the minors could store the water flowing in during the night and release it during the day to the Abu Ishreens. During the night, water levels in minor canals would rise to a maximum level in the morning, when field irrigation would start again. During the day, water levels would go down, as the irrigated fields would use more than the inflow into a minor. At the end of the day, the cycle would start anew.

Obviously, this approach did require changing the minors. Original Gezira minor canal banks were about 50 cm higher than the so-called Full Supply Level (FSL—maximum water level). Butcher proposed to increase the allowable FSL to 20 cm below bank level, yielding some 30 cm of storage space to be filled at night. If necessary, canal dimensions could be adapted to accommodate the storage volumes. The night storage would decrease the safety margins in minor canals, as less room for fluctuations was available and higher water levels would increase the risk of breaches. Nevertheless, Butcher concluded that "the risk must be taken if anything like adequate storage is to be provided."[31]

Given the many minors in Gezira, Butcher preferred "some more automatic system" to make the necessary shifts twice a day. Therefore, he divided each minor canal into sections, to space the storage evenly over the canal and allow closer control in the sections. Each section had a structure at its downstream end consisting of a weir with a pipe. During the day, the pipe would be open and pass water to the next section. During the night, the pipe was closed; the water level upstream would rise with the section filling up. Once full, water would flow over the weir to the next section—keeping the water level in the upstream section constant. The only action needed in this system was closing the pipes in the minor and to the Abu Ishreens. This would work well, "provided the ghaffir carries out the simple duties allotted to him," and start upstream.[32]

As water-level changes in minors would influence water levels in the larger canals upstream of the minor, Butcher designed a flow control system for the entire Gezira canal infrastructure. He drafted a system of control points, dividing the Gezira into blocks between 17,000 and 20,000 feddans. Water flows to these blocks would be continuously monitored. To make the system even more water-tight, Butcher advised that his control areas should coincide with "the divisions of agricultural control," in other words the SPS Blocks.[33] Monitoring and measuring water flows was vital for Butcher—not really strange given that engineers readily claim that measuring is needed[34]—but in the night-storage system it was even more important, as margins of error were smaller. Furthermore, night storage could not use water levels when monitoring discharges, as a standard relation between discharge and level in a minor would be impossible to establish.[35]

As he was asked to solve the problem, Butcher wanted to refrain from judging the need for night storage, but he did stress that his proposal should be weighed against the advantages of avoiding night irrigation. Night storage meant more work for the irrigation staff, and even Butcher could not resist stressing again who was to blame for this. He "emphasized" that "proposals, designed to relieve the cultivator of some part of his full responsibility," were putting "additional strain on the machinery of water distribution." Butcher considered night storage "as a temporary solution," to be discontinued as soon as the tenants were "adapted to the new conditions of free flow supply."[36] The Gezira canal system could be changed back to continuous flow any time, even after implementing the measures he proposed.

A Hobson's Choice

Temporary or not, when Smith entered the Gezira as a young engineer, he and his colleagues must have been busy with changing the Gezira minor canals. Although we have no direct evidence of Smith's work from the sources, the conversion must have asked much of his attention. Implementing the night-storage system with its additional weirs in already constructed canals was a challenge. The year before the opening of the Gezira system was one of "agitated construction," to such an extent that the contractor did not always pay the attention needed. Complaining about this was difficult, as everyone was busy with "the extra work thrown on them by the introduction of the Butcher weirs."[37]

MacGregor kept Schuster informed on progress of irrigation in the first months of the 1925 season; canal performance was one of the major topics.

In the first months after opening, night irrigation was actually practiced, as "it would have been risky to have attempted the Butcher system with absolutely new banks."[38] A major risk of night storage was breaching of canal banks. Night irrigation meant that the outlets to the fields were open at night, but the tenants themselves did not remain on the fields—they went to bed. This labor-extensive irrigation method was what the SPS wanted to avoid. MacGregor acknowledged that the SPS was "perhaps right," but (again) he could not resist mentioning that he had "seen less mess as the result of night watering than one was led to expect."[39]

The night-storage system itself was implemented gradually. First, the night-storage system was implemented on selected channels; once it worked on those, it was tested on other canals. Newly constructed banks had to be strengthened and had to settle—which was sometimes done by arranging tenants trampling on the banks. In September 1925, MacGregor could report that the night-storage arrangements worked satisfactorily. Only in a few cases night irrigation was still used for durra and lubia. He emphasized that night storage caused backwater effects on canals upstream, requiring additional control actions and hiring additional ghaffirs, but thought that the situation would improve with time. One major advantage of the intense first months was that "all ranks" were "gaining practical experience."[40]

Indeed, complaints about the water control required by the system do not return in later SID documents—apparently the engineers started getting used to the actions required in Gezira. In the 1930s, however, an interesting new complication for the irrigation engineers concerning night storage appears in the sources, at a time when everyone would have been used to night storage and new minor canals were designed for storage. Apparently, drawing a suitable longitudinal section for a night-storage minor was not easy, as "so many considerations" were required and a night storage canal was "entirely different to any orthodox canal."[41] Even after more than five years, the SID was still struggling to consider all relevant issues for designing minors and including these in a design drawing.

Working with standardized representations of minors was important. Without them, the night-storage system was "apt to be vague in the minds of the section engineers and younger officials."[42] Even as late as the 1950s, we encounter the apparent necessity from the side of the SID to standardize information on design procedures for night storage. In 1952, a proposal was made to use tables allowing engineers to "easily compute" water levels and storage volumes in minor canals. These tables would include the length of a reach, its maximum storage depth, bed slopes, and the depth of flow for a given discharge. Each combination of these parameters would yield different storage volumes and many different combinations were potentially possible. Luckily, the Gezira plain had relatively low and regular slopes and much of the area could be standardized. The results could be kept at "three full-page tables."[43]

Getting used to the routines of night storage took time for the engineers, but they never would agree with it. Throughout the lifetime of Gezira, engineers complained about, struggled with, and dealt with night storage. As late as 1944, an experiment with night irrigation (see chapter 4 for its context) allowed the Director of Irrigation to claim that night storage "had been controversial for years." At the same time, "both parties knew each other's arguments pretty thoroughly" and could only "agree to differ." Nevertheless, for him it was clear that Gezira worked against "the natural laws (hydraulics, gravity etc.) to a degree that cannot be justified as a permanent measure."[44] The amount of work required, the sediment problems, and the complications night storage brought for water distribution could all be—and were all—solved, but what the engineers kept on stressing was that night storage was fundamentally wrong. In 1925, it was "not in accordance with sound practice" and "a distinct challenge to the laws of nature;"[45] ten years later night storage was "the wrong principle" and certainly not "a 'for ever and ever' arrangement." Night storage was a "Hobson's choice" for the irrigation engineers.[46]

We can safely conclude that a 1990 World Bank report was not entirely correct when stating that night storage in Gezira was selected "when it was realized that tenants were opposed to irrigation at night." [47] Indeed, night storage was not anticipated upon in the design at all, but tenants were not the ones asking for night storage. Introducing night storage in Gezira was not a smooth and uncontested operation, but the resulting Gezira canal layouts and management arrangements show that the engineering efforts to avoid night storage were not terribly successful. Night storage remained to be implemented in all extensions of Gezira, be it in the late 1920s, the 1930s— and even in the 1950s and 1960s as we will discuss further in chapter 6.

Standards

After the first challenge to change the infrastructure under construction, Smith and his fellow irrigation engineers had only little time to enjoy stability before they had to work on the changes needed in Gezira infrastructure in the late 1920s and early 1930s. We already discussed the need for an extra branch canal to the western extensions in the 1920s; other changes that needed to be incorporated were the change in crop rotation in the early 1930s and the resulting different water distribution. Finally, the main canal was enlarged a few times, which was made possible with the 1929 Nile Agreements with Egypt. Once the cotton extensions would be ready, the Sudan Government—the SID—had to provide 450,000 SPS feddans plus 45,000 KCC feddans with water from Sennar. An additional 15,000 feddans of the SPS were supplied with water with the former Had Abdulla pumps. In a normal October month, however, the main canal could provide just enough water for the SPS area. The SID had no "legal obligation" to enlarge the Main Canal but did feel "a certain amount of moral obligation." Furthermore, if the Main Canal was not enlarged, durra could no longer be irrigated in October. This might be "agriculturally sound" but would "not be popular with the tenants." [48]

The original main canal could carry 80 m³/s, measured at K57, the starting point of the irrigated area—84 m³/s was diverted at Sennar. The agreement with Egypt allowed 96 m³/s in the canal from Sennar. How many feddans could be irrigated with this water was subject for debate, as it depended on one's starting point. Just considering a total area would not work, as any arrangement in the canal system could only take "a very wide margin" given the possible "artificial rather than natural emergencies"—which referred to the changes in crop rotations a few years before. A three-year rotation required more water for the same gross area compared to a four-year one. With the original rotation changed to a "less concentrated system of cropping," the major canal system carried more water than needed. A possible answer to this over-design was extending the irrigable area, which was only possible because such a change was in the convenient direction. A change from four to three years would have required limiting maximum flows in the main canal or to enlarge—remodel—the canal.[49]

Other, more local adaptations to the canal system were not uncommon either. In 1935, after some correspondence between Syndicate and government, the Efeina Block was transferred from the SPS area to that of the KCC. The SPS was given an extension of 14,000 feddans at Wad Rabi and some 7,000 feddans at Mezeiglia East, which required new canals and adapting existing ones.[50] In other areas, additional drainage was required, for example, in 1936, when the area to the west of the Main Canal between K41 and K55 was prepared for irrigation in the next year. As the effects of rain on the cotton had become better known compared to the 1920s, "a strong request" had come from the Syndicate to provide this drainage.[51]

In an attempt to plan potential changes in the irrigation infrastructure—after all, how does one design a system knowing it is likely to be changed in any future—the SID designed and constructed canals and control structures differently. From the 1930s onward, canals were designed for a four-year rotation.[52] Canals, however, were built in earth and as such could be widened if needed. Control structures, which were constructed in masonry and more difficult to change, kept being designed for a three-year rotation. As the dimensions of control structures depended on their water discharge capacity per unit of area downstream, their dimensions could be standardized.

The first efforts for standardization were already made by Butcher. After his work on the night storage, he designed a moveable weir for water regulation, which was standardized in three series.[53] The smaller Series I weirs were used at heads of minors to regulate incoming flow; larger ones were used for regulators and majors. Even for these standard structures, however, allowance had to be made for the exact location of the weir. In 1927, gates had to be constructed in the Northern Extension in such a way that Butcher's regulation structures could "be installed without any difficulty" in the future. In some cases, however, the water levels required for a structure to function could not be met completely. In such cases, additional "horizontal timbers in subsidiary grooves" were used to manipulate water levels.[54]

Figure 3.2 Syndicate diesel engine. Reproduced by permission of Durham University Library (F.B. Hunt; SAD.692/19/1)

These and many other standards for design and construction of irrigation infrastructure were drafted by the SID as so-called design sheets. One of the most typical standards in Gezira must be Design Sheet 106, which describes the "Syndicate Plough Turning Cycle" (figure 3.2). The sheet prescribes that the minimum radius on the outmost edge of the turning cycle, at the border of a field, is 14.63 m (and not 14.62)[55]. Other design sheets include standards for canal slopes, which depended on length and area served; for pipe diameters, which depended on the area served; for discharges per meter of weir; for weir levels above the full supply level; etc. Design standards were drafted to save on computational time when designing infrastructure—not a real luxury when computations needed to be done by hand—but also standardized the actions of engineers themselves. No engineer should be allowed to work "without first giving him an example of what is or is not required."[56]

Standardization may suggest that design procedures were fixed, but design sheets changed over time; they had to, as Gezira changed. Changing rotations brought changes in water amounts per area and canals had to be adapted to that. An example of adaptation to structures is Design Sheet 32, which specified the width needed for night-storage weirs. In the original Sheet 32, the standard width is 4 m.[57] In Design Sheet 32a from January 1928, the width has been increased to 5.5 m.[58] There was even a design standard for nonstandard night-storage situations. Design Sheet 17, Paragraph 1, specifies what to do when a pipe regulator would serve less than five Abu Ishreens. In such conditions, none of the Abu Ishreens might require water during the following day, making stored water useless but also potentially impossible to remove from the canal. In a case like that, a night-storage weir had to be avoided.[59]

A final design standard worth mentioning is one anticipating changes due to the natural conditions in Gezira. Design Data Sheet A.21 explains that (light) control structures built on Gezira soil would usually rise after irrigation had commenced, as the clay soil would swell when wet and the structures would rise with it. The way to deal with this—according to the sheet—was to construct the structures below final design level. Series I structures were to be built 10 cm lower, the heavier Series II 5 cm.[60] In case the weir after soil swelling became too high, experience from practice recommended "chipping ou [sic] a little concrete" from the weir.[61]

At all these weirs, water flows were measured. In 1928, Smith was still a junior engineer within the SID, but he already knew that archiving data in a correct way was vital; if not doing that, why record all these numbers in the first place? With the lower Sudanese staff actually doing the measurements and archiving the data, Smith considered "it essential that the daily gauges be recorded in Arabic." To comfort the English-speaking engineers—one assumes—he added that "Arabic gauge books are quite as easy to read as English." Most British could read Arabic figures, and only after "a week or two," the name of a regulator had become irrelevant, as anyone would know "its place in the book."[62] These and other statistics were well kept. All was on paper, obviously, and the enormous amount of paper collected over the years made some data disappear.

That is at least the impression coming from a note from Smith in 1943, in which he discussed one of the basics of Gezira irrigation design, the water factor. Gezira canals had water factors "according to files," expressed as m³ per feddan served. Smith considered that the original F13.5 (which stood for 13.5 m³ per feddan) was based on the "maximum factor of 13.6 at least" given by Butcher in his 1924 note, but that same note mentioned F15 for the main canal. Smith thought he remembered a value of F13.4, but notes he had made in 1930 mentioned 13.8. A Mr. J. L. Chapman claimed that the original design had used F24 gross for the Main Canal and F20 gross for distributaries, but Smith could not "trace the reference."[63] Drowning by numbers was not just possible in Gezira fields, it was a real option in the SID offices too.

Surveys Have Camels

In case numbers did not provide enough drowning possibilities, daily realities might have. After all, keeping paperwork on design and construction is one thing, but keeping up reality in relation to paperwork is something else. Take something like ensuring that measurements were done with a similar reference to ensure that drawing canals on a map would be possible and the correct spot in the field to actually dig the canal could be found. For this purpose, the Gezira area was covered with a grid of permanent marks, in a process that was described in 1928 as "painfully," with practice different from theory "in some unrecorded manner." The marks had to be put at the correct places by surveyors. However, "[s]urveys have camels," which could damage the marks. As if that was not enough, sometimes a plough "waddle[d] along...."[64]

These and other details of daily reality on the ground were reported by the responsible engineers in Handover Notes, in which engineers explained the particulars of their area to their successor when going on leave. Every location in the canal system where canals would branch off had its own target water level. The main canal at K169 had a design level of 6.35 m—relative to reference (remember the camels changing some of the references)—but daily experience had shown that it could go up to 6.50 m without harm. Actually, that was a good thing, as it allowed keeping a little more water to divert if a shortage was expected. The buffer might just be enough to have the shortage "tide over until more water arrives from the dam."[65]

Certain points in the system might not behave as intended. Some canals had to bring water to areas that were relatively high compared to water levels in the canal. Especially with low flows—and thus levels—such canals would have problems in bringing the required flow. In 1934, such a situation was reported for the Wad Shannan canal. Despite the problems being "repeatedly investigated since the start of the scheme," there was not much one could do, and the engineer in charge concluded that "the present layout is probably the best."[66]

Given all these adaptations and daily improvisations, it is not surprising to find suggestions from engineers that being too accurate on water levels in canal management did not really work. Instead of keeping "too rigid a level," it was much better to work on balancing flows. Knowing the time it took for a change in water discharge made at Sennar to arrive 57 km downstream, would be quite enough in managing changes required in the control settings at K57. The time lag could be taken as 6 hours for a first notable effect, and it took some 19 to 20 hours for a final effect. This made it "safe to start moving at K-57 six hours after the Dam" and control whether a full effect was noticed some 20 hours after the change.[67] Even in years as late as 1942, when Gezira management logic would have been pretty clear, layout and management at location K127 of the main canal was changed. The new fixed outlet opening should avoid "continual fiddling with the gates."[68]

All in all, all these SID activities shaped a more or less elaborate system, in which daily actions on the ground, standardization in design and construction, and responsibilities for actions were related. Making sure correct actions were taken was vital, and correcting those lower in rank was part of daily routine. Training new engineers was the responsibility of the more experienced ones. New engineers should develop real engagement and "Divisional Engineers should foster this interests."[69] In February 1937, the Divisional Engineer (DE) at Wad Medani wrote to his Sub-Divisional Engineer in Messellemia that "you and your Assistant Engineers need guidance as to the design to be adopted." After all, the DE had modified the design "in the light of experience." To make sure that the message would come through, he enclosed five copies for the Assistant Engineers as well.[70] Control was loosened somewhat once certain standards were established. In

1939, Director of Irrigation F. M. Chinn indicated that he did not need to see all the detailed canal designs anymore. Until then these had been sent to him, but checking was "lost in a cloud of details" anyway and times had changed. Most DEs had "seen upwards of 14 years' service and should be able to approve a major canal design."[71]

Apart from constructing extensions, remodeling canals, and managing water flows, there was plenty to do in Gezira to keep the system going. Gezira required continuous maintenance, as weed and silt competed with water for space in its canals. The SID fought "a perpetuated war against weeds."[72] Even this major task, necessary to keep the system working without changing anything, was not free of negotiations. Many maintenance tasks had to be performed in minor canals, where stagnant water during the night resulted in major silt and weed problems. The minors were also the level where SID and SPS responsibilities were closest together and very often mixed. In the original planning, the minor level would be under SID responsibility, but allowing the SPS inspectors a larger say in water management made sense from an irrigation perspective. Rational as the new division of tasks might have been, for canal maintenance, it created problems, as there were no clear arrangements between SPS and SID on responsibilities for tasks like clearing weeds and repairing small breaches.

The issue was "a diplomatic one." Officially, the SID was responsible for maintainning all canals and everything in or on them, up to the downstream side of the Field Outlet Pipes in the Abu Ishreens. Operating the pipes, however, had become the responsibility of Syndicate inspectors.[73] The ghaffirs that managed the canals, including removing weeds, were officially under the SID but "Syndicate ghaffirs" in practice. As a minor canal was typically several kilometers long, its maintenance was quite a task. To compensate for SPS trouble, the SID made a budget available to the Syndicate for additional efforts where needed. The SID feared, however, that SPS inspectors would show a "reluctance to apply" for these funds with SPS management, as this might be taken as a sign that they could not do their work properly—in chapter 5 we will find out more about the internal Syndicate hierarchy. The SID could not put too much pressure on inspectors either, as the answer could be, "Well if you don't like the way we do it, do it yourselves, all of it."[74]

Concerning the many small repairs in minors, practice made it again much more convenient to have inspectors arranging matters with tenant labor, as it was much cheaper. That model did not work very well during busy periods. In the cotton-picking season, SID laborers still needed to come in. As a result, it remained challenging to create a clear-cut division of responsibilities between SID and SPS. The SID engineers needed to act "with discretion."[75] Basically, daily maintenance was planned out, but engineers and inspectors made it up as they went along. Within all those formal and informal negotiations, something like regularity did develop. Day-to-day decisions might need some improvisation and flexibility; new standards might have to be drafted every now and then, but all in all, Gezira irrigation with its associated tasks became quite predictable after the start-up years and the major changes of the early 1930s. Flexibility was part of the design process, but it was a rather predictable flexibility, especially after the first busy years were done with.

Nostalgia about the Carefree Days

As one can imagine, the Second World War put much stress on the machinery of the SID, with a lack of personnel, budget, and equipment. The difficulty of keeping up maintenance standards during World War II, with its "drive for economy," shortage of labor, use of SID excavators by the Army, and scarcity of spares, had considerably increased silt and weeds in the Gezira canals.[76] In addition, disposing of excavated silt became a problem. The original design had reserved space for depositing silt, but it was becoming increasingly difficult to find space for the sediments. Roads were sometimes blocked, which blocked the draglines that had to clean the canals. Expropriating strips of cultivated land along the canals was a temporary answer, but redesigning each local canal system was seen as the better option. The post–World War II silt problem was reason for the engineers to stress again that the "fundamental cause" for it was the night-storage system.[77]

The postwar years, however, would not offer the "transition from the austerity and difficulty of war to the plenty of peace" so desired by the SID.[78] One of the first items on the development agenda was adding another 100,000 feddans made "possible within the resources of the Sudan's present entitlement of water."[79] The post–World War II years were full of new programs requiring "constantly changing planning, designing, estimating, and improvising"[80] in a spirit of planned development. Given the key role of Gezira for the financial position of Sudan, Gezira would be a central element in any development plan. In an attempt to answer to the British and international call for a broader concept of colonialism—one that actually made an attempt to develop colonies for the colonial population itself[81]—five-year development plans were made by the Sudan Government.

The long-term goal of the five-year plans was to "survey the natural resources of Sudan (including man-power)," resulting in "a general long-term plan for the conservation and coordinated development of these resources," with the goal to yield "the maximum practicable improvement in the standard of living of the inhabitants of the country." In addition "minor supplementary development schemes" could be executed.[82] Whatever was planned, the perceived need to "look for the money" remained strongly in the center of debate. Although education and health would be productive in a wider sense on the longer term, "revenue producing projects must be given the highest priority."[83] Strictly speaking, it was not necessary that all projects should be financed by the available government budget. After all, "sound development schemes"—those that would increase the national income—could be taken on board even when the government could not finance them directly from the current budget. Table 3.2 shows that the "well-planned, productive projects" were infrastructural projects.[84] Even the—undetermined—allocation

Table 3.2 List of investments[85]

Central Government	Estimated Capital Costs of Program in £E
Health	716,000
Education	986,500
Agriculture and Forests	1,846,170
Irrigation	1,575,000
Veterinary Services	84,800
Public Works	1,722,500
Post, Telegraph, Telephones and Wireless	665,000
Broadcasting	71,000
Aviation	250,000
Sudan Railways	2,500,000
Miscellaneous	13,500
Total	**10,430,470**
Local Governments	**1,050,000**
Total All Government	**11,480,070**

under "Local Governments" included 400,000 and 175,000 pounds for the Khartoum and Port Sudan sewage system respectively.

Another significant development after 1945 was the growing political activity in Sudan. In the summer of 1946, Sudanese workers—including SID ghaffirs—went on strike. Engineers and inspectors had to find ways to keep the system working. For example, canal discharges were lowered to a safe figure.[86] To show their dissatisfaction about the distribution of profits, and stressing their demand for a larger share, Gezira tenants went on strike as well.[87] As a result of both actions, water demand in the irrigation season of 1946 "was extremely light."[88] Without these new developments, it was already a major challenge for the SID to maintain normal irrigation standards on the regular 900,000 Gezira feddans.[89] Something as routine-like as replacing worn-out steel pipes with concrete pipes proved to be a huge task. With a rate of 300 sets of pipes per year and over 10,000 pipe outlets in Gezira, the "magnitude and importance of this matter" was clear.[90] The expected year of completion for the entire operation was 1957,[91] but the program was delayed because of the Managil Extension (see chapter 7).[92]

North by Northwest II

Despite the changes in colonial policy of the British and despite the difficulties in keeping up normal routine in Gezira irrigation, new extensions of the irrigated area were high on the post–World War II agenda. Extending Gezira had always proven to "greatly contribute towards the financial strength of the country."[93] As mentioned in chapter 2, the potential irrigated area of the Gezira Scheme was set around 1.7 million feddans in 1938, with possible further extensions only feasible with increasing population pressure.[94] In that same year, plans were prepared to extend the Gezira area with an additional 90,000 feddans in the Northwest—or even 180,000 feddans when water use on these feddans would be limited. Another option was to realize a much bigger irrigated area in the Southwest, but that would need negotiations with Egypt.[95] The Northwest Extension could be realized within the allotted Sudanese water share. Enlarging the main canal to the full 168 m^3/s would allow a canalized area of 976,000 feddans.[96]

At a meeting on February 2, 1946, the possibility to realize the additional 100,000 feddans to the west of the existing Gezira area in the Managil region was discussed. More detailed plans for this extension should allow comparing costs and benefits with the Northwest option.[97] Both extensions required enlarging the main canal, but the Northwest project only required enlargement of the existing canal between K57 and K169, whereas the Managil project required a new (Managil) branch. This new branch would initially serve the 100,000 feddans only, but it had to be constructed in its final size given the much larger potential area. This made the Northwest project cheaper to realize on the short term.[98] What made the extensions cheaper as well compared with the pre–World War II ones was that land in the Northwest was expropriated and not leased. The lease model in the Gezira Land Ordinance had been designed to assure that if "the scheme were to fail," land property could be easily restored. With success guaranteed, national development high on the agenda, and "no longer any need to buy [Sudanese] goodwill," the expensive land-holding lease arrangements were no longer needed.[99]

Constructing the Northwest Extension started in January 1949. The SID started the project with rehabilitating its own "appallingly congested" Headquarters in Wad Medani.[100] The first 50,097 feddans received water in July 1951, a second set of 49,625 feddans was ready in July 1952. The enlarged main canal carried the additional discharge required. Work on the main canal continued with the heavy equipment that was available to anticipate "the possibility of 50 per cent cropping in the Gezira" and other (small) extensions "possible in the future."[101] Although the newest extension, several branch and major regulators in the Northwest area were equipped with outdated gates that "were in stock"; it was "easier and quicker" to use them as standard "in the short term available." Despite the project to change the field pipes in Gezira, most of the Northwest area had steel pipes for the same reasons. The Sennar concrete pipe-making machine could not produce sufficient numbers, although concrete pipes were used in the Northwest when steel pipes could not be delivered in sufficient numbers either.[102]

The area was planned as a Syndicate area, with one change from normal practice. The inspectors' houses were positioned closer to each other than in the original Gezira area, in the interests of their social welfare. The main decision to be made for the Northwest area was—as to be expected—the water distribution strategy. The SPS concession would end in 1950, and as such the need for night storage could possibly change. As early as 1941, the option to start night irrigation in future extensions was mentioned. In anticipation of this change, irrigation designers were instructed to design larger minor irrigation areas with minor canals further apart.[103] In 1946, a Special

Committee advised to start night irrigation but requested compensation for tenants as these would need boots and clothing.[104] The Financial Secretary replied to this demand that the issue would be tested.

And indeed, in 1947, SID Director Smith proposed to use part of the new Northwest Extension as an experimental facility. One double Block would be operated under a continuous watering system and another double Block on night storage.[105] A third—smaller—area was to be used for a special test: alternate canals would be on different watering systems, with each cotton number considered a replicate.[106] Because of "administrative difficulties" being "too formidable," this third area was later included in the night-storage area, on the request of the successor of the SPS, the Sudan Gezira Board (SGB).[107] The tenant's land holding on either the night storage or continuous system was the standard 40 feddans with 50 percent fallow.

To Waste Water Is a Sin

With Gezira reaching its limits within the Sudan share of the Nile water in the late 1940s, the "need for economy in water was pressing." New extensions or increased crop intensities would require either a larger share of Nile water or another way of using the available water. In the tenancy model after the Syndicate concession had ended, tenants were to become small-holder farmers. Those farmers would not be forced to grow cotton, but had to be convinced. At the same time, "most careful thought" on how to ensure those small-holders would use water economically was needed. Introducing water rates was considered an option "to convince the cultivator that to waste water is a sin."[108] Another option was to impose quota on water use.

In the 1950–1951 season, the huge cotton crop made it impossible to give as much water as usual during the second half of March. The cotton harvest was still considerable, only 10 percent lower, which suggested that cutting water at the end of the season was a "probable" future strategy.[109] To be able to make a sound decision on the effects of decreasing water use on crop yields, trials were designed. For these trials, it was necessary to measure the total volume of water applied to a field. Several field channels were fitted with so-called Dethridge wheels.[110] The first results showed that irrigation rates were often twice or three times higher in the morning than in the evening of the same day—a result not that strange in a night-storage system with higher water levels in the morning. However, some plots occasionally received twice as much water as others,[111] confirming "the surprising variations" in field irrigation practices.[112] Even after decades of experience, field realities could surprise SID and SGB employees.

Saving water was vital in the total "outline of attack" of post–World War II SID plans.[113] Any further enlargement of irrigated area depended on the availability of water—something not likely to come for several years, however. Nevertheless, SID plans showed quite a "field of possibilities," including converting the Gezira area back to its original rotation from a 50 percent maximum cropping intensity within a four-year rotation.

Remodeling was expensive, however, and its outcomes unclear; it was not put very high on the agenda. Other options were realizing irrigated areas elsewhere, for example, the Kenana system of some 250,000 feddans, but it was clear that extension and rehabilitation in or close to the Gezira area had priority. The "first stage of future development" promised to be the Managil Extension of 700,000 feddans, with "the great advantage of experience of the existing" Gezira Scheme. All to know was known; Sennar Dam could be used, giving Managil the promise of "the quickest and most assured return." The potential area of at least 700,000 feddans was too large to develop in one go, partially because of costs, but also because water availability was not enough yet for the entire area. The idea was to develop areas of 100,000 to 200,000 feddans each time: "growth by installments."[114]

Managil will return in chapter 7, but for now Managil serves as the perfect example of "planning it all out and making it up as one goes along." When the SID started planning Managil, much was not decided upon yet, including the basic layout and typical holding size. The SID engineers assumed that Managil would copy the Gezira model with unit plots of 10 feddans. The ever-important issue of continuous watering or night storage popped up again. In the first designs, minor canals were spaced at "double-night-watering" intervals. With this idea, the general outline for the major canal design was drafted, as this was less likely to change. If night storage would be selected after all, intermediate minor canals could be inserted. Finally, the designers simply assumed the highest cropping intensity possible when drafting the preliminary design. All these assumptions allowed the engineers to proceed with preparatory work "with the knowledge that it will not be wasted." [115]

In 1955, it became clear that Managil was indeed to be designed as an extension of the Gezira; "no fundamental changes" were proposed, as Gezira had "proved so successful in the past." Minor canals were designed for night storage. The only major change was the return of three-year rotation, this time with irrigation ending on January 31 each year. In the existing Gezira area, irrigation could continue after that date, although it was expected that irrigation would also stop "earlier and earlier" in that area.[116] In 1955, the main canal was still being enlarged to 168 m^3/s to supply the existing Gezira area plus the first 200,000 feddans of the Managil Extension. A second main canal parallel to the first one would be needed to supply the full Managil Extension.

All in the Family

Development and construction of the extensions could be more or less covered with existing senior SID engineers, but there was quite some pressure on the SID in post–World War II Sudan in keeping its staff and filling all positions. In addition to the general policy of Sudanization of government posts—replacing British staff with Sudanese staff, discussed in chapter 6—the SID seemed to have a continuous shortage of engineers of every kind. Hiring

British engineers was obviously not an answer, if that would have helped anyway with career perspectives being low for British in a Sudanizing country. Some new British staff was hired, as development and construction was suited for short-term contracts; British engineers were "on the whole considerably more efficient from the technical point of view" as well. Only few Sudanese could "hold their own without our expatriate engineers."[117] However, the new Sudanese engineers who were sent to the United Kingdom to gain experience earned "high praise."[118] There was simply an "acute shortage" of new Sudanese engineers. Other engineering professions—such as the railways—were more popular, as an irrigation engineer spent "a smaller portion of his working life in towns."[119] Already in 1947, this was seen as "a grave handicap" to the future SID and the "development, by Sudanese, of irrigation in the Sudan."[120]

The SID developed a working structure in which the British "back-room boys" could work on designing projects. Some of the staff could then focus on longer-term planning, allowing the SID "occasionally to look into other peoples' projects instead of, as now, being forced to slam the door."[121] Much of preliminary designing was routine work, which could be done by staff with less experience without full-time supervision. The Sudanese staff needed to gain more practice on actual design, "while there are still those left who can guide" as well.[122] After all, the idea was that Sudan would continue constructing larger-scale irrigation systems. For operating and maintaining the existing Gezira system, the situation was less dramatic. Skills and experience needed for such routine work were less, "long years of experience and an intimate knowledge of the country" were more important.[123] Sudanese recruits from before World War II were "running the Gezira Scheme," even though they might have "attained a somewhat lower academic standard" than postwar recruits.[124]

Some British had been "full of the news" that Gezira would have "no British irrigation engineers,"[125] but indeed, in the 1953 season, the SID organization in Gezira was "fully Sudanised."[126] To give an idea of numbers, in 1954, the SID included 40 expatriate staff members (22 civil engineers) and 55 Sudanese staff (53 civil engineers, 42 from before the war). These numbers were short of the 79 Sudanese engineers actually needed but seem to have been enough to run Gezira. Some positions were not Sudanized as quickly as others, though. Apart from development and construction, expatriates—the new term for "British"—were used in posts that were seen as too crucial to risk changes, such as the Resident Engineer at Sennar. This was a "post of great responsibility," with potential for disaster in Gezira and problems with Egypt if flood season control went wrong. The Assistant Resident Engineer at Sennar could be Sudanized "within a year or so," providing an excellent "training post for the first Sudanese occupant."[127]

At headquarters, in contrast to field posts, Sudanization was seen as "in large measure to meet a popular demand," as "political."[128] Obviously, the work at HQ was much more related to political discussions in Sudan. The British engineers tried to plan the Sudanization process of the SID, but the quick developments in the political arena seem to have forced them to make it up as they went along, too. In 1951, it had been assumed that there would be no Minister for Irrigation for a few years to come. Just one year later, a Minister for Irrigation, a complete Sudanese Cabinet, and self-determination in 1953 were foreseen. "These rapid steps have led to very real uncertainty amongst Departmental staff."[129] Sudanese hoped for promotion, and British were uncertain of their future.

The issue concerning which posts should be for British and which for Sudanese was a continuous balance between nationalities and persons. "It is expected that the post of Deputy Director (or Director [D]) will be filled either by Mr. Bootheway, or by Faris Eff. on a Sudanisation directive. In the latter case the post of A.D.I. (Admin) should revert to an expatriate and be occupied by Mr. Bootheway."[130] For each Sudanese post, the discussions usually included an element of (lack of) expertise. In 1953, two Sudanese were to be put "under definite training—not figureheads but taking their full share." This would allow them to learn the job and be ready if "the march of events call for a Sudanese titular head."[131]

In 1955, 34 British engineers had left the SID (20 of them civil and 14 mechanical engineers), leaving six British engineers in the department.[132] That same year a Sudanese Director of Irrigation was appointed. Recruiting new engineers was a most important task, possibly from Egypt. When this was less successful, attempts were made to recruit staff from India and Pakistan.[133] In 1956, the SID recruited 20 engineers from Pakistan and India (plus eight from Egypt).[134] Only another two from India were recruited one year later, plus one Greek, 16 from Egypt, and 13 new engineers from Sudan itself. In 1958, nine Sudanese and ten Egyptian engineers were recruited,[135] but these numbers did not solve the shortage of engineers. In the Western Gezira only 50 percent of the posts were filled in 1957.[136]

The Edge of the Precipice

The pre–World War II years may not have been the carefree years Smith considered them to have been, but given the "formidable difficulties" the SID faced in the 1940s and 1950s, the "magnitude of the achievements" of keeping Gezira working and extending the area with the Northwest system can hardly be overestimated.[137] At the same time, SID stress was man-made by rather ambitious postwar development plans, especially given SID capacity. With British engineers retiring and not enough Sudanese engineers to fill the empty positions, pressure on the SID staff was high, and in the early 1950s, there was a "substantial back-log" in realizing the 1946–1951 program in the early 1950s.[138]

In April 1953, the SID had "lost" R. J. Smith, after he had been the SID head for seven years. What was particularly remembered about Smith was his "outstanding organising ability."[139] As we have seen, being able to organize the SID and its work was not a superfluous luxury in the years that Smith was directing the department. Many tasks had to be fulfilled, with an ever-increasing pressure on staff numbers, with the ultimate goal—in his own words—"rescuing the Gezira from the edge of a precipice."[140] The neglect of maintenance in the war years, the additional "fringe" crops during and after the war, the general pressure on resources, material, and people after the war, in addition to the "progress in development at expense of adequate progress in maintenance," made this quite a task.[141] At the end of his long career with the SID, Smith continued using similar phrases as his colleagues Roberts and MacGregor had done in the 1920s when discussing the night storage system. The "struggle with weeds and silt" was directly linked to the "peculiar night-storage system."[142] Each time, however, night storage remained the choice, and the SPS continued to manage the extensions.

After World War II, the larger projects in Gezira—especially the Managil extensions—could only be realized because much of the work was done by contractors "from outside the country"—with quite a few being British.[143] Involving outside agencies was "the only practicable way" to ensure that "a stock-pile of known possible projects" would be realized.[144] Indeed, more and more international consulting engineers became actively involved in planning and making up Gezira as they went along, as we will find out in chapter 7. What they typically shared among each other and with their predecessors in the first half of the twentieth century is ignoring those that did most of the actual work in Gezira. We turn our attention to these thousands of tenants working the fields of cotton and durra in the next chapter.

Chapter 4

Making the Best of a Rotten Deal: Tenant Realities and Resistance

New Syndicate inspectors received detailed information about the rhythms of Gezira in the shape of a booklet with notes. The booklet served as their first encounter with Gezira tenants, "the most important factor contributing to the success of the Gezira Scheme," as well. Knowing tenants' "background and characteristics" was considered "essential to ensure smooth collaboration between Inspector and Tenant." As much as the notes given to inspectors "would be incomplete without some remarks"[1] on the tenants, this book needs to pay dedicated attention to the tenants and laborers in Gezira. Obviously, as with any other story on the colonized that has to be largely reconstructed from the perspectives left in the sources by the colonizer, my discussion on tenants is heavily influenced by those sources. Gezira is not different from other colonial histories.

More often than not, the British complained that "the average Arab dislikes continual manual work," especially "the exacting work of continual irrigation and seasonal operations" so "irksome to him."[2] Close supervision, or the necessity for "drive," is usually explained in terms of failure of tenants. When colonial sources discuss anything else than failure of tenants, we actually do encounter bits and pieces of deliberate strategies of tenants within their socioeconomic context. As tenants "only received 40 per cent of the profits of the cotton crop" they were less interested in that crop. Many tenants would have preferred growing irrigated durra and only became tenants "because of the 5 feddans of tax free irrigated" durra in each tenancy. Nevertheless, with more and more tenants "who have never known an Agricultural system other than the Irrigation Scheme" over time, British hope that new tenants would "take to it better than did their fathers" was never lost. The new tenant would "appreciate his rightful place in the Scheme and realise that the new ways are better than the old."[3]

These ideas and ideals on planned irrigation and profit in Gezira had to be realized as much by colonial officials as by Sudanese farmers. The British were concerned about the influence of the Gezira project on Gezira social life. Keeping the Sudanese social structure intact was a major goal of British policy, especially before World War II, but modernizing the country was a goal as well. We will discuss the British claims on the social reality encountered before and at the start of the Gezira Scheme and the British ideas on the ideal tenant as a yeoman farmer. Once recognizing that realizing those ideals was not easy, the British responded with increasing their planned efforts. We will also discover what the first person actually interested in the Sudanese tenants and laborers had to say about them.

The chapter will make clear that tenants did not necessarily agree with the ideals planned for them. Sudanese farmers responded in their own ways to the new realities of Gezira. They "engaged with, bargained, deflected, or resisted the demands of colonial power" and its representatives.[4] Tenants' preference for food-producing options in Gezira was a source for continuous conflicts between SPS staff and tenants. Despite the yeoman ideal, tenants used large numbers of hired laborers from other regions in Sudan and even West Africa. The desired family farms did develop in quite other ways than the British intended. Resistance to the cotton drive or yeomen claims is probably best described as the tenants' version of using their weapons of the weak in order to maximize the goals and profits they themselves wanted to achieve.[5] In addition, through discussing Gezira tenants, this chapter makes it very clear that even those with a genuine interest in them usually still construct a model of how tenant reality should be before discussing tenants' strategies.[6] When actions of Sudanese are discussed by others, (implicit) claims how these Sudanese should behave are never far away.

A Spider's Web

One of the more enduring myths is Gezira as an "empty wasteland" before the Irrigation Scheme. In "its natural state," Gezira had nothing to offer "but yellow grass and scrubby trees" that had "to do without water." Those "native proprietors" that lived in Gezira could hardly grow enough grain to "keep them alive." In the old days, even good rainy seasons would have required only a few months of hard work. In the other months, Gezira inhabitants led a "very cheap and lazy" life of "attending weddings and funerals" and doing "odd jobs" like trading in oil or tobacco.[7] Therefore, most Gezira landholders would have been active elsewhere.[8] Going against the image of emptiness, Garstin had described Wad Medani as "the most prosperous town in the whole of Sudan."[9] In 1917, the Board of Trade provided a pretty accurate image of Gezira; it might have been empty, but wet years allowed "large crops of native maize." Although in dry years only "patches" of crops were found, Gezira farmers did even grow some cotton when possible—the quality was just not high enough for export.[10] Rainfall in the Gezira was erratic indeed, but yet the plain was an important granary for Khartoum in years with good rainfall, when the durra crop would provide good yields. In years with little rainfall, waterwheels in the Blue Nile provided water for narrow strips on the banks.

Obviously, too much Sudanese prosperity in Gezira would not fit the idea of British initiative required to make the area prosperous again. The "theory of decline and dislocation" was much better suited when proposing expensive irrigation works.[11] However, the area should not be too desolate and empty either. When proposing economic activities, having a population around to provide labor and/or consume the results was convenient. In 1912, a committee was appointed to collect "evidence, especially of that bearing on economic considerations involved in the execution of the Gezira Irrigation Scheme." It counted 79,647 men, 98,369 women, and 77,813 children in the potential area of the scheme; "in adjacent districts" even more people were found: 100,077 men, 124,835 women, and 126,429 children.[12] The conclusion was that "by Sudan standards," Gezira had a "considerable" population of "an intelligent type."[13]

Desolate the economic perspective might have been from British perspective, most of Gezira land had an actual owner. In 1913, about 13,300 Sudanese owners were registered. What being registered actually meant, however, was less clear. It was estimated that half of the properties were registered as of "heirs of so and so," without clarifying arrangements in sharing the land between individuals. The average size of a holding was 37½ feddans, with smaller plots near the Blue Nile and in the eastern, more populated areas of Messelemia and Wad Medani. Areas further west had less rain and less plot divisions among family members.[14] As a general rule, the British considered that Gezira land management left much to be desired. Plot boundaries ran "in all directions like a spider's web."[15] This "puzzle-plots territory" with its "crazy-pavement plan" of landownership was not suited for rational land use. Rational development required "regular chequer-board holdings" through government action.[16]

To deal with the situation, the Sudan Government designed a land tenure system that was rolled out over the existing property distribution, as an extra layer. Land was neither expropriated nor bought from owners—except in some cases when land for permanent works was concerned. The land in the scheme area was leased from the owners: the Sudan Government—compulsory—rented land from registered holders in the scheme area at a fixed annual rental for a period of 40 years. During the lease, subdivision of registered holdings was not allowed. Selling land "to foreigners or to African usurers" was forbidden as well. Other sales were only allowed under specific conditions. Land could be sold to Gezira tenants, "close relatives," and members of the local administration, like sheikhs, as long as these owned less than 720 feddans of land. Selling land to "bona fide local cultivators" with less than 40 feddans could be approved as well, as could selling land to the government.[17]

With the land under government control, tenancies were allocated to three different groups. "Right-holders"—landowners and their sons—could claim tenancies first. Landowners with 40 feddans or more could ask for multiple tenancies, with the idea that the actual landowner would cultivate one tenancy himself and nominate others for the remaining ones. This second group, "Nominees," included cultivators and relatives of right-holders who did not own land. The third group consisted of "preferential tenants," those landowners holding 20 feddans or more "but not enough to qualify as right-holders." Right-holders and nominees had automatic rights to tenancies, unless they had "a bad cultivation record." Preferential tenants had no automatic right but had—as the title suggests—preference over "landless men" and others.[18]

This system of lease certainly transformed the existing social reality in terms of landownership and associated economic power but did not make it disappear from Gezira. A major reason to lease Gezira lands, and not expropriate them, was indeed that a lease system would interfere only minimally with existing property rights. In case Gezira would not turn out as a success, land could be handed back to the original owners with relative ease.[19]

For Political Reasons

Next to concerns about Gezira success, however, the government considered its main role protecting the agricultural society—whatever that was—from the harmful effects of modernization. Especially Syndicate supervision and drilling, necessary as it may have been to produce cotton and profit, would be potentially harmful to Sudanese society. As discussed in chapter 2, a major reason for the Sudan Government to consider increasing the irrigated area outside the Gezira Scheme without involving the SPS was the potentially negative effect Syndicate methods would have on the relation between British colonial rule and the Sudanese.[20] In 1926, when the Northern Extension was agreed upon, the Governor-General's council wanted to include specifically—"mainly for political reasons"—that the Syndicate would manage the area "in a manner conducive to the general well-being and contentment of the native cultivators."[21] The Syndicate was granted its "desire to make a financial success of their undertaking," but this had the danger of minimizing attention for "the native agricultural community." Therefore, the Sudan Government had to be "alive to the importance of establishing the agricultural community on a social system which will conduce to tranquility and contentment."[22]

These noble words aside, there was a certain pragmatism in government policy too. When, for practical and/or financial reasons—like not being able to manage extensions because of a lack of people or expertise—methods of intensive control were to be employed, it was better that a company would do so instead of the official government.[23] With higher cotton profits not only securing Syndicate income, but also that of the Sudanese tenants, the government did see some advantages in SPS methods. Protecting the Sudanese against the Syndicate would have been quite a task anyway. In 1924, government staff in the province Gezira was located in was limited to 12. There were two government offices within Gezira itself, whereas from 1924 onward the SPS employed more than 100 British field inspectors—with a maximum of about 180 in 1930.[24]

Occasionally, jurisdiction—if one could use the word—between government officials and SPS inspectors overlapped. Inspectors interacted directly with tenants, which was not even close to a model of indirect colonial rule of keeping distance between colonial rulers and ordinary Sudanese. Although without formal judicial authority, an inspector could evict unsatisfactory tenants. The SPS model was built on direct interference with tenants to secure cotton yields and profits, but in daily practice, inspectors did involve the local sheikhs to some extent in their actions. A sheikh could be "a bad cultivator," but as the Sudanese were "used to being dealt with" by them, sheikhs were involved,[25] for example, when allocating tenancies.[26] In an attempt to strengthen local Sudanese leadership, courts under the responsibility of sheikhs were established in 1926. Tenant problems would be treated in court not by inspectors. However, in 1930, an inspector could still order (!) a local leader to fine people that had stolen grass from tenancies. The case "caused great consternation among government officials."[27]

This inspector had at least approached the local leader; many inspectors simply bypassed governmental structures when arranging matters. Setting fire to his durra because a tenant neglected the cotton did occur. Such actions could be subject to complaint by tenants and were treated by government officials, who were often uncomfortable with the situation, as became clear when inspector Morgan—in SPS service between 1928 and (at least) 1953 and the author of the instructions for new inspectors—burnt a tenant's hut "in order to impress on them that they must either obey his orders or leave his area." A government official observed that "the failure to pick cotton or carry out correctly some agricultural operation" was not an offense against any law in Sudan, but that "[b]urning a hut most certainly is."[28]

Government officials did not necessarily share views, though. The Sudan Government was certainly not the all-agreeing, monolithic block of British opinion that the term "government" so easily suggests. Gezira plans were not seen as problematic by everyone. In 1923, the Governor of Blue Nile Province claimed "that village life is not being destroyed by the Scheme."[29] Later that year, another government official suggested that the Punjab irrigation colonies in India were "likely to furnish useful guidance" for Gezira plans.[30] The India Office was approached for documents on the Punjab.[31] Governor-General Stack, however, did not see any need to compare Gezira with the Punjab. Gezira was not a colonization project, and village organization would remain strong. Suddenly, Gezira was not the empty desolated area it was so often portrayed to be. On the contrary, the Punjab colonies had been developed "in uninhabited tracts" of land.[32]

Interestingly enough, starting in empty land did not mean that Punjab colonies required Gezira control. In Punjab, villages were constructed with different holding sizes. Larger holdings were "to be held by men suitable to act as village headmen." Irrigation was planned on the basis of the holdings' pattern; in other words, village and irrigation were directly related. Stack argued that this relation made Gezira incomparable with the Punjab. Growing cotton was "a very delicate operation" and Gezira inhabitants had "far less natural aptitude for agricultural operations" than Punjabi. Therefore, "extremely close supervision" was vital, "at least for the first years." Once the cotton increased Gezira welfare, new social values would emerge. Keeping track of these "existing developments" was the key to better policies in future. Tenants would become good countrymen with Syndicate support, given its "past attitude in these matters."[33]

Yeomen Farmers

In the early 1920s, Syndicate officials expressed their more than great satisfaction about Gezira. The "arid plain" hosted "fresh green extensions" of cotton fields. Each of the pump areas within the future Gezira was a "nucleus" of Sudanese families who were trained to grow cotton.[34] Such a Gezira family was planned to include about eight people per tenancy, excluding small children.[35] Obviously, many tenants had "to be taught how to cultivate on up-to-date lines," but once this was successful, success was huge for everyone, even for camels. "The old lady, whose useful days were assumed to be over, finds herself much sought after and able to earn ample money by picking cotton; the young man and maid find it a lucrative field to earn their marriage portion before returning to their own homes, and the old mare camel, whose productivity had ceased owing to poor feeding, again starts to reproduce when fed on succulent fodder."[36]

In the same year these happy words were expressed, Gaitskell—then a simple field inspector within the Syndicate—was courageous enough to express his deep dissatisfaction to the SPS management about progress in building "an industrious cultivator type of peasantry" in Gezira. The idea(l) of co-partnerhip—after all an important aspect of legitimizing the Gezira model—seemed to be real to Gaitskell. It was "the essence of the scheme," crucial to create "a sense of responsibility & cooperation in the native" as well as "a unity of interest between Government & Syndicate." It was the copartnership that would encourage all three parties to efficiency. However, despite "this favourable treatment of the native," things went wrong. This had started, Gaitskell complained, when the "very status of tenant" had been "cheapened" by granting plots "to all sorts of young boys & slaves."[37]

The major problem, however, was that families had too many tenancies and as such were overloaded with land. A family—"a man & his family working with him"—was only assumed to hire outside labor when it was "imperative." The average family, however, simply did not have the family labor available to run all plots satisfactory. Illness of a family member was a huge problem, the option to leave the tenancy every now and then hardly available as the Gezira schedule demanded tenants' presence every day. There was "no peace in the cultivation"; the "very monotony" in Gezira destroyed "true interests" and made tenants look for "furtive devices of escape."[38]

In addition to the—perhaps slightly more objective—issue of available labor, Gaitskell complained that tenants were far from industrious. They behaved like "a collection of idle rich," by not working their fields themselves and celebrating the "coveted position" that success meant hiring to do the work. Tenants had an "exaggerated standard of living," with the risk—still according to Gaitskell—that in years with lower cotton prices, "widespread discontent" among tenants was to be expected. Discontent in a group "lacking in responsibility & self control" with a "veneer of Western civilization" would only lead to "future political trouble." If this was not bad enough, financial losses were to be expected because of tenants' lack of interest. Gaitskell considered the higher losses in cotton picking in the old Blocks as a sign for "retrogression in efficiency as time goes on."[39]

What was to be done? Continuously "driving" the tenants was not enough and would only make tenants become more dependent on the inspector. Many would "readily relapses into [...] former laziness." Gaitskell had good hope, however, as many other tenants were already "fairly competent." They were the hope for the future and should train others. Anyone else whose "circumstances or character" continued to be "incompetent" should be evicted. Obviously, tenancies should be allocated "more in proportion to the efficiency expected." Tenants from elsewhere could be encouraged to join, to relieve the pressure on the existing population. Northern Sudanese should make Gezira a "Sudan scheme" instead of a "provincial" one.[40]

The answer from the SPS management to Gaitskell suggests more than a difference in opinion; the letter reads like a clash between generations in the Syndicate. In chapter 5 this clash will be taken up further, but now I will focus on the ideas on Gezira social development and the role of tenants. As a first line of defense against Gaitskell's accusations, Senior Inspector William Porter Palgrave Archdale pointed to the "material and population" available. Tenants had "extremely strong views about old customs." The quick expansion of the scheme had made many inspectors forget that basic fact—according to Archdale. Furthermore, the many new British inspectors without "real experience" had made supervision less effective, but "two or three years of tightening up control things" would do the trick. Training the "industrious cultivator type of peasantry" required years. After all, "one must walk before one runs." With time, the "utopian" ideas would materialize.[41]

Part of Archdale's optimism was his own experience at Zeidab, his first position within the Syndicate in 1911. In the early years, very few Zeidab tenants did manual labor, but in 1928 tenants and their families worked hard on their fields. Even sons of tenants "were out at sunrise"; Gezira would see the same development over the years. The population shortage might have made allocating tenancies slightly less ideal, but this would change. People from outside Gezira were as welcome as ever; in Tayiba and Barakat, early settlers had "established large numbers of huts." Archdale saw the need to allow tenants some freedom every now and then. It was his own "invariable rule to let each tenant away for 2 to 3 days" after irrigation turns "as long as his work was satisfactory."[42] Hubert Poyntz-Wright, Syndicate director, took up the line developed by Archdale and pointed out that the Gezira Scheme was accomplished "in spite of the opposition" when trying to realize Gezira, "from both the Government & the native himself."

Referring to the changes in the Tayiba tenant arrangements, he explained why animosity against the Gezira Scheme "prevailed for years."[43] Gaitskell was simply too impatient.

Benefits of Tribal Discipline

Gaitskell was also a forerunner (and quite possibly creator) of discussions in (as early as) 1931, when Sudan Government officials started mentioning the option to transfer managerial responsibilities from British Inspectors to Sudanese personnel. With the Gezira cotton crisis, the moment may have been less suited for such new policies, but later in the decade, the Gezira did become "an arena of 'devolutionary' experimentation." Interestingly, this was perhaps even more so because the presence of the Syndicate actually did provide conditions in which devolution could flourish.[44] In February 1937, an informal advisory body was established in Gezira with all 13 sheikhs from the area as members. They were to be the link between government and Gezira population. For the first time, the Gezira Scheme was seen as a separate administrative unit. This experimental advisory body became the Gezira Local Authority in 1939. Another government act in 1937 was creating model villages in the Fawar and Gondal Blocks, under the heading of social development. As much as the experiments should build viable village communities, individual tenants should be able to profit and go their own way. Local sheikhs had to be trained as leaders.

The government realized that cooperation from the SPS (inspectors) was crucial. The Sudan Government saw options for increased tenants control over their own tenancies and increased Sudanese administrators control over village affairs, but the Syndicate was not necessarily convinced of the benefits of such an approach. The issue was not so much that close cooperation with local Sudanese authorities was useless, as became clear in a meeting between government and SPS in 1935, where Archdale—now Assistant Manager of the Syndicate—"suddenly 'blew up' and said that of course the SPS used sheikhs and anyone else who was of any value."[45] Even experiments with other types of management could be acceptable in principle, but the bottom line was that cotton growing had to be guaranteed. The SPS was not really in a position to block government policies either and had to accommodate and/or support them somehow. The resulting SPS strategy of restricted resistance will return below.

There was one part of the larger Gezira project where the Sudan Government could develop its own community model including the irrigation system and management model allowing the Sudanese tenant to develop along his own desires. The Abdel Magid area was part of the canal infrastructure of Gezira—it received its water from Sennar Dam—but it was not part of Gezira management, as it was managed by the White Nile Board. Abdel Magid was one of the Alternative Livelihood Schemes created by the government to compensate those Sudanese who had to leave their lands when the Jebel Aulia Dam on the White Nile was completed in 1937.

Egypt paid Sudan 750,000 pounds because Jebel Aulia served Egyptian interests on Sudanese land. Most compensation projects were small pumping schemes along the White Nile, but the Egyptian money allowed the larger Abdel Magid area to be opened as well. White Nile Schemes were a symbol of the government policy to stimulate and/or improve social development in Sudan. The Gezira model of tenancy was kept, but some other Gezira features were not copied.

In contrast to the original Gezira area, where villages were not included in the plans, White Nile Schemes did include planned villages where tenants would live within or close to the cultivated area. White Nile tenants were sorted into tribal groups, after which they were placed in new villages. Using to "the full the benefits of tribal discipline and the corporate life of a compact village community under a chosen head," Canal Courts were set up to deal with tenants not following the rules. A Tenants' Council consisting of sheikhs and some tenants, chaired by the Inspector of Agriculture, was established as well.[46] In Abdel Magid, a District Commissioner was appointed to arrange the 1938–1939 crop season. Although much of the work was about the technicalities of agriculture, the Commissioner was a member of the Political Service. The Commissioner used the "Syndicate Block or Estate" as a model, but the "primary conception of the scheme" was to run through "village communities working under their selected leaders."[47]

Syndicate management was replaced by management by the sheikhs, although still "subject to general supervision by the British staff"—a British Inspector of Agriculture or Irrigation Engineer. A "higher standard of agricultural efficiency" was not a main goal.[48] It is not really surprising then that night irrigation was practiced in Abdel Magid, although the actual extent is not entirely clear. Night irrigation seems to have been discontinued in 1945 with canals becoming weed infected and effective watering not "within the tenant's capabilities,"[49] but to what extent the more stressed conditions of World War II played a role in this is not clear. A push for controlled agriculture was not absent either. Abdel Magid may have been controlled in a less strict way, but it was—as other White Nile systems—functioning along general Gezira lines. Annual tenancies from July 1 to June 30 were given out "at the absolute discretion" of the White Nile Board.[50] Most tenancies included 5 feddans of cotton, about 3 feddans in durra, plus some additional wheat and lubia, but the Board decided on rotations. As in Gezira, cotton seeds were supplied by the Board to tenants at their own expense. Sharecropping was not allowed. Frequent meetings with the sheikhs were held, at which representatives from the Agricultural and Irrigation Departments explained what was desired.[51]

The X-Files

Despite its status of "experimental ground for ideas," Abdel Magid was very much a centrally planned reality too. Experiences must have been positive, though, because in 1940 "Schedule X" was started,[52] a policy experiment to allow growing other crops and training tenants to run their own affairs. The instructions sent to the SPS field staff were jointly signed by George Richard Frederick Bredin, the Governor of Blue Nile Province, and Gaitskell, who had become SPS Manager.[53] Schedule X prepared Gezira for what would happen after the "concessions to the Companies are ended," although in 1940 the decision on renewal of the SPS concession had not been officially taken yet—see chapter 6. In order to "make the best use of the permanent irrigation system established in the Gezira" in the post-Syndicate period, "an orderly organisation of village communities" was to be achieved. The village became the most important social and political unit within the Gezira.[54]

The first villages to enjoy the new arrangements were in the Hosh Block. From its thirteen villages, eight did establish councils right away. A favorite subject discussed in the councils was the similarity (or not) of Hosh compared to Abdel Magid, especially whether Hosh would see similarly high cotton payments for tenants as Abdel Magid. Farmers were clearly aware of what happened elsewhere. In February 1941, a Hosh Block Council was established to coordinate between councils and smoothen communication between inspector and villages.[55] Each village and Block[56] in Gezira would control cultivation and manage agricultural operations.[57]

Had World War II not happened, policies like in Hosh might have started in other "suitable places" too, but "owing to the scarcity of British Inspectors" after 1940, extending the experiment could "not be rapid."[58] As replacement for the many British staff joining the armed forces, 350 Sudanese samads were appointed to supervise village cultivation. Running small areas of 700–800 feddans under British staff supervision, samads could be "a great help to his Inspector and the tenants."[59] Samads were directly accountable to the inspector, bypassing village councils. Samads were assumed to be temporary, but the Sudan Government did consider giving "official recognition and remuneration" to them given the "good service" they provided. Bredin saw enthusiastic and capable samads as the answer to the many sheihks unable "to cope with the peculiar conditions created by the Gezira cotton Scheme." The illiterate and aging generation had to be replaced with young Sudanese relating to a modern, forward-looking local government.[60]

Until the glorious moment of orderly village communities, including agricultural control under a village council, supervised by an increasing number of Sudanese inspectors, becoming reality, British oversight was seen as vital as ever and British inspectors as indispensable as they knew all about their Blocks and tenants. After all, the British had to ensure that the new councils and sheikhs operated in the right spirit. Sudanese leaders should realize that they were not appointed "for their own personal advantage or profit," nor just for profit of government and Syndicate. The new leaders were responsible for "the common welfare of the tenants put in their charge."[61] Village councils had to balance the power of samads. The quick growth of samads—who were not elected by tenants nor controlled by village councils in most cases—posed the danger that samads would "become petty, but tyrannical, bureaucrats."[62]

After the turbulent times of World War II, during which the Sudan Government had decided not to renew the Syndicate concession after 1950 (as explained further in chapter 6), the policy of village councils only became more important. Devolution may have been a firm new policy, but the shift from an executive to an advisory position for British officials in Gezira—whether from government or Syndicate—was expected to take years—by the British. In the meantime, building stronger personal relationships between British and Sudanese was encouraged, as it would show Sudanese that the British were serious about devolution. Only when the Sudanese saw this, they would become "receptive learners of British ideas."[63] British should avoid showing the Sudanese "lack of sympathy" or refusing "the legitimate aspirations" of the Sudanese "for whom we are trustees." Without Sudanese faith, the entire colonial policy of devolution would collapse even before it started.[64] The general unrest in the Sudan in October 1946 only confirmed the need for building stronger relations between Sudanese and British. It was also a sign to stress the need for continuing British rule, as every "reasonable person" knew that without the British "anarchy and chaos [were] inevitable."[65]

Cooperative Socialism

The summer of 1946 started with tenants striking against—what the tenants thought—unfair use of the Tenants' Reserve Fund, the collective arrangement from the late 1930s in response to the problems Gezira had in making money in the early 1930s. As chapter 2 has shown, British ideas on tenant arrangements in Gezira came to include more and more collective elements. Defending the resulting collective policies as the way forward became more sophisticated over time as well. With the wisdom of hindsight, bright pictures of a grand future were painted, as if collective policies had been planned from the start and yielded immense promise and success.

As their English colleagues had done for England, fine Sudanese yeomen should become the "backbone in the formative years" of Sudan.[66] Freed from "the mercy of precarious rainfall," Gezira tenants had grown as farmers showing more interest in "the financial details of the scheme." This interest was fine, as long as the tenants—and everyone else—understood that "individual enterprise" in Gezira was "subject to certain cooperative features" that were "essential to its success."[67] One of these features was protecting the yeoman tenant against slavery by a "minority of the big men."[68] Gaitskell in particular never failed to emphasize that the collectivism of Gezira was not an answer to a financial crisis in the 1930s, but a key solution for a static society with wealth concentrated in the hands of a few large landowners and merchants.[69]

In 1943, actually one year before the unhappy message of nonrenewal would be delivered to the Syndicate, Gaitskell gave his optimistic view of the world to come, in which money no longer was "the only thing we think about." The keyword was "cooperative socialism." To Gaitskell, the "most interesting of all features" of Gezira were the "co-operative advantages" for tenants. A tenant did not need to provide capital, he did not pay rent for land or water, he could ask for loans, and he had the benefit of scientific research. Furthermore, "[i]nstruction, supervision, tools and materials are all provided." The collective Reserve Fund helped tenants through bad years. And indeed, "all these advantages of co-operative Socialism" did not prevent a tenant from making his own profits. His cotton profits were paid to him based on his own yield, and his grain crop was his own.[70]

Gezira had affected its society enormously, both "for good and for bad." Slavery had disappeared and a cooperative society had been created, but this all had happened at the cost of the breakdown of old family authority—we again encounter the family ideal. "Personal gain" had become too important for tenants, an interesting remark coming from a SPS manager. To counter personal interests, a spirit of community service and social morality was needed. Better standards of sanitation, health, elementary education, and recreation were needed as well, as this would make Gezira life "less brutish and these country folk more spirited and capable." Learning "self-management in their own affairs"[71] was required for tenants, and not straightforward at all.

This view on tenants as being simple and still focusing on old values was persistent in British colonial circles. Despite the huge progress, tenants would still have trouble making the shift from nomadism—which was "still so close as a neighbour to the inhabitants of the Gezira"—to irrigated farming. The need for "routine in one spot" instead of the "far less exacting" nomadism was still demanding.[72] After making the shift, though, it was clear that Gezira tenants had to remain comfortable with being a cotton grower—or any other crop for that matter as long as done within clearly prescribed conditions. Sudanese tenants were not supposed to have personal ambitions that were too high. This rather limited future was not an "irksome duty" but "the door to a wider and richer life" once villages would function as social units. If and when the tenants could manage their own affairs within their village, cooperative society could prosper.[73]

In order to ensure that irrigation and social life could prosper together, specific irrigated areas should be attached to a village and its tenants. This was new, as in the original Gezira, administrative and irrigation boundaries were not correlated. When the gravity system started, the scheme was part of three districts, each having areas within and outside the scheme. In an attempt to bring some administrative unity, a "Commissioner of the Gezira Area" was appointed in 1928, but District Commissioners kept their powers. Syndicate authority and native authorities overlapped, but it was the native administration that was split across several agricultural blocks. The agricultural infrastructure had become the "dominant spatial influence upon the region," which was at least partially materialized in the canals. A simple walk to another village was complicated because only a few bridges crossed the canals. The Sudan Government adapted to the existing physical space when reshaping its administration. "The dominance of the Scheme upon administration was complete."[74]

The irrigation engineers were the last ones having problems with such a policy. For them, a "fundamental requirement" of irrigation was "co-operation" within "a community served by a communal water-course." In Gezira, once communities were organized on Abu Ishreen and Minor Canal levels, Syndicate control could be relaxed.[75] This may appear as a technocratic argument related to effective water control—which it partially was of course—but it turned into something close to a political position when the SID argued that an irrigation-based community could only develop with "reasonable permanency of tenure." This should allow farmers and their children to improve their land.[76] In developing proper tenants, the many wrong indents from inspectors did not offer "good training for the cultivator."[77] A key issue for the engineers was—obviously—the night-storage system. In 1944, an experiment with night irrigation in the Abdel Hakimb Block allowed the SID representative once again to emphasize that, contrary to the ideas of the SPS, it was "possible to gradually train the cultivator to normal irrigation methods"—read: irrigate during the night.[78]

Creating advantageous infrastructure, training tenants, and changing the SPS approach was obviously "very slow work." It had to be the Sudan Government to "do the work at the start" ensuring the communities would emerge.[79] The new Gezira communities would be different from those elsewhere in Sudan, as a "sophisticated rural area" like Gezira did not fit in the typical rural/urban dichotomy of Sudan Government policies.[80] Gezira became the forerunner and its tenants the elite of devolution, a concept that lingered long within the official discourse. As late as 1967 we read that a tenant should "care personally about the land he works" and create "an organic, self-renewing society." Instead of sticking to "the tradition of their society," they should be proud producers of the cotton to "benefit the country as a whole."[81]

In 1947, 60 percent of the Gezira Scheme was covered by village councils.[82] Joint SPS and government directives in 1945 and 1947 explained the role of social development, including new services and potentially increased tenant initiative, in relation to the increasingly important role of village councils. Despite the beautiful words and pretty concepts in the joint instructions to SPS personnel, the decision not to renew the concession created a problem with the new Gezira dual mandate of cotton production and tenant education. Given the obvious differences between short-term SPS benefits and longer-term government goals, how could an inspector work to the best advantage of his employer? What was the best advantage exactly?

The first years after 1945 did not allow for the standard level of Syndicate control as before—with lower number of inspectors and tenant unrest one would not have expected anything else. The basic rhythm of control, however, was quickly regained. In October 1945, SPS inspector F. Bertram Hunt—whom we will meet again in the next chapter—thought that the cotton was "looking lovely" with "truly enormous" plants.[83] A few weeks later he saw that the SPS ploughs had "done the two odd corners I had requested the engineer to plough,"[84] and in March 1946 he just went on as usual with payment day. Completely in line with regular procedures, he "paid out only half" to punish "bad picking." He concluded, however, that it was "a poor crop this year."[85]

With the desire of the Syndicate to go back to business as usual (emphasis on "business") after World War II, it was only to be expected that the SPS and its employees "had to limit their views to some extent to the financial loyalties and responsibilities to their shareholders." This was only more logical given the amount of work that was asked from SPS staff,[86] including improving livestock and milk supply, mechanization, or stimulating growth of fruits and vegetables, to name only a few items on a list from 1948 that included an enormous amount of hints, projects, and tasks that could be done by an SPS inspector.[87] There was simply too much to be done "to be satisfied by the haphazard enthusiasm of a few unassisted individuals."[88]

The 1947 version of the shared directive from SPS and government had called for "a unified approach" in the development of village councils to ensure "the best agricultural results" achived by "healthy, contented and self-reliant" tenants' communities. In an effort to ensure unification of activities, the Block inspector was made responsible for the councils of his area. Councils should be involved in daily decisions, but any "final confirmation" on issues like allotment of tenancies, dismissal of tenants, and arranging "tulba" and "wakils" was to remain with the inspector "a long time to come." The inspector was even supposed to have a strong say in the formation of the village councils themselves. The ultimate goal might be to organize council elections, but in postwar colonial Sudan, it was apparently "too early to resort to elections by secret ballot." Instead, members were appointed by the Sudan Government. The Block Inspector could advise on "important sections and interests" and suggest "influential and respected" council members.[89]

SPS inspectors may not have been governmental staff, but as their work was semipublic—and had become even more so with the shift in policy after World War II—they too had to arrange their duties within a new perspective. In line with these new duties, another set of instructions was sent to the SPS field staff in 1948, jointly signed by the SPS Manager and the Governor of Blue Nile Province, confirming what was already policy and practice.[90] Assuming that Gezira local government could only be based on the village group and that SPS inspectors had very detailed knowledge of the people in his Block, each inspector was assumed to have a key position in stimulating village councils. His "almost daily contacts" with tenants gave him an "unrivaled position" to assist the British administration. Inspectors had "to encourage the people along the path of progress."[91]

Village councils were not expected to take on all new social development issues at once. Inspectors had some freedom to suggest activities that "their" councils could work on, including more attention to education, recreation, and health. It was for the Block inspectors to stimulate "village conscience" when making those choices. The 1948 instructions gave inspectors quite a bit of room to make their own selections. After all, one could not expect Block inspectors to be willing or able to "handle all these ideas at once." The best policy was to be developed slowly, with inspectors working on issues of their own interest. Increased attention for councils should not lead to a "fall-off in efficiency," but stimulate the opposite. The need for efficiency was taken for granted, but the way to ensure it had to change from "the simple order and obedience" approach into stimulating "a better understanding" why such orders were given for the "well-being of the tenant and his crops."[92] As mentioned earlier, the larger goal for the tenants was transforming them from subcontractors to yeomen farmers with the proper mindset.

Given this whole setup, it is not that strange that a later observer—who had been an actor at the time—concluded village councils were "much more autocratic than was theoretically desirable."[93] One could, however, doubt whether the outcome could have been any different: the policy of devolution in Gezira kept the tension between controlled cotton production and increased tenant freedom and as such did not prevent regular SPS policy to prevail in practice. For a contemporary observer, however, it was clear that the "ultimate aim of self-work" was still a dream because of "the apathy" of the Sudanese "to improve their lives by working harder."[94]

A Little Lip Service

To many British, there was not much else to expect; they fully agreed with Gaitskell that many tenants were "still simply children."[95] In his other writings, Gaitskell emphasized the need to move from "the 'parental' stage" in which decisions were made for the tenants "just as parents decide for their children" to "the 'school-teacher' stage," in which tenants had to be taught to become happy. In trying to bridge the Syndicate and government interests, Gaitskell treated both parties continuously with new texts to ensure Gezira developments would take the right direction. Balancing between Syndicate and government ideas, he could both define Gezira as "one huge plantation" without room for individual initiative from tenants and maintain that there were great options to develop the desired "new, stable, social, and cultural life."[96]

Government officials were more likely to agree with Gaitskell that discussions on devolution and more tenant influence were needed if alone "to avoid at least part of our troubles when the war is over."[97] On the SPS side, he did not necessarily meet much goodwill. For Syndicate management, bringing too much of all these debates out in the open was useless anyway. Those Sudanese who could actually read would not be "capable of understanding the meaning" of any document anyway,[98] but the effect of those ideas on the "uneducated mass of very backward nas [natives]" would be dangerous right away.[99] Gaitskell was unhappy with the results of devolution policies and regularly complained that there was "insufficient guidance" for

SPS staff;[100] the issue was of course that the guidance had to come from an equally overloaded Provincial government. However, true as this observation from a rather pragmatic angle may be, there is more to say about the reasons of less than desired results from devolution.

Gaitskell would have done well to check upon the internal standards of his own Syndicate before complaining that others did not provide enough support for the inspectors. The internal SPS instructions given to new staff stressed that using village councils was to be restricted primarily to agricultural issues, although "social matters of benefit to the local people are not excluded." Councils could not be efficient "without regular supervision," but inspectors should not spend too much time on them, as otherwise the inspector could "lose touch with his 'maroor.'"[101] It is very clear that—notwithstanding several individual inspectors being much more engaged with the issue—it was standard postwar SPS policy to go on with business as usual, "except for a little lip service" to devolution.[102] The Syndicate simply had never been on board for any policy of devolution. Already in 1938, MacIntyre wrote to Asquith—still on the SPS Board—that he did not think that the SPS "should start to support any of the social services" discussed by the Sudan Government. This would only "subscribe money" to something that would not make "our tenants benefitting any more than they do now."[103]

When the Sudan Gezira Board took over Gezira management in 1950, social development and devolution were made obligatory throughout the Gezira. Despite the many years of experiment with devolutionary activities—in Abdel Magid, in Hosh—the average Gezira inhabitant was still assumed to be "blissfully ignorant" of devolution. Even in "the more sophisticated areas" in Gezira "among the more intelligent tenants" who did start to get the point of what their British educators and advisers were trying to get at, "the multiplicity of councils, committees and associations must be very confusing." Apart from measures on health and basic infrastructure—that could be standardized to a certain extent—it was standard SGB policy "to find out by experiment the right answer to social problems."[104]

Among the better defined activities was a stronger focus on issues like health. Every Thursday, canals close to houses had to be "completely dry"[105] to control mosquito growth; on the same day the main feeder channel would be oiled for the same reason.[106] In addition to the major work keeping the irrigation infrastructure running without weeds, silts, and bugs—and constructing extensions—the SID encountered new work in the "so-called roads of the Gezira" suffering "under increasing heavy traffic."[107] In Syndicate times, decisions on roads, or bridges over canals, used to be taken in meetings between SID and SPS. Now, village councils were responsible for roads and bridges. The SID understood perfectly well that many councils asked for more bridges. Crossing canals was not only handy, but a "new bridge stands as a visible monument to something done." With many possible sources for financing bridges available—local government, national government, the social welfare budget of the SGB—payment was heavily debated. Unfortunately, these arguments did "not, however, produce bridges."[108]

As a result, the SGB did have quite a few projects and activities, but instructions for personnel on what to do remained rather restrained. This was partially a choice by necessity as financial resources available would not be enough to meet all potential social development projects, but the ideological defense of the Board was at least as important. The SGB did not want to take all the initiative in a post-Syndicate Gezira. The SGB did "not accept a policy of spoon feeding"—a clear indication that the SGB wanted to be different from the Syndicate.[109]

Not Much Fun for the Inspector

However, there was one aspect where the SGB was not different at all from the Syndicate. Throughout Gezira sources, whether they are from the Syndicate or the SGB, tenants remain described in terms of "natives" not being "over-industrious."[110] The complaints about tenants being lazy or still needing supervision continue well into the times of the Sudan Gezira Board. In the early days, the Sudanese were often compared to their Egyptian colleagues—not in favor of the Sudanese who were "not capable of doing the heavy manual work involved"[111] or had "experience nor capital" to develop their own field canals.[112] Sudanese tenants were not seen as incapable of learning, as several did indeed develop a "more intelligent interest" in cotton cultivation after being a tenant for many years.[113] Once tenants did make a choice, however, based on their interest, this was not always recognized. As late as in 1965, under the SGB, the financial successes of tenants growing vegetables in the northern blocks close to Khartoum were acknowledged, but the tone of surprise that tenants were actually able to do so by themselves is clear.[114] Needless to say the vegetable growing had to become the responsibility of the SGB, in order to improve it.

Despite avalanches of ideas and plans for Gezira tenants, what they were like and how they should behave, accounts from tenants themselves or people who could report about tenants from firsthand experience are rare. British people in Gezira lived "encased in English suburbia" without contact with villages on either bank of the Blue Nile. Living in Gezira might have meant being "physically in the midst of other people," but most Syndicate people appear to have stayed "utterly ignorant of how [tenants] really live." Not too many of "the Lordly Syndicate people" did bother to engage with the tenants.[115] The information that new SPS inspectors did receive about tenants was mostly stated in terms of what the tenants did not do well. The average tenant was granted knowledge on durra cultivation, although "carelessness in cultivation" of the grain was directly associated with "the average Arab" disliking "continual manual work." Interestingly enough, tenants were seen as rational: inspectors were informed that tenants did not like to work on cotton as they only received 40 percent of the profits.[116] Indeed, we have clear evidence of the economic motivations of tenants when most Tayiba tenants refused to accept changes in the tenancy agreement in 1913 (discussed in chapter 2).

The Gezira documents tell a story of lacking skills, with tenants unable "to get a job done by a fixed date," or "managing their financial affairs" as they fail "to look ahead." Tenants lacked "the qualities of attention to detail and perseverance." Strategic behavior toward SPS inspectors was recognized, but dismissed as "a childish tendency" to "'bluff' the Inspector."[117] Tenants did not work "easily after rain"—something that holds for inspectors as well as we will discover in the next chapter. We read about tenants summoned to work extra but who "seemed soon to melt away."[118] In periods like Ramadan—for obvious reasons—"not much work" was going on, which was "not much fun for the Inspector."[119] Interest in something else than cotton was not really on any SPS horizon, as such interest was "useless from the point of view of getting things done."[120] Nevertheless, inspectors were encouraged to learn tenants' names, their social relations, and their "circumstances," as this made tenants "more responsive."[121] Whether it would have added to the fun for inspectors remains unclear, but corporal punishment was officially forbidden.[122]

In 1949, Culwick came to Gezira as a government employee to study food and alimentation in the scheme;[123] she returned in 1951 to work for the SGB on social development. Apart from being the first noninspector working in Gezira (although not for the SPS, but later for the SGB), and the first woman in a nonadministrative position, Culwick is the first Brit who appears to be genuinely interested in the Sudanese inhabitants of Gezira. She was full of admiration for Gaitskell—"a brilliant and charming man with a real interest in social development"[124] and "with real vision"[125]—but it is also clear that she wanted to come much closer to the Sudanese than Gaitskell ever wanted to be. In order to do her work, she arranged housing in a Gezira village to "see all sorts of interesting episodes as an unnoticed spectator."[126] The house was luxury for a Gezira village, as it was a guest house from a wealthy local Sudanese, and Culwick may have continued to visit the Gezira clubs and inspectors houses for a "civilised bath" every now and then (in addition to her "trek bath" in the village house), but she did come as close to village life as probably possible in colonial Sudan, including the "flies in the village which have to be experienced to be believed."[127]

Culwick even joined the fasting during Ramadan, partially because she wanted to see the effects, partially for more strategic reasons as it went "down extremely well with the people."[128] Indeed, fasting did work really well: to her "own or everybody else's astonishment," all 98 households were running food records and weighed and measured themselves within weeks after Culwick's arrival in the village.[129] Obviously, her contact with the village women was key here, even though these women could not "think why [Culwick] should be interested."[130] Building on her early observations that the social status of women in Gezira were "about as dark"[131] as it could possibly be, she returned to the issue in her work in the early 1950s. The "duty of personal appearance in the field" was a clear obstacle for women.[132] In addition, women could not easily attend public events like the massive (collective!) payment days. Another, more general, but major obstacle was that women were not recognized as possible tenants, let alone as good tenants, by Sudanese and British (men) alike.[133]

The Mist of Lost Illusions

In addition to her sharp observations about SPS inspectors—something we will see again in the next chapter—and the position of Sudanese women, Culwick was pretty clear as well about the—imaginary—idea of the family as building block for Gezira agriculture. For her, the idea that Gezira tenants acted like small-holder or peasant families on their holdings had "never been more than an illusion."[134] Earlier, similar observations had been made, in the sense that one could doubt that a family unit of labor had "ever existed."[135] Outside labor in the peak seasons was not the only factor; Culwick argued that "its association with slavery" made "field work a despised occupation," making tenants use outside labor to the "maximum extent possible."[136]

The labor issue and the claimed laziness of tenants were two sides of the same coin. Using labor of slaves (including poorer Sudanese) for all possible manual labor was seen as the original way of doing things. To what extent this image is correct is not that easy to say, but it seems to be clear that although only a small percentage of the population could have hired labor and accumulated individual wealth before the irrigation system, the associated values and aspirations were apparently a source of inspiration—a dream—for the wider population or simply the model to work with. Becoming a tenant opened possibilities to become a person who could actually hire external labor. Hiring the labor was not only a way to raise social status, but actually necessary as well. The British in Gezira—Syndicate and government alike—preferred to view Gezira tenants as lazy and self-serving, not as people with aspirations.[137]

Many of the workers in the picking season came from villages outside Gezira. Especially the remote villages were "really grim," and some even "completely deserted," when the inhabitants had gone to Gezira to pick cotton (figure 4.1).[138] In addition to pickers from close by, many seasonal laborers ("Westerners," as they came from Western Sudan, French Equatorial Africa, or Nigeria) came in from areas further away to stay in the Gezira in picker camps. During the cotton picking season, families could have their meals "all over the place," depending on whether the woman responsible for cooking was picking cotton or not.[139] Already in 1934, there were 8,000 Westerners living in "Western settlements" across the scheme. Several others lived in the original villages. In 1940, there were 62,000 Western laborers within the scheme, on a total population of 380,000.[140] Over time, laborers started cultivating tenancies with poor land far from a village—as they has aspirations too.[141] In the economic crisis of the 1930s, vacant tenancies were offered to Westerners, who were more willing to work along SPS standards.

Figure 4.1 Cotton picking (Australia National Library). Reproduced by permission of the National Library of Australia (nla.pic-an23565551)

With all the changes brought in Gezira by the British cotton project, the original hierarchical society in the area did—obviously—not disappear completely. Larger Sudanese landowners dominated in the scheme and were member of a body like the Rural District Council of South Gezira District. The village councils that were established in the 1940s in an attempt to strengthen local Sudanese rule were only open to tenants or ex-tenants. The growing number of non-Sudanese tenants in Gezira became an issue after World War II with increasing nationalistic idea(l)s in Sudan. Westerners' communities were seen as undesirable, both on economic and ethnic grounds. In 1945, for example, the Rural District Council of South Gezira District passed a resolution recommending that all Westerners with more than three previous convictions for crimes of theft and violence should be repatriated.[142]

All in all, the typical Gezira tenant did not exist; underneath such a term, large socioeconomic differences were hidden.[143] Nevertheless, for many Sudanese and non-Sudanese tenants, the Gezira scheme was more than an "exploitative and unwelcome economic structure," as they did use Gezira "to achieve culturally embedded aspirations."[144] Through their actions, the tenants made the "imaginary peasant farmer" disappear "into the mist of lost illusions."[145]

Tenant Tarteels

In a way, this chapter is a preparation for the discussion on "development" I will turn to in the epilogue. Even in the first three decades of the twentieth century, when economic profits were the main driver for building Gezira, the need to develop backward Sudanese and lead them to progress was claimed by government and Syndicate officials alike. I could imagine that SPS representatives had a clear interest to include those considerations in their public communications to please the government, but the even more interesting possibility is that an actor like MacIntyre—not the first to defend social development—actually had ideas about the Gezira project guiding the tenants to a desirable social status. He seemed to have strong ideas that making sure that "the youths of the Gezira" would become "accustomed to the Tenant 'Tarteels'" was actually a good thing.[146] To ensure young Sudanese (boys) would become "agriculturally minded," small farm schools had to be established in empty inspector houses. Gezira boys could grow two feddans of cotton. Paying them would make them feel they were "becoming a real tenant."[147]

The SGB—and its first chairman Gaitskell—had a much stronger claim and urge to "develop," which was at the same time a good reason for most British to stay in Sudan. After all, the British were the required impartial guide to help the divided Sudanese nation. As we will develop further in chapters 6 and 7, and frame as a larger debate in the epilogue, such a position of "guide" can only be kept up if something like an independent "measure" or "standard" of development exists—and that Britain itself had already reached the required notch on that standard. In chapter 7, I explore how the Gezira development history could be (and was) turned into a development model. In chapter 6, we will find out how the post-SPS Gezira was to be arranged. Before those chapters, however, we turn our attention to those who as employees of the Syndicate were supposed to do the actual spoon-feeding of the tenants. We will discover that quite a few Syndicate inspectors did not necessarily like using spoons in such a way—even though they did like good food and plenty of drinks!

Chapter 5

Another's Week's Toil: British SPS Inspectors and Their Idea(l)s

When visiting the Gezira in 1942 after a few years on voluntary duty in the Sudan Defense Forces (SDF), Syndicate inspector F. Bertram Hunt entrusted to his diary that he had decided to leave the SPS. He did "not like the atmosphere," as there was too much "[p]etty self-seeking," including "jockeying for position and trying to catch the bosses' eye." Catching the bosses' eye was crucial, however, as they had "great powers," basically through "one's earnings through unpredictable salary increments." With their "stingy attitude towards housing and equipment," however, the bosses did make the work "unattractive."[1] Obviously, Hunt's—extensive—diary is a little colored when it comes to interpreting Gezira realities; comparison with other material, including letters and diaries from others working for the SPS and/or in Gezira, however, suggests that the general setting is very well pictured by Hunt. It's just that he did like it less.

An Oxford graduate, Hunt had joined the SPS in September 1934. He joined the SDF at the outbreak of World War II.[2] Despite his decision in 1942 to leave the Syndicate, he returned to Gezira in April 1945 to take up an inspector position in the Northern area. Hunt did not observe any "change in the attitude" of the SPS, and the manager was "a complete dictator."[3] The first season after the war had a poor cotton crop, but the return to the "old chase-chase again" was planned for the 1946 season.[4] Hunt tried to move to the Department of Education in Sudan, and indeed on September 30, 1945, he was offered a contract of £660 per year (plus 26.5% as there was no pension involved). Hunt wanted a little more, but even when he was offered £900 a month later, he continued to doubt and finally turned it down on January 22, 1946. Hunt did not want to become a teacher but decided to return to England. In the meantime, he continued working as an SPS inspector, until he could go on leave. In April, Hunt went on leave, to arrive in Liverpool in May; he formally resigned from the Syndicate in June 1946.

Figure 5.1 Sudan Plantations Syndicate cotton inspector standing next to a motorcar. Reproduced by permission of Durham University Library (F. B. Hunt; SAD.692/19/5)

In Hunt we encounter an inspector with negative views on Gezira work and life. As we will see in this chapter, there were enough inspectors who enjoyed their work in Gezira, or at least stayed on as inspectors for long periods. We will encounter Edward Philip Gibbons, who started working for the SPS in 1923 and retired from the service of the Sudan Gezira Board in 1959 (!). Ethelbert Adrian Patrick Taylor was an SPS inspector between 1926 and 1946. His colleague Thewlis Clarkson Johnson joined in 1930 and retired in 1953, again from SGB service. Their careers shed light on the ideas and reflections of the British SPS field staff on working and living in the Gezira area. This will offer a perspective that has only very rarely entered discussions on Gezira—let alone on colonial life and work in Sudan or colonial areas in general—namely the perspective of the British people that were supposed to implement the colonial agenda. SPS inspectors were not part of the Sudan Government, but they mingled with engineers and civil servants in clubs, house dinners, and picnics, sharing their Britishness in the Sudan.

In the following paragraphs, I will focus on the working conditions of SPS inspectors (figure 5.1), including their responsibilities, their routines, and their responses. I start with an overview of the way the British SPS field staff was organized, the number of inspectors in SPS service, and their average time of employment. Who were these inspectors, and where did they come from? What were their conditions of employment? This will be followed by my attempts to sketch daily routines that are hidden behind these numbers. The British SPS inspectors—in the early years individual men, later men and their families—lived in Gezira and worked within a strict daily schedule. The result of their work—cotton—was closely monitored by SPS management; decisions on yearly bonuses were eagerly awaited by inspectors.

Outside work, inspectors were often bored, as there was not much entertainment in Gezira with its monotonous landscape and very strict routines. Why did many inspectors become great polo players, and pretty much all of them experts in off-road—even amphibious—driving? Finally, I discuss how the British SPS staff responded to the changes that came upon Gezira—and their daily work—after 1945.

Almost One Hobby

Inspectors were recruited from Britain—with quite a few, though not all, having degrees in agriculture from the better universities.[5] Figure 5.2 shows the total number of employed people by the SPS (or the SGB after 1950) over time, with quick growth in the early 1920s and later stability. In the years before the opening of the larger Gezira area, anticipating the larger number of field staff required, the SPS employed more field staff than strictly necessary. The existing smaller pump areas were used as "training ground" for inspectors.[6] These numbers hide that many of inspectors worked relatively briefly for the SPS, as figure 5.3 shows. Even if we correct for the fact that British inspectors starting to work for the Syndicate or SGB after World War II could not have worked long for each entity either, the large majority of field staff was employed only up to five years in Gezira. Living in Gezira was "by no means to everybody's liking," which made "a fairly big turnover of British Field Staff" normal.[7]

When looking at the numbers of inspectors and their employment records, one should realize that leaving the SPS was only practically feasible during one of the biannual leaves one would have. Being back in Britain when resigning avoided the need to pay the return trip from Sudan oneself. After two years of completed service, an inspector was entitled to four months leave plus an additional £100 travelling allowance. This leave was taken from the end of March to the end of July. Fourteen months later, field staff was entitled to another home leave—this time of three months from end of September to end of December. As one had to be back at the end of December, this meant leaving Britain just before Christmas.[8] During July and August, the first two months of the growing season, and January to March, the picking season, leave was out of the question. Practical opportunities to resign were limited to every second year. Despite the large number of people working only briefly for the SPS, it is also clear from 5.3 that there were inspectors with an entire career within the SPS—Gaitskell being among them.

In February 1923, Edward Philip Gibbons joined the ranks of the SPS, for what turned out to become an impressively long career—longer than "Mr. Gezira" himself, Gaitskell. Born on April 16, 1901, Gibbons went to Eton between 1914 and 1918. He continued with the Degree Course in Agriculture at Magdalene College in Cambridge in 1919. After his final exams in June 1922, he applied for a position in the Colonial Agricultural Service, not really strange for someone whose "almost one hobby, all his life,

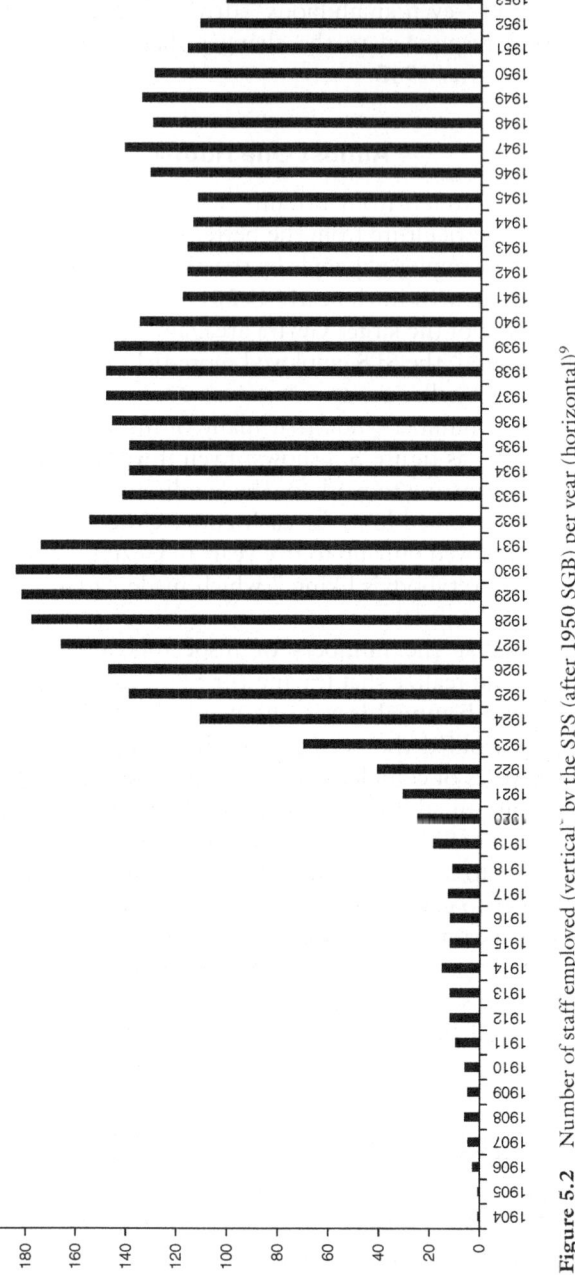

Figure 5.2 Number of staff employed (vertical) by the SPS (after 1950 SGB) per year (horizontal)[9]

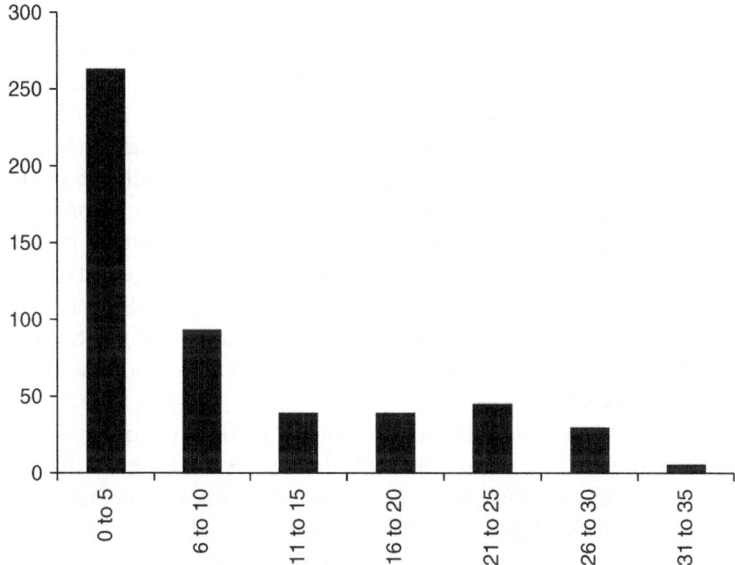

Figure 5.3 Five-year interval distribution of total years of employment of SPS staff[10]

has been farming and Agriculture."[11] Before his degree in Cambridge, he had been closely involved with agriculture at his father's farm. His combined experiences "gave little idea of the future job" in Sudan, however, as the information given by the SPS after recruitment were only "vague and incomplete." Together with another new inspector—a graduate from Oxford—he travelled to Egypt and took the train from Luxor to Shellal, "a good introduction to dust and flies." A Nile Steamer brought them to Khartoum. Finally, they arrived in Barakat, where the Syndicate Headquarters was located. In 1923, Barakat counted five houses for British staff and a village for the Sudanese.[12]

Gibbons started as a junior inspector in 1923; most likely he was promoted to Block Inspector in 1925 in the new Gezira area. In 1930, he was given a larger Block. On Christmas day in 1931, he was promoted to the position of Group Inspector for the Northern Group of 180,000 feddans (some eight Blocks) in Turabi.[13] In 1950, Gibbons continued to work for the successor of the SPS, the SGB, in different positions. He retired on January 1, 1959. In his long career, Gibbons was employed in most positions a British SPS inspector could have in the Gezira. As junior inspector, he learned the job; as Block Inspector he was responsible for his own work and a few new junior inspectors. All inspectors in a Block were supervised by a Group Inspector, which he became himself in 1931. His positions under the SGB are representative for a typical career of a British inspector turning into an expatriate expert. Gibbons first stayed on as Group Inspector, was then asked to guide new extensions, and was finally appointed to advise on new extensions. After 1950, Gibbons gradually moved away from fieldwork.

Although the archival material is silent on the details of his appointment, it is likely that Gibbons was appointed in a similar way and on similar terms as his colleague Taylor, who joined the SPS in November 1926. In the letter of employment, the secretary of the SPS announced that Taylor's starting salary would be 400 Egyptian Pounds (EP) per year, starting on the day that he would sail to Egypt. His travel allowance of 70 EP would be paid upon joining; his passage was booked by the SPS, and he would be informed when to sail. The contract could be terminated at any time by either party, with one month's previous notice, but in case of resignation at the initiative of an inspector, no travel allowance was paid. Furthermore, no SPS staff member was allowed to become involved in irrigated agriculture in the Sudan for a three-year period after retirement from the Syndicate without written permission from the SPS.[14]

Taylor was told to be vaccinated, inoculated against typhoid, and have his teeth attended to before travelling.[15] Health was something of importance, as the "incidence" of " certain diseases" was "at times" very heavy among Syndicate inspectors. Malaria was practically unavoidable given the wet conditions and large-spread stagnant water in Gezira. What was suggested was "careful supervision" and treatment of servants and neighbors, "particularly the former," as these were assumed to be infected anyway. Avoiding dysentery was a matter of drinking only clean water and discouraging "[d]efacation in canals by natives." Contact with Sudanese women was discouraged, as "every native woman should be considered as infected" with venereal diseases. The "only adequate protection" was "abstinence." Using alcohol was discouraged as well, as it was under its influence that "caution is disregarded." Two medium whiskies per day were fine, but "anything in excess of this must be looked on as definitely injurious."[16] We will discover below that this advice was somehow lost in Gezira life. Alcohol could occasionally make inspectors very wet too, as we will find out later in this chapter.

Just Go and Look

Upon arrival in Gezira, inspectors were supposed to have or arrange their own kit, which included anything from riding boots to medicine against snake bites, and from a folding camp table to light clothes of "white or khaki color."[17] To what extent the kit was useful is unclear; records on that issue from SPS inspectors are unavailable, but civil servant James Wilson Robertson, who was recruited in the early 1920s as well and whose first post was in the Gezira area, reports that those lists were "usually all rot. I haven't needed half the clothes I brought out."[18] The Syndicate provided (unfurnished) housing, which during the early period of Gezira still being constructed—which extended well into the 1920s—meant that field staff lived in tents or other temporary quarters.[19] We can only assume that the experience of being in a Syndicate tent in Gezira came close to the experiences of surveyor Crompton in the early twentieth century. The habubs (sandstorms) would not have made camping much easier. Throughout Syndicate times, accounts on living quarters full of dust—similar to the tablecloth with the circle by Crompton[20]—remained pretty normal.

Once available, SPS houses came in standard series: the "A" pattern houses were reserved for Block Inspectors (and presumably Group Inspectors), the smaller "B" pattern houses for the other inspectors. Hunt started in Gezira as third inspector in a Northern Division Block under Block Inspector M. E. Webb and Group Inspector Gibbons. Hunt lived in a house about two miles north of Webb's house. The house of second inspector W. H. Robertson-Aikman was toward the northern end of the Block, some three miles north of Hunt's quarters. Near the Block Inspector's house was the Block office, basically a storage building plus a (smaller) pattern "D" house for the Bash-Katib (Chief clerk), who was usually a Greek or Syrian.

A "B" house was a single story, red brick dwelling with a flat roof, with two rooms, a bathroom, a store room, and a veranda on three sides (figure 5.4). Especially during the hot season, one lived on the veranda. Floors were made of smooth red cement tiles. Inside room walls and rafters supporting roof and veranda were painted every few years by the Engineering Department of the Syndicate. Color choice included "cream" or "stone grey," although the choice between the two would not have mattered that much. When an inspector was phoned by the SPS Engineering Department to check if a paint job had gone well, he was asked what color had been put on the walls. "The inspector replied 'Although they have been working here for a week, I am blowed if I know, I will just go and look.'"[21] Hunt assures us that such response was pretty common in Gezira.

Figure 5.4 Sudan Plantations Syndicate cotton inspector's house, probably B-type. Reproduced by permission of Durham University Library (F. B. Hunt; SAD.692/19/6)

In the 1930s, SPS houses had no electricity, gas, or running water, which offered some challenges for working and living under a "grilling hot sun"[22] without fans or refrigerators. Water, both for drinking and bathing, was carried daily from the canal outside the house compound, and stored in "zeers"—large earthenware vessels on wooden stands on the back veranda. Hunt kept his butter in a jam jar that was suspended in the zeer to prevent the butter from melting. Some of his colleagues had hand-operated cream separators, but only if one kept a cow. In the early days, so-called "shamadans" (candles with a glass cover) provided light in the houses, later pressure lamps took over.[23] All SPS houses were set in the middle of a 10 feddan plot. Many inspectors made attempts to cultivate a small garden, with apparent mixed results.[24] Some grew fodder for their ponies, for which each house had two stables.

Tasks like gardening and especially fetching water would be done by the servants, up to six for an inspector. The cook was usually the senior servant, but the "suffrage" ("house boy") did most of the work. The suffrage cleaned the house, washed and laid out cloths—including "a clean linen watch strap every day"—and served all meals, from bringing early morning tea to pouring out drinks or serving dinner at any time between 9 p.m. and 3 a.m. to the inspector and any guests. The "car boy" cleaned the car—once available—and fetched it from the garage; the gardener (obviously) took care of the garden, one servant "emptied the bucket" from the lavatory set some 25 yards away from the house itself and brought all the water from the canal, and finally a "syce" would look after the ponies—if any, for which the SPS provided an additional allowance. All servants lived in an outhouse, which also contained the kitchen,[25] but the SPS did not build those outhouses, which had to be provided for by the inspector himself.[26]

This inspector house(hold) was set in one of the flattest landscapes to be found anywhere. All accounts agree on this, including adding a negative tone. The Gezira landscape features as "the least inspiring you could possibly imagine."[27] Civil servant Stanhope Rowton Simpson found that "the general environment itself, once the novelty had worn off, was really rather depressing." After a while he "began to get quite a craving for the sight of a hill."[28] Gezira flatness was "criss-crossed by the most bewildering network of canals, field channels and drains" with the added problem that travelling in Gezira was like "a sort of jigsaw puzzle to the stranger, as one dodges this way and that to find a bridge." It was "most irritating at times" being near the destination "and yet be hours finding your way to it."[29] Even experienced drivers could easily lose their way "in this incredible maze of irrigation"[30]—especially at night. Quite a few cars ended up in the canals—as we will also discover in more detail below.

Within the flat, canal-dominated landscape, British society in Gezira appears as pretty confined. The evidence is rather overwhelming on this aspect, but the judgment on it differs per person. Hunt was pretty clear about his disliking it many times—even though he did enjoy the many parties. In 1938, after a visit to Khartoum had allowed him to be "away from it all," he complained about Gezira life being "mentally pretty narrow," always meeting people "with a very similar outlook on life here."[31] In the 1930s, there was no newspaper circulating in the Gezira. Some British staff members had a battery-powered radio for entertainment and news from the outside world. In 1938, Hunt "heard the wireless news at Glennie's (an irrigation Engineer)" about the "international storm raised by Germany's entry into Austria."[32] Somewhere later in 1938, Hunt arranged a Philips set for himself for such news, especially after dark, when the BBC signal came through much better. His daytime was filled with his inspector duties anyway.

One of the other major links with the larger outside world was the mail. Letters from the United Kingdom came with Imperial Airways Empire Flying boats, which took four to five days. Stamps for these letters were similarly priced as for home postage. Letters were delivered weekly to each Block office by van from SPS Headquarters in Barakat, together with other (official or company) letters. In addition, there were post offices in Medani and Hasi Heissa, plus one on each of the two main trains between Khartoum and El Obeid each week. Mail addressed to one of the local stations could be collected from the station master's office. For parcels, the story is a little different. Parcel receipts were handed out for signature when the train stopped, which had to be signed. Parcels remained on the train as long as they were not collected by the inspector himself or his servant. In practice, this made collecting parcels an interesting little bureaucracy of its own. First one had to collect the receipt, then sign it, and then go oneself or get a servant to meet the next train. Apparently, some parcels traveled up and down the line for several weeks.[33]

Divide and Rule

SPS Inspectors were recruited to supervise the tenants cultivating the cotton. Each working day, an inspector made three tours through the area under his supervision—his "maroor" or "marour." After a cup of tea around 6 a.m., the first inspection tour was between 6:30 and 9. After breakfast, the second tour was from 10 until about 12 p.m., with a third tour—after a siesta—from about 4 until 6 in the afternoon. Inspectors were expected to possess at least two ponies upon arrival, plus saddles. Junior inspectors were expected to do their inspection tours on horseback well into the 1930s.[34] Hunt was the third inspector on his Block and did his work on horseback. In warm weather, the inspection tours became "more and more of a sweat."[35] Cars must have made their entrance already in the late 1920s, as in 1930 the SPS management decided to go back to horses instead of cars, as one measure to cut costs. Typically, Block Inspectors as well as second inspectors in a Block were given a company car for work—and recreation. In the early days, these cars were T-Fords; in the early 1930s, these were replaced with Austin 7s, and from about 1937, Morris 8 models were used. Occasionally, a Block Inspector would give an inspector without a car a lift or even lend his car to an inspector, for example, to drive to Khartoum on a weekend. Furthermore, with two leave periods per year, there was frequent change of inspection tours, which could include using car and house of the inspector on leave. In the late 1930s, when a small car would be in the order of some £100, many inspectors bought one. Owning a car saved one from going to dinner on a pony or waiting for the Block Inspector to take one "home from polo."[36]

The different houses and differential accessibility of means of transport were clear expressions of hierarchy in the Syndicate. Next to houses, cars, and ponies, a vital instrument expressing that hierarchy and structuring the inspector's working day was the telephone in each house—an Ericsson handset connected to a central Exchange. After ringing to the Exchange, the—English-speaking—Sudanese employer would connect one to the required number, actually more often to a name. The telephone was the way of passing instructions or items that had been noticed and should be dealt with. For an inspector, a very early morning call from the Block Inspector, or sometimes the Group Inspector, was normal. Within this rather rigid hierarchy of brick, telephones, and transport options, a junior inspector occasionally "felt ashamed and resentful of criticism, especially at so early an hour."[37]

Inspectors had to be careful with too much criticism, however, as every year the Group Inspectors would evaluate all inspectors under his command. Based on these evaluations, raises in salary or bonuses were awarded, of up to £60. A Group Inspector would make an unannounced tour of each inspector's area some two weeks before raises were due. One of the Directors from London—sometimes an SPS manager from Barakat—would join. Their big motor cars—Fords—would leave large tracks clearly visible for several days. Even car tracks can be a sign of hierarchy...Once the raises and bonuses were told to each individual inspector, the telephone network was used to find out about everyone's payments, and "one was soon able to discern where oneself stood, compared to one's fellows, in the estimate of the Company." The telephone operator could often already tell the results; when one inspector asked to be put through to a colleague, the operator informed him that the colleague had been given £30.[38]

The annual evaluations seem to have been serious and pretty direct on occasion. In 1943, the evaluator—Gaitskell, Assistant Manager at the time—advised that Zeidab manager F. W. Wall should be replaced with the second in command, G. B. Tame, who seemed "to have found his niche." In the Gezira, Tame was "not outstanding among his contemporaries," as "his personal interest in the natives" was "rather too excessive." At Zeidab, this "particular trait" was valuable. Wall does indeed disappear from the SPS employment list in 1946, after being in its service since 1926! Tame had started working for the SPS in 1928 and was still in service in 1953. It is interesting that Gaitskell takes the tone of a full-blown SPS manager here, focusing on drive and not too much contact with natives, whereas he later becomes the champion of the new approach in Gezira (as we will see in chapter 6). It is also interesting that even in 1943, dealing with experienced inspectors, the evaluations are rather harsh. The Zeidab phrasing would reflect the general tone, as the same SPS management was involved. Junior

Zeidab inspector E. A. O'Dowd apparently had an older brother in Gezira in service since 1924, but Gaitskell considered O'Dowd senior as "a liability to us" for "lack of drive and effectiveness." The younger O'Dowd had to prove that he did not have "the same defect,"[39] which he probably did as he was still in service in 1953. J. H. B. O'Dowd, the older brother, moved out in 1945.

Officially, SPS management in Barakat and the Group Inspectors could be consulted at any time, but in practice communications may have been less straightforward. On March 2, 1938, after realizing that the picking in his maroor might appear very unorganized, Hunt wanted to know whether this state was exceptional, as he never saw other areas. As "none of those in authority seem to worry much," with "Gibbo" (Gibbons) only occasionally passing by but not saying much, Hunt did not have much of a clue.[40] On March 9, however, it became clear that both his superiors had seen the picking in part of his area and had "pronounced it a shamble" with cotton "lying all over the place."[41] It meant that Hunt had to put in much effort in that area and possibly neglect other parts of his maroor. To what extent superior inspectors could be approached and would communicate depended on the superior himself as well. In Hunt's Northern Block in the late 1930s, the Block Inspector did "not collaborate in fixing your rise,"[42] which suggests that other Block Inspectors did. In 1940, Hunt worked in a Block supervised by "Robbo," who had "a reputation for backing one up."[43]

Given the people involved, it seems like different areas did develop slightly different routines in daily management and exchanges between the different ranks, but there is no doubt that the overarching SPS model went for "value for money." This motto is strongly symbolized in Syndicate Chairman MacIntyre—"a very austere character"[44]—who stands for the "prevalent Scottish attitude of parsimoning and the value for salary insistence of the management," leaving "a lasting impression" on Hunt.[45] The "Divide and Rule" method of the Syndicate toward its staff[46] encouraged Hunt to join the SDF in May 1940 after an Italian attack on Kassala. However, even despite his clear dislike of Gezira, Hunt shows a certain pride in his SPS work. His maroor was bordered by the railway taking out the cotton each afternoon during the picking season; each time the bales of cotton were loaded Hunt was very satisfied "to observe the fruits of our labour being taken away for sale."[47]

The Most Monotonous Job on Earth

Syndicate hierarchy and arrangements did not remain undisputed over the years—even though actual chances in arrangements seem not to have been made by the Syndicate. Already in 1928, an inspector had drafted a letter to the SPS Manager on the position of junior inspectors and the way the inspections were arranged. Inspector Gaitskell argued that fewer inspectors were required when Gezira would be organized in larger areas per inspector. Larger responsibilities and a car would give inspectors a better position toward the tenants; tours on horseback were a "discouragement"[48] and actually challenging the authority of the inspector. In a reaction to this letter, senior inspector Archdale stated that touring an area three times a day was not monotonous, as long as one focused on different parts each time. Riding on horseback would actually give "a great chance to a Junior man to get to know his tenants and learn their ways," something that could not be done "from a car."[49] No changes were made in 1928. In 1932, concern for junior inspectors was focusing on the economic problems not really increasing their career prospects. Many junior inspectors were looking for jobs elsewhere to escape from the "most monotonous job on earth" without clear options for promotion.[50]

In the same letter that mentioned the juniors in 1932, the position of Group Inspectors within the Syndicate hierarchy was discussed. It is likely that possible budget cuts by removing the most expensive staff was at least partially the reason to suggest changes, but the idea was defended in another way. Group Inspectors might have been crucial in the early days, when Block Inspectors were still "untried men ignorant of their duties" and developing a common routine was needed,[51] but after some years, the Block Inspector had become the key in Gezira day-to-day management. The relations between Block Inspectors and Headquarters in Barakat were important. Group Inspectors might not be needed anymore, because a good Block Inspector did not require a Group Inspector. A Group Inspector might even conceal bad inspectors from the view of Barakat.

Block Inspectors should still be controlled somehow. Block Inspectors should not become too independent from HQ, which required HQ "to have more ears." Replacing Group Inspectors with two Assistant Senior Inspector who would check one Block a day—and be supervised by the already existing Senior Inspector—was proposed. In such a way, HQ would still have a good idea of what happened, but a Block Inspector "would be otherwise a free agent to run his own show" responsible for his own area. This would result in higher motivation of Block Inspectors. At least twice a year a meeting with all inspectors would be held for "a feeling of unity" and to inform them about progress in the Gezira.[52] The new arrangements would allow cutting the number of field staff. In 1932, the SPS employed 98 inspectors within Blocks, excluding Group Inspectors. The new organization would require 90 inspectors; eight men plus at least two of the four Group Inspectors were not needed anymore. Although I did not find any official statement about the 1932 proposals, it is pretty clear that SPS management did not reorganize the staff. The system of Block and Group Inspectors continued to be used.[53]

When after World War II the hierarchy between British and Sudanese became an issue, the hierarchy within the SPS and the position of young inspectors was discussed again. Many SPS employees favored maintaining SPS hierarchy between experienced and young inspectors. After asking quite a few questions during a meeting, a new inspector was summoned by his superior SPS inspector, who explained that "one didn't open ones mouth like that in the Gezirah until one had been here much longer."[54] Another junior staff member was told that in Gezira "young Inspectors should be seen & not heard!!."[55] For at least one inspector, this was reason to go directly to Gaitskell and hand in his resignation. Gaitskell, Syndicate manager in Gezira since 1945, did not accept the inspector's resignation and asked him to "go on opening his mouth."[56]

Balls and Parties Every Night

Inspector Hunt was one of those inspectors who did not want to stay long in Gezira. An element of Gezira life, however, that Hunt definitely did seem to like was the social life, especially parties. The extent of parties and visits must have increased with the increase in the number of cars, but in the early days British social life in clubs and private houses was already important in Gezira. The SPS was the only nongovernmental institution employing a number of British in the area, but the type of British it did employ, with degrees from Cambridge and Oxford, was pretty similar compared to the members of the Sudan Political Service. Quite a few SPS inspectors had applied first for the Colonial Service, including Gibbons. As such, the two groups mixed easily at events, clubs, and private parties.[57] Public events like the Race Meetings in Medani in winter, tennis tournaments, and the Polo Cup were important meeting venues.

The importance of horses ensured that polo and horse racing were major leisure activities in Gezira (figure 5.5). With SPS headquarters nearby and several Syndicate inspectors within easy reach, Wad Medani was the place to enjoy "good polo—perhaps the best in the whole country" for those "who could not only afford it but had the requisite ball-hitting ability."[58] Many inspectors were "first class" polo players;[59] Gibbons himself was "a keen horseman" owing much to "the success of the Wad Medani Race

Figure 5.5 Gezira polo. Reproduced by permission of Durham University Library (F. B. Hunt; SAD.692/19/19)

Club."[60] Civil servant Simpson—who joined the Sudan Government Service in the 1920s—was not a good polo player, a continuous source of "considerable frustration and disappointment."[61] With cars becoming more important in the 1930s, polo popularity declined a little, and attention for tennis increased.

Next to cars, "wives and families" created a larger focus on tennis.[62] Officially, and certainly in the early years, inspectors were not supposed to marry, although in the early 1920s, quite a few inspectors were actually married.[63] The letter of employment stated clearly that an inspector could not marry "until development work is completed and then only if he can show sufficient means to support a wife."[64] As late as the 1940s, references are made to the unsuitability of Gezira life for women "who have nothing to do and have no entry to their husbands' jobs." Ninety percent of the wives of SPS inspectors would "hate" or at least "dislike the place." For the inspectors it was "no fun" that their wives were not happy.[65]

To cheer them up, the SPS maintained two clubs for its personnel. There was the 88 Club (obviously) situated at Kilometer 88 on the Main Canal; the other was the Messalemia Club, some 20 miles north of Wad Medani. Both clubs had a polo ground, tennis courts, and buildings with billiards, bar, showers, and sleeping options to stay for the night. When the parties were special, even the Northern inspectors would go to the clubs. Gibbons was "a prominent part in the social life of the country" and "a most useful member on his local Club Committee,"[66] which is the more remarkable given that Clubs 88 and Messalemia were quite far from the North. Hunt was stationed on a Northern Block, about halfway to Khartoum from Medani, and he often drove to Khartoum on Sundays or holidays. Next to going to Khartoum, so-called Soda&Mezza Clubs could replace the official clubs. After polo or tennis, everyone simply went to one of the houses in the vicinity, where hosts—well, the servants would have served it—provided tea, sodas, sandwiches, "whilst everyone brought his own flask."[67] Apparently, it was not uncommon to have a Soda&Mezza Club with the host absent. When coming home in April 1949, one Bill "discovered he had had a lunch-party, which had drunk his gin and eaten his food and left an appreciative note for the absent host. Such seems to be the way of life in the Gezira."[68]

Heavy drinking was clearly a part of Gezira life. When Hunt visited inspector Webb to borrow the car, he drank "two whisky and sodas and a Gin and bitters with him"[69] before driving off. Not surprisingly, cars could end up in canals. Once, an inspector in the backseat was "heard to shout 'Shut the door the fish are coming in.'" At another moment, a car was found in a full canal with the lights on. Even then, hierarchy was kept: the car belonged to a Group Inspector and was driven into the Main Canal.[70] In May 1947, a Brit lost his way between his office and house; in the rain and mud he managed to drive the car into a canal near his own front gate. That same night, "one of the cotton syndicate chaps ditched his car" in the same place, even though the road had "pretty well dried up." The inspector must have had "a good evening!"[71]

Traveling through Gezira was already challenging when sober, even with motorized transport more readily available. In the rainy season, powerful showers accompanied by sand storms transformed narrow Gezira roads into pools of mud. The Gezira clay was "perfectly fiendish just after rain."[72] Even light rains would turn the inches of dust to mud,[73] clogging car wheels and having anyone traveling onward run the risk of becoming stuck in the mud.[74] There might be willing helpers in a near village, pushing out cars—but obviously not during the night. One was never sure where it had rained or whether it would rain along the way. One could either try and find out by telephone or just muddle through. The clubs functioned as hubs for those who were stuck because of rain; often it was better to spend the night at the club instead of spending 6 hours for 30 miles to get home.[75] In case of delays, taking lunch at the club or using the phone to inform others about one's whereabouts was common, including assuring one had not ended up in a canal.

Several big parties were held each year, when "[n]o trouble was spared" to ensure "a huge success." For a fancy dress dance at the Messalemia Club, the "order went out, 'Come as you would to a dance in Paris' and the preparations were so thorough that shortly after leaving home I passed a road sign which said 'To Paris 40 km.'"[76] Inspectors did not always feel super after such a party. Luckily, servants could offer medicine, while occasionally "observing that Syndicate inspectors were apt to drink a good deal."[77] After a party, it was not uncommon to sleep until early afternoon, have some lunch, and go back to sleep again, especially if it was hot. At the end of the afternoon, one could go for some polo or tennis and then spend the evening at an inspector's house for some drinks, darts, and cards. On Sundays, one should not be too late at home and in bed, as one had to be up "again all too soon to start another's week's toil."[78]

The End of Slavery

The weekend parties—whether they were special events like the Paris party or the general club nights—were not only important social events but also venues where SPS inspectors could ventilate their ideas, frustrations, and other feelings about the work, the Syndicate, and Gezira. On a February weekend in 1938, Hunt joined a "first-rate party" with many participants bursting into song. Fellow inspector P. E. M. Mellor sang his two self-composed songs; the first was entitled "A little B house in the Bhor," and the second song described the career of bad tenant Ahmed Abdulla Sakran (figure 5.6).[79] Inspector Jake contributed to the general happiness with his song "They say that Gaitskell's potty over me, and so he ought to be, as you can plainly see; for what I owe the SPS is nothing to what the Syndicate owes to me."[80] At a party in 1946 at the Medani Club, Macdonald—"the works & buildings man"—made fun of Harold Wooding, one of the Directors, "the root of all discomfort out here, in fact a Syndicate Ogre!"[81] At a party in the late 1930s, one of the pub quiz questions was on the year slavery ended.

Figure 5.6 SPS poster with the career of bad tenant Ahmed Abdulla Sakran. Reproduced by permission of Durham University Library (SAD.842/11/1)

One of the inspectors quickly and loudly answered "1950," referring to the expected end of the SPS concession period; that inspector was not referring to ending slavery for the tenants, I may have to add.[82]

This possible ending of SPS involvement in Sudan and the associated end of employment in the company—whether considered positive or negative—was already discussed by inspectors before World War II. World War II itself brought more involvement of Sudanese in Gezira's daily affairs. Many of the SPS inspectors, as their colleagues had been in 1914, were eager to join the armed forces to fight for their country—the United Kingdom that is.[83] A few inspectors—mainly the younger inspectors—joined the British forces after resigning from their job in Gezira. Many inspectors continued with their work in the irrigated area.[84] Rationing and shortage of supplies was daily routine, and public transport facilities other than the railway were nonexistent.[85] Gibbons asked for leave in 1943, arguing that he had not seen his wife and family since June 1939; his father had passed away in March 1942, and he needed to arrange the formalities for the inheritance. The handwritten comment on the letter reads rather sadly, "Never acknowledged, never answered."[86]

It is not hard to imagine that much debate was created in 1944 when the Sudan Government announced that the concession would indeed be ended in 1950. World War II changed the British perspective in and on the Sudan. As in other colonial areas, Britain was to change its policy in Sudan to include the Sudanese themselves and their wishes in development plans and the colonial project. British in Sudan, government employees, and SPS inspectors alike had to come on much friendlier—but not totally equal—terms with the Sudanese, who would take over British positions in a gradual process of "Sudanisation." What that meant for Gezira was not immediately clear. It was anticipated that many of the British field staff would stay in the new set-up, as it was certainly the idea to keep Gezira as the money machine of Sudan and it was clearly impossible to train enough Sudanese inspectors at such short notice. Whether the inspectors that stayed on would become government servants or employees of some new entity was not clear as late as 1949.[87] There was a rumour that all inspectors over 50 would have to go when the Sudan Government took over in 1950.

Before the war, the British field staff in the SPS had been 107, but in 1945 only 65 were left. With several resignations expected of those inspectors that would try and find another job[88] or simply wanted to retire, two inspectors would be under 35 in 1950. Finding new and young inspectors was not that easy, with the perspective of longer-term employment in the SPS gone and the uncertainty of the British position in Sudan in general. The general atmosphere in Gezira among SPS field staff will not have helped either. Some claimed that "90% of the chaps" were "disgruntled & looking for jobs elsewhere" or would get out "as soon as they have got back the money they've had to spend on equipping themselves here."[89] The rising prices immediately after the war would not have made a compelling argument to come to Sudan either.

The End of an Era

In addition to uncertainty about their own future, SPS inspectors were uncertain about how to deal with the Sudanese after World War II. Opinions were divided what a changed social and political context actually meant. As a British subject, a Syndicate inspector was supposed to engage with Sudanese, take them seriously and listen to their opinions. That was a big change for many British persons, not only in the SPS. In July 1940, Ms. Aglen, wife of civil servant Aglen, complained that her garden was full with weeds, but that her gardener had "one idea only and that very strongly—to do as little as possible." She continued complaining that this was "the all pervading idea in this country—they are all just about the laziest set of people out—which does not make it any easier to get things done."[90] She was certainly not alone in her reserved views on the Sudanese. SPS inspector Johnson and his wife—to whom we owe a set of fascinating diaries on the last years of Gezira under the SPS concession—wanted to engage more with Sudanese, but even Ms. Johnson remarked that the Sudanese were "like children." Perhaps "one cannot stop children growing up," and the "surging movement for freedom" was not to be stopped either, but the Sudanese needed "more time to think what they want to do."[91]

To Culwick, some of the SPS inspectors had "a real Herrenvolk outlook" who tried to continue living "inside blinkers."[92] Many British SPS inspectors "of the old school"[93] refused "to have anything to do with 'devolution' and village councils and what-have-you." Maintaining "an efficient cotton-producing machine by an iron hand and direct rule" was their goal.[94] Other inspectors were more engaged with the changes brought upon Gezira and could be labeled "social pioneers in addition to their normal S.P.S. duties."[95] Others might have a somewhat mixed attitude toward the whole matter of involving and liking Sudanese, as Inspector Roddie Dew, who apparently was "a good devolutionist on paper but when pressed to mix with the Sudanese does not respond!"[96] It may have been difficult for many SPS inspectors to change their way of running their work; perhaps many simply "saw which way their bread was buttered." For them "understanding or really being interested in the Nas [. . .] was not in the contract!."[97]

Differences between Sudanese and British lead to "[h]eated arguments" between different SPS inspectors on "Sudanese manners" even when only "exemplified by the Bakht El Ruda crowd at a local football match" in 1949. An issue with high practical and symbolic value was club entrance. Only British citizens were allowed in the many clubs in Sudan, although in Medani Club, "after a row," a Russian and his wife were allowed in. In trying to change this, the Johnsons were actually having a little "reputation of rebel leaders." They simply considered it rather strange that in a context of planned coming together of British and Sudanese, the clubs would remain separated.[98] Only in 1950, the Governor-General gave the directive to the Medani Club Committee that they had to permit Sudanese as members, "whether they liked it or not."[99] Other clubs had to follow.

On the cover of her diary of 1950, as usually written in a school notebook with a tag on the front to fill out the name of the school and the pupil, plus the class and the subject, Ms. Johnson duly filled out the details: "School—The Sudan; Name—Thewlis & Winifred Johnson; Class—Unknown; Subject—The end of an Era."[100] In that same year, five Sudanese inspectors were hired by the SGB; Ms. Johnson could not help to see the humor in this fact, as just a few years earlier "people were horrified at the thought of Sudanese inspectors & thought us mad when we said it would happen & that they would naturally share the club with us."[101]

A Personal Arrangement for a Particular Situation

For Hunt, the continuous SPS mentality of "Take the cash and damn the staff"[102] was reason to resign from Syndicate duty in 1946. Inspector Johnson and Ms. Johnson left Gezira and Sudan in 1953, one year after Gaitskell had left. The other of the two Gs (Gaitskell and Gibbons) stayed. Gibbons continued working in Gezira until 1959. Directly after the change of Gezira management in 1950, he was asked to supervise the Northwest Extension, in addition to his work as Group Inspector of the North Group. He would receive an additional £500 per year, as "a personal arrangement to deal with a particular situation"—the extra salary was not permanent and would cease once the responsibility would change.[103] In 1953, with many of his colleagues leaving or preparing to leave the Sudan, Gibbons was also asked about his "details of retirement," as in the process of Sudanization a balance had to be found between Sudanese prospects and British desires to continue or leave. The standard letter sent to him explained that financially "the most advantageous time" for his retirement would be December 30, 1953, especially as he would "qualify for the full special bonus as per the agreement with the Financial Secretary in 1950"; staying on meant forfeiting the bonus. Furthermore, to stimulate his thought process, Gibbons was offered free passage and six months final leave if he would retire.[104]

In the personal letter that every British employee with more than 28 years of working experience received as well, Gibbons was explained that the offer was based "on the assumption that 28 years in this country was a reasonable length of time to allow a chap"; they expected him not willing to stay longer. However, the choice to retire was "entirely" up to Gibbons himself and "optional," even though the hint was clear enough to take the offer.[105] There is no formal response from Gibbons, but on the back of the letter his draft answer is clear enough: "As a first thought the idea of arriving in England on 1st of January does not appeal to me." Gibbons wanted more information, as he foresaw problems with Gezira work, especially in case "the present low standard of Sudanese recruits prevails." He seriously doubted who would ensure things would go well "if seniors go."[106] So, Gibbons stayed.

Almost immediately, useful employment was found for him. The SGB had set its eyes on managing the new Managil area and in the summer of 1953, the Managing Director was "anxious" that the SGB should start planning almost at once to ensure that Managil would come under the Sudan Gezira Board. Managing the Managil area—or at least claiming that the SGB should manage Managil—would require experienced staff and as such offered "a highly suitable opening for our present British staff who are willing to undertake the pioneering of such a scheme." Gibbons was thought to have "the enterprise, guts and 'know-how'" to assist with realizing the area. Therefore, he was to have "little fear of being Sudanised," as there would be no suitable Sudanese for quite a few years to come.[107] A committee was to prepare the necessary plans, and Gibbons was to be the chairman of the committee.[108]

When he was finally appointed for the work, Gibbons was offered a contract from January 1954 "to act in an advisory capacity in further extension schemes."[109] Gibbons accepted the offer,[110] which also meant he was no longer Group Inspector of the North Group. The terms of service are interesting and show some continuity with SPS rules. Employees with lower salaries (less than £600 a year) still had to obtain the Board's consent before they could marry and spouses were not supposed to accompany any new employee during his first tour of duty "save with the Board's consent." An innovation was that new employees (clearly supposed to be men) had to actively support Sudanization, devolution, and social development, as this was "an essential part of the employee's duty."[111]

What Gibbons was to do exactly seems to have been negotiated on a year-to-year basis; at least, in the spring of 1955 Gibbons was "anxious to get some more definite information about what I am to do in the coming year." A main concern was that he did "not want to spend a complete hot weather in 1956 in this country."[112] In a response to this request, he was informed that his post was planned to be Sudanized in October 1955, but that he was requested "to serve the Board" for one year after October. His work would be "purely advisory, not executive," as he would support and advise Group Inspectors and HQ "when called upon to do so."[113] Gibbons became the Development Adviser for the new Guneid Pump Scheme.[114] All these changes did bring up the question whether Gibbons would still qualify for the retirement bonuses available to British staff, as quitting one job and starting another was not really retiring from a job anymore. Gibbons' career was "in a somewhat different category from 'final' retirement."[115]

The issue was solved in March 1956, when another two years were added to his work for the SGB after October 1956 "with the possibility of extension by mutual agreement." Gibbons was to become the Senior Development Officer in the Managil Extension Project, but he kept his option "to resign under the Sudanisation Scheme."[116] He took this option in October 1958, when he left the Sudan for final leave, to retire officially on January 1, 1959. There was some formal hassle about his bonuses, but the SGB was thankful to him, as he had "cheerfully and loyally done [his] duty" and had "helped to build up the tradition of integrity and service always associated with the staff of this Scheme."[117] Gibbons requested to be allowed to hand over his shot gun to his "thoroughly reliable" driver.[118] He also gave detailed instructions to his bank to transfer the balance of his account "less LS.1 at the end of October, November, December and January to Barclays Bank Ltd., P.O. Box 8, St. Helier Jersey, Channel Islands. Any balance provided that it is less than LS.10,- is to be transferred to the Wad Medani Church Fund on the 28th February 1959. Any amount in excess of LS.10 is to be transferred to Barclays Bank Ltd. Jersey."[119]

Directly after retiring, Gibbons was approached by Gaitskell for "an 18 months job in Pakistan with the International Bank." Gaitskell had been asked himself, but as he was not interested he was asked to suggest anyone else.[120] Gibbons refused, as he was already not feeling well. He passed away on July 4, 1959, only half a year after retiring. Five years after poor Gibbons's death, on July 11, 1964, his wife Alison Gibbons invited former SPS inspectors for a reunion at her house at Elm Hill. Those who went to these reunions clearly had wonderful times, were delighted about how "the spirit of days gone by is kept alive,"[121] and fully enjoyed meeting their former colleagues and remembering Gibbons. "It is of course delightful to meet ones old friends although unfortunately they get fewer and fewer. Gibbo had been out—about a year before I got [...]. He gave me my first instructions in growing cotton."[122] Former inspector Hood had not seen his former colleague Wilcox since 1936![123]

The reunions were held every two years for "Gezira people—because (unlike the government people!) there was no official gathering."[124] On July 5, 1981,[125] the last meeting was held. The invitation called upon all to make an effort to join, as it would be the last one. Specific attention was asked for ensuring that children and grandchildren would come as well. The former inspectors might have looked like "real old fogies to the young ones,"[126] but they did at least manage to keep "the Sudan spirit" alive, for example, for Sara Taffinder, who not only enjoyed her happy memories of her Sudan childhood, but who also stressed that "[w]henever there have been discussions about the British colonialists [...] I have always been very proud to point to the Sudan and its colonists." The Sudanese people "had been well guided" toward their independence. "And the wives who put up with the pretty uncivilised conditions and separation from husband and children made a great contribution."[127]

Men of Good Personality

The last project in which Gibbons was involved, the Managil Extension, was later presented as an "achievement of the Republic of Sudan after independence,"[128] which it obviously was. At the same time, Managil was also the last effort of British colonialism in the Sudan, given the huge involvement of British staff—people like Gibbons and the senior irrigation engineers—both before and after Sudanese independence. We could consider Managil a very strong example of the transition between colonial development and development aid, with several changes in political responsibilities, but strong continuities in planning, engineering, and staff involvement. In the context of development aid, the Gezira example and the experience of its British employees became valued (re)sources for other projects, both in- and outside Sudan.

In realizing their potential value and finding the new jobs they needed in the 1950s, British governmental employees in Sudan were offered support by the Sudan Government. The British staff of the Sudan Gezira Board—formerly SPS—were employed in the private sector; the Gezira Board Bill explicitly stated that SGB staff was not government staff.[129] This also meant that in the process of Sudanization, the standard support available to British government employees was not available to the SPS/SGB staff. The SGB—in the person of Gaitskell's successor Raby—we will see much more of him in the next chapter—did make some attempts to change this, but the Sudan Employment Bureau did not consider SGB staff as government officials. The Bureau saw the SGB staff "in very much the same position as any British commercial firm in the Sudan, and they did not really come within the Bureau's terms of reference." More than "all possible advice on re-settlement problems" was impossible.[130] Going to the same clubs might have been the norm in colonial Sudan, but after Sudanese independence the different positions of business and government were emphasized anew.

One of the problems the Sudan Employment Bureau had with SGB staff was their "somewhat unique experience and knowledge."[131] Many of the British field staff were between 20 and 40, "young ex-Service men of good personality," mainly of "School Certificate-standard," and clearly not easy to place.[132] They might qualify for the practical work done with the new Colonial Development Corporation and a few might be "acceptable as Rubber or Tea Planters."[133] Despite their differential access to support finding new positions and their very specific job experience, quite a few former SPS inspectors actually did find jobs elsewhere, particularly within the post–World War II development aid programs. Many of the British that left the Sudan as colonial oppressors returned a few years later as international development experts—and went on to develop other countries as well. But let us develop that story in chapter 7 and now move to the issue concerning what was to be done with the Gezira Scheme and its management once the Syndicate concession had ended. Obviously, we already know that part of the outcome was the Sudan Gezira Board, but our focus in the next chapter will be on the negotiations and debates on the question itself.

Chapter 6

Move from the Old Grooves: Gezira Continuity and Change after World War II

On July 1, 1950, the Sudan Plantations Syndicate no longer managed the Gezira Scheme. The Sudanese "became owners of one of the most spectacular agricultural experiments in the Middle East and Africa, certainly the largest agricultural unit of its kind in the world, and one of the most profitable." With the Sudan Gezira Board taking over scheme management, the "cotton-growing machine" promised to change into an "educative movement." Cotton-growing would not disappear, but tenants would "be taught to become a mixed farmer," growing other crops and holding cattle. As such, Gezira would only grow in its role as a model for "all agricultural development" in Sudan, with the SGB setting "an example to those working for social development elsewhere in the country." The dual goals for the new Gezira were to "prove to be the heart of an enlightened Sudan" but at the same time to stay "its economic backbone."[1]

Sudan Governor-General Hubert Jervoise Huddleston informed the Syndicate in a letter of March 25, 1944, about the decision to end the SPS concession in 1950, the formal end date set in the agreements of 1926.[2] The Syndicate had initiated negotiations on concession renewal after 1950 in 1938, but these were suspended at the start of the World War II. At that time, however, both the government and SPS were quite confident that a renewal of the concession would be possible, although the conditions might be changed. Therefore, the 1944 letter surprised and slightly angered the SPS, even though it had to acknowledge that the changes in Sudan policies toward "increased native administration" had altered "prospects as a commercial undertaking" too much anyway.[3]

Postwar Sudan would have to show improved British colonial rule, in line with general changes in British colonial policy and elsewhere—explained in chapter 4. British rule should do more than take care of business; it should also stimulate social development and (more) equal political representation.[4] Discussions on Gezira—the flagship of development in Sudan—regarding how to include more emphasis on human development and new relations with the Sudanese had started when the SPS was still managing the scheme, but World War II accelerated the whole process. We already know that the SPS was never really fanatically going along with the new policies to increase Sudanese engagement with Gezira affairs. After the March 1944 announcement SPS enthusiasm did obviously not increase.

In the decade immediately after World War II, at least as much paper on Gezira and decolonizing Sudan as on cotton in general has been produced in Sudan. One of the main producers of paper was Gaitskell, who was extremely enthusiastic about the new course—which in fact he had encouraged right from the late 1920s. In 1944, he defended both the SPS profit-based course and the new focus on social development. "One does not blame a camel because he cannot pull a goods train. One looks around for an engine to do this. Our Camel was only designed to carry Cotton, but it is a goods train that we have got to pull in future in the Gezira Scheme if we are to achieve those moral and social objects, as well as the economic ones."[5] Gaitskell, who saw himself capable of changing cotton camels to development engines, was asked by the Sudan Government to become the new Gezira manager after the end of the concession. When the SGB took over from the SPS, Gaitskell was its first Chairman and Managing Director with the mandate to grow cotton and stimulate social development in the Gezira area—the latter to be financed with part of the former SPS share of cotton profits.

This chapter will trace the implications of discussions on changes in colonial policy in the Sudan after World War II for Gezira and its associated organizations SPS, SGB, and SID. After the exchanges on the concession, the discussions on what to do after 1950 within the Sudan Government are traced. What should Gezira management look like, and what should it do? Next to increasing the scope of Gezira policies toward social development, how to increase Sudanese influence, and the numbers of Sudanese working in management positions were major issues. This chapter focuses on the discussions directly related to Gezira; in the next chapter the broader Sudanese and international development context will be discussed.

Boulevard of Broken Dreams

In 1938, the SPS management decided that "the time was now opportune" to discuss the possibility of extending its Gezira concession after 1950 with the Sudan Government.[6] MacIntyre was authorized to enter into negotiations with the government. He approached Governor-General Sir Steward Symes to explain "the inconvenience" of the short period remaining for the SPS for its management—for example when recruiting staff. MacIntyre reported that "both the Governor-General and the Financial Secretary had been sympathetic," but he and his directors realized that certain concessions from Syndicate side would be required. Perhaps a merger between SPS and Kassala Cotton Company (KCC), or a change in the distribution of profits as in the 1920s, was required. Another possible issue would be how to arrange the loans to the tenants, as it was likely "the Government wished to appear as direct lenders to the Tenants."[7] The Syndicate did realize its role had to change, but remained convinced it could play a role in producing the cotton that Sudan so much wanted.

One year later, in 1939, the prospect was still fine. Financial Secretary Francis Rugman "appeared to be keen to continue," and the positions of government and Syndicate "were not very far apart except for the settlement of financial details."[8] The start of World War II interrupted formal negotiations, as another World War had done decades earlier in 1914, but discussions between government and SPS continued. Wooding—who took over from MacIntyre, who had fallen ill—continued with the merger between KCC and SPS to "pave the way" for a new agreement "in due course."[9] In 1943, the Assistant Civil Secretary presented a glorious future for the Sudan with its new and shiny local government, but warned as well that local self-government would not be "the cure for all ills." After all, the "art of self-government" was not something that would fly just on theory, it had to be practiced and one could not practice "without mistakes being made."[10]

Gaitskell saw the same great opportunities for new development ideas, including a strong emphasis on devolution and Sudanese involvement, but his own Board did not agree. Until 1944, the SPS management had no reason to think that its concession would not be renewed; there is certainly no evidence for any concern in the archival material. The Board minutes of meetings in 1940 mention all kinds of progress mentioned above. After this, only very brief minutes of very scarce meetings are available, caused by restrictions imposed "not only by the Defence of the Realm Regulation but by the Paper Control."[11] However, even in 1943, apparently a time of scarce paper and ample secrets, SPS Minutes would have shown a response to any rumor or news that the concession would not be renewed.

The letter in March 1944 was an extremely unpleasant surprise to SPS management.[12] In the letter, the Governor-General explained that the changes in political context had made the government decide not to renew the SPS concession. A combination of new options for irrigation development "arising from the elimination of Italian East Africa," "great improvement" in governmental finances, and the differences of opinion between SPS and government were the reasons underneath these changes. The year 1950 would "see the end of a stage of development in the Gezira."[13] A few months later, the Governor-General suggested that in 1939 the idea had indeed been to renew the concession for another ten years. Since then, however, it had become clear that "the Government's long-term policy of social evolution" would clash with "the companies' necessarily short-term commercial policy." Cooperation based on the "goodwill of both parties" would not be sufficient anymore. The pressure on the Sudan Government to become much more involved in Gezira—"to take over the whole concern"—was growing in Sudan; continuing any type of private concession, even a modified one, would be politically difficult.[14]

Directly after the announcement to end the concession, an exchange of telegrams and letters (similar to 1926 on the new, extended concession) discussed the exact text to be sent out to the general public and Syndicate stakeholders. On August 13, 1944, the SPS Board could finally approve the text of the Circular to Stakeholders, in which they were informed about the termination of the Syndicate's concession in the Gezira.[15] MacIntyre reminded the stakeholders that this nonrenewal "was not anticipated by your Director." There was the obvious need to talk business as well. MacIntyre informed his audience that the SPS would "continue to receive its share in the proceeds of Crops grown during the remainder of the Concession period."[16]

At the General Meeting in spring 1945, MacIntyre repeated his surprise about the unexpected developments. He could not believe that an enterprise so positively described by the Sudan Government itself—for example in that same Circular to Stakeholders—could end this way. The government had always shown "its high appreciation of those great services which these Companies have rendered to the Sudan." The SPS had established "a great agricultural undertaking" in just 25 years, in a way showing that "they had the welfare of the population at least as much at heart as their own immediate interests."[17] After such a positive testimony, MacIntyre asked for understanding that "some surprise may be felt" that such partnership—"frequently been quoted as a model for others"—had to end.[18]

Obviously, the government was simply exercising its right within the rules of the agreement and did so with the changes of policy "following, if not leading" the general change in British colonial policy in mind. MacIntyre could simply not grasp that the government was prepared to slay the goose with the golden eggs. How many Ministers of Finance would "willingly contemplate the withdrawal of between five and six million pounds of public money from fruitful employment in the country which they administer"?[19] MacIntyre could only grumble to his stakeholders that it was the financial success of Gezira—which was very much the result of SPS efforts—that gave the government the budget to develop Sudan and Gezira by itself.[20]

In London, the SPS and KCC stocks went down after the announcement for nonrenewal. Most English newspapers were not exactly thrilled either. Even though a perfectly legal decision, the nonrenewal was shocking and sad at the same time; "enlightened private enterprise" was "engulfed in the impersonal maw of Government control."[21] The Sudanese press on the other hand was thrilled by the "delightful surprise" of nonrenewal and demanded immediate clarity on what would happen with rents, profits, and "the number of Sudanese" to be employed in the new administration. Would there be a full nationalization of the scheme?[22]

For the SPS, the end of the concession was a sign of personal changes too. In September 1945, Archdale, the SPS Manager in Gezira, resigned and went to South Africa; Gaitskell succeeded him as Manager. MacIntyre resigned as Managing Director early in 1946; Wooding replaced him. In May 1949, Wooding succeeded MacIntyre as Chairman of both SPS and KCC. That same month, the Zeidab Estate was sold to a private Sudanese investor.[23] On March 28, 1951, the Declaration of Solvency of the SPS showed a positive sum of £9,285,198.²⁴ On May 19, 1952, all business affairs were settled at the General Meeting of Members of the Sudan Plantations Syndicate Limited in Liquidation. Stakeholders received payments up to April 1953.²⁵

Realizing an Ideal Arrangement

Once the decision not to renew the concession was made public, the issue of what would have to be different in Gezira in comparison to SPS management became the center of debate—as discussed in chapter 4 and to an extent in chapter 5. A closely connected issue was which type of entity should manage the new, extended Gezira affairs, the focus of this chapter. Should Gezira be managed by an entity outside direct government influence, or should it be a governmental body running the cotton machine—either within a department or separate? Gaitskell never doubted what to do: the scheme had to be run with "complete neutrality from the political point of view." His ideal was an organization along the line of the Tennessee Valley Authority (TVA) in the United States, an entity outside regular governmental organizational lines responsible for a public utility. Not everyone agreed, however; there were strong voices—both British and Sudanese—to bring the post-1950 Gezira Scheme directly under the government.²⁶

Given the importance of the topic, the procedures that ensured everyone's voice was heard, and the busy period for government officials after World War II, it took a while before the final shape of the successor of the Syndicate was clear. The Gezira Scheme Bill that finally settled the issue was accepted on April 19, 1950, just a little over two months before SPS management in Gezira would end. The apparent lack of progress in reaching a decision itself was cause for speculations on the direction the decision might take. In 1949, the Sudanese newspaper *Rai El Amm* expressed fear that the government might even ask the SPS to continue for another few years, or—potentially worse perhaps—that the scheme would be given to "a few Sudanese capitalists." This last worry was enhanced by the sale of the Zeidab Scheme to a Sudanese business man.²⁷

In his recollection of the process, former Civil Secretary James Wilson Robertson—a key player in the negotiation process leading to the Sudan Gezira Board and the same civil servant who wrote in chapter 4 about how useless equipment lists were in the 1920s—framed the basic dilemma pretty accurately. The Gezira Scheme had become an established economic institution on which the Sudan heavily depended for revenue and food. Whatever one might think about Syndicate management, it had certainly produced profitable cotton and a group of inspectors who knew what to do and who apparently "were generally popular with the tenants" as well. At the same time, any type of private, let alone foreign, "control over a national asset of this kind" would no longer be acceptable in Sudan.²⁸

In the United Kingdom, the new Labor government stimulated the formation of public corporations to manage railways, coal mines, and energy supplies. With the discussion on the possibilities of what type of government control should take over Gezira from the Syndicate, it is not strange that the option for such a corporation sprang to many minds in the Sudan.[29] The direct connection one can assume between one Gaitskell brother in the UK Labor Party (and after 1947 in the government, namely Hugh Gaitskell) and the other being a major figure in Sudan and Gezira would not have worked against that option either. However, a public corporation was not the favorite idea of everyone engaged in the discussions. It was suddenly not the first option mentioned in the string of governmental committees and reports on the subject between 1944 and 1950.

A Special Committee from the Advisory Council for the Northern Sudan was appointed in 1944 to offer suggestions on how to deal with Gezira after 1950. The first meeting was on December 11, 1944. On September 17, 1945, the committee visited the Block of Inspector Hunt.[30] In July 1947, the committee presented its report: it proposed to organize Gezira "as an independent self-contained unit of the Agriculture and Forest Department known as 'The Gezira Scheme Agricultural Division' or the like." In addition to such a unit, a Gezira Scheme Higher Board of Management plus a Local Gezira Scheme Committee should be established.[31] The Committee suggested that Gezira land should not be owned by the State. The "ideal arrangement" would be that the land should be registered in the names of "the cultivators as a whole"—as long as they were Sudanese. Such collective registration did not imply any kind of free choice in farming for the tenants, as the Committee continued to explain that such landownership did not mean at all that tenants "should be given complete freedom" what to do with the land. Gezira was a national scheme with Sudan's wealth depending on it. Therefore, tenants were "discharging a trust to his country" when properly working according to their "duties," including centralized crop rotations and collective cultivation arrangements.[32]

Continued centralized agricultural management did not mean either that only cotton should be grown in the new Gezira. After all, a new Gezira policy should not be "confined at financial benefits only." New ways should be found to ensure Gezira would serve "the welfare of the country as a whole," but certainly that of the "community which live within the boundaries of the scheme" as well. Cotton might not be the best crop to reach both goals. Marketing cotton might become more difficult in the future; furthermore, there should be more room for tenants to grow lubia and vegetables. In a true Gezira spirit, such agricultural issues were to be coordinated by a Committee, which would "study the methods of co-operative and social farming." Tenants became "farmers" in the Committee's report, but they still were to be "instructed" in modern methods of cultivation. Experimental plots throughout the Gezira area would allow teaching the farmers such methods, "of course [...] under the supervision of a specialist."[33] In other words, the new Gezira might have farmers instead of tenants, who might actually grow more than just cotton, but regular Gezira supervision and standardization was to be maintained.

In his (undated) response to the Special Committee's report, the Financial Secretary stressed how much the change of Gezira management in 1950 was "a delicate proceeding"; it was vital "to maintain the momentum and earning power of the Scheme," but this had to be done "with the intention of carrying out any reforms recommended." Obviously, Gezira earning power—which gave Sudan "a foundation of great security"—was a main issue for any Financial Secretary. He supported the idea of a Board responsible for policy development, but stressed that it would be necessary "to entrust business management to a small executive body" with members from the larger Board. The small executive body would have to be chaired by the Financial Secretary. The two bodies would report directly to the Governor-General's Council. A Gezira manager "with wide and clearly defined powers" would manage the scheme and report to the executive body.[34]

For the Financial Secretary, "the primary function" of any Gezira organization was "successful management of an agricultural enterprise" "too vast" for a simple division of the Agriculture and Forests Department.[35] Obviously, new Gezira management should have access to "the best possible advice" and support from government departments, including Agriculture and Forests. Furthermore, although the Financial Secretary agreed that it was important to offer security and "inducement to cultivate" to tenants—as Gezira could simply not "afford to risk a temporary or partial abandonment by the tenants"—security could also go too far. Land ownership was not what the Secretary had in mind. The dominant position of cotton as a main crop was not to be changed either, even though some diversity of agricultural production was desirable. In terms of supervision, the technical departments should provide tenants with "the best possible advice and help" but not "dominate the individual farmer." Having "too few agricultural officials" was as dangerous as "having too many."[36]

Copying a Penguin

Now it was the turn of Civil Secretary Robertson to provide his ideas on future Gezira management. He proposed an organization along the lines of the TVA. Actually, Robertson rather gave his own version of Gaitskell's ideas, as he was not sure about "how such a T.V.A. authority will work in with departments re staff, money, etc." But it was "worth having a little ballon d'essai once in a while." Apparently Gaitskell had been convincing enough on the topic, as Robertson sent his note for comments to Gaitskell, in order to "brighten up" the "rather bold and colourless" note. This strong connection with Gaitskell was not supposed to become too public, as Robertson proposed to send the final note to the various departments without mentioning Gaitskell's name. Knowing that the Financial Secretary would not support a TVA model, Robertson was "not sanguine of much success" anyway.[37] Robertson had done some reading on the TVA; he had borrowed the Penguin Special on the TVA from Deputy Governor of the Blue Nile Province D. Hawkesworth.[38] Although Robertson confessed he had only "read parts of it and glanced at the rest," he felt confident enough about the possibility of managing Gezira on TVA lines.[39] At the time of writing to Hawkesworth, Robertson had not received Gaitskell's answer yet, but he must have received it a little later. Gaitskell was "delighted" that Robertson had been able "to note down so clearly the gist of my conversation with you" on the TVA. Gaitskell just had the minor suggestion to stress the "political importance" of deciding on the Gezira Scheme "and of doing it now." Furthermore, Gaitskell advised a more positive reference to the cooperation with the Blue Nile Governor; Robertson's phrasing "might cause some feelings to be hurt" in that office.[40]

For Robertson, the existing cooperation between the Governor of the Blue Nile, the Gezira Advisory Board, the SPS, and many other groups had been "spasmodic" at best and should be improved. There were no shared plans for the "integration of progress in all spheres of economic, administrative and cultural advance." The Gezira area had to be seen as a whole, which was the strongest argument for bringing all governmental activities within the area inside one authority able to deal with "the diverse difficulties as parts of one general problem." Such authority could work "on the lines of the T.V.A. having final authority under the Governor-General," with a separate budget and its own staff. A TVA model would "inspire a greater enthusiasm and efficiency than the present departmental arrangements."[41]

A TVA setup differed considerably from the model of the Advisory Council for the Northern Sudan and the Financial Secretary. Robertson thought that a Board under a charter would put too much emphasis on "business and technical interests." Gezira was more than "a machine for the production of cotton and money"; he continued in the real Gaitskell style that Gezira might be "the scene of a real experiment" in which education, social improvements, cooperative enterprise, and democratic local administration could be combined within a major agricultural scheme. Robertson considered it "most unlikely" that anything but a single authority responsible for the whole scheme, with adequate legal and financial power, could do the job. A single TVA-style Board was simply the only way "to cut across departmental red-tapes."[42]

Robertson's final note was sent out on September 24, 1947. Bredin, Governor of the Blue Nile, and the Director of Agriculture and Forests (Director AF) each wrote a response to the Robertson note. In his note from February 1948, Bredin quoted "in extensio" an earlier note from Gaitskell himself, something Gaitskell was "particularly delighted" about.[43] The Governor saw a need to keep Gezira's agricultural efficiency, but a necessity to move away from the "drive and initiative of the British field staff" asked for another approach to avoid "inefficiency and corruption." For the Governor—and Gaitskell—the way forward would be a "a deeper sense of their responsibility to the nation" with the tenants. In addition, the Sudanese who were to succeed the British staff had to develop a genuine interest in realizing a public service.[44]

At the same time, any political complication between future management—especially government—and tenants was to be avoided. A strong Gezira Authority would be able to keep focus in Gezira on local issues, which would save the government "from the embarrassment" of doing so itself. After a reference to Labor in the United Kingdom, which would have solved a similar problem through decentralized boards controlling (nationalized) industries, the Governor continued with remarks Gaitskell himself could have made—and actually had made in his note. "[B]old and constructive policy" had to be agreed upon immediately, given the "world events outside this country," including the need to "discharge of our responsibilities to our African territories." Action was needed with the world "torn between the American creed of free enterprise and the Russian creed of communism."[45] We will return to these more political—and rhetorical—claims in the epilogue; for now it is enough to focus on the proposed central authority for Gezira.

The Director AF had a slightly different view on that matter. Obviously, he supported the need to broaden the scope of future Gezira policies to issues beyond cotton, but his main concern was which type of Gezira management would "preserve most safely the outstanding economic importance of the Scheme." It was a "cold economic fact" that any new Gezira setup had to guarantee "efficient business." He then moved to the TVA and the new British Transport Commission, both examples of public bodies with important overarching responsibilities. The Director AF did not see a big difference with a Gezira Board, as proposed by the Gezira Advisory Board. Furthermore, Gezira was "already semi-nationalized." Any future Gezira management had to focus on agricultural efficiency and the welfare of its workers and tenants. The welfare of the general population in the area was to remain the duty of the government.[46]

Three Little Piglets

These different notes suggest all involved in the debate neither doubted nor disagreed on what needed to be managed: the Gezira area had the double duty of agricultural production bringing the money for the Sudanese state budget and of stimulating social development. The disagreement was about how to arrange that, for whom exactly in the Gezira area, and the relative importance of the two duties for future Gezira management. Basically, there were three positions, each with different emphasis. The Financial Secretary proposed a separate Gezira Board responsible for agricultural production—something the Gezira Advisory Board had also suggested. The Director AF proposed a similar Board within the Department of Agriculture. For both, social development was a task for other, already existing governmental institutions. The Governor of the Blue Nile, the Civil Secretary, and—with (through) them—Gaitskell proposed a separate overarching Board responsible for both Gezira duties—with the TVA as model.[47]

On April 8, 1948, the people proposing these different models met within the Gezira Advisory Board to discuss the future of the Gezira Scheme. The discussion was framed along the note of Robertson, who had criticized the original Gezira management proposal by the same board in November 1946. As Gaitskell had played "a major part" in the debate and had helped to prepare the Governor's note, Gaitskell was attending the meeting as an observer—with the consent of SPS Managing Director Wooding. Financial Secretary John Wilson Edington Miller, Chairman of the Advisory Board, started with expressing the "most earnest hope" that Gaitskell would be prepared to "manage for the Government the Scheme to which he has given so much under the Companies" after 1950. With "his worth and his ability," there was no other that had "so completely the confidence of Government, tenancy and the Country."[48]

Miller started the formal discussion in the meeting with summarizing the original idea of the Advisory Board. There would be a Board, operating under a charter, with representatives of main interests in Gezira to control its policy. Gezira management would be done by a small committee of the Board, consisting of "those whose interests were primarily business and technical." Finally, there would be a Local Committee to represent the tenants. To Miller, this setup was to avoid the "fundamental error" in Schedule X, which had mixed up the scope of local government and management of an agricultural enterprise. Obviously, he was not "against the Governor's suggestions"; on the contrary, "many of the objectives suggested" were within the model as suggested by the Advisory Board.[49] Miller saw no real differences of opinion.

Miller agreed that management should pay attention to social services, but claimed that it should limit its services to tenants. In his response, Robertson argued that "in an agricultural scheme of this nature the social and political aspects could not be divorced from the management." The Governor of the Blue Nile, after assuring all that "both he and the management" realized that submerging the economic and agricultural aspects of the Gezira scheme with social goals was not desirable, agreed that an organization with agricultural responsibilities was needed. A general Board responsible for policy was required too, but what was really needed was a really powerful Managing Director, with "wide experience" and "free to travel abroad and keep in touch with modern developments." The Director AF agreed with most remarks, with one exception: there was a need to separate the agricultural (economic) aspects from the social aspects. The Managing Director should be "impressed with the need to resist un-proven change," as a focus on revenue was vital. The Managing Director should be directly accountable to the Financial Secretary or the Minister of Finance "over the head of the Policy Board" when it came to Gezira revenue.[50]

Being the last of the respondents, all that Gaitskell had to do was drop the ball in the open goal. Obviously, he "wished to associate himself" with the ideas of the Governor. A Managing Director and Manager were needed to avoid the Gezira scheme would "drift on without proper guidance and introduction of modern methods." The Managing Director would be responsible for planning and expansion of the scheme and develop contacts around the world. When asked by the Financial Secretary, he agreed that the Managing Director should be a member of the Management Committee and not replace it. Basically, Gaitskell argued for continuing something as close as possible to the SPS. He saw the field inspectors as key to combine the two Gezira duties. The British staff should not be isolated from the "civic side of the Scheme," since they acted as "the feudal father" ensuring that any "aspirations and hopes, both agricultural and otherwise" of the tenants were made known to the "appropriate authorities."[51]

In an impressive compromise, the Advisory Board finally proposed a Policy Board, a Management Committee, and a Local Committee, with a Managing Director being member of both Policy Board and Managing Committee. It is tempting to conclude that this recommendation was made knowing that Gaitskell would be that Managing Director—possibly the post was specifically created to ensure Gaitskell would stay. Village-level coordination on the issues of agriculture and social services should continue, "pending the results of an enquiry as to whether or not separation of these interests at the village level was necessary"—apparently a concession to the Director AF. All in all, the proposed structure kept all responsibilities intact, did not create a separate TVA-like entity, but did allow for more initiative on social development by the future Board and Managing Director than originally anticipated in the 1946 ideas. After drafting such important and—as it turned out to be—determining advice, the Advisory Board went on with business as usual, including recommending that food security in the years to come required that "all possible fringe areas should be rented" to tenants.[52]

Freedom to Exercise the Necessary Authority

In July 1949, just one year before the SPS would hand over Gezira management to the new entity, a proposal for that entity was sent to the Legislative Assembly in Sudan. The assembly discussed the note on October 26 that same year. After stressing again the importance of Gezira for the finances of the whole country, and any "further social advancement," Financial Secretary A. L. Chick stressed that the new organization should all and foremost be capable of continuing to manage the "essential functions" of Gezira "in order to safeguard the revenue from the Scheme." These essential functions were similar to the SPS responsibilities.[53] As the government had to provide the capital for taking over the scheme, the new Board had to reserve part of its revenues to pay interest. The new Board would have to pay tax on its profits as well—which was new, as the SPS did not pay those taxes.[54]

The new Gezira Board would have six directors: the Financial Secretary (or a representative), a Managing Director to be appointed by the Governor-General, the Governor of Blue Nile Province (or a representative), and three other members appointed by the Governor-General, of whom at least two should be Sudanese. The chairman of the Board would be appointed by the Governor-General from among the directors. To allow social development and keeping the scheme a business, the Board was to be "left free to exercise" its authority, "as is the practice in other countries" as a separate legal entity with contractual powers and perpetual succession. At the same time, the Gezira Board would be fully responsible to the Executive Council, represented by the Financial Secretary.[55] Such considerable control of the Executive Council over the Gezira Board—"greater than that exercised by the U.K. Government over State Corporations such as the Coal Board and the British Transport Commission"—was claimed to be essential to guarantee government revenue from the Gezira Scheme.[56]

Social development in Gezira was to be coordinated by a new Social Development Committee, consisting of members from the Board with "wide responsibilities." A Gezira Local Committee consisting of "those living and working in the Gezira" would advise Board and Social Development Committee "on all matters affecting the welfare of the inhabitants." The Local Committee would include representatives of tenants, other local inhabitants, the Administration, the Board itself, and Government Departments. The already existing Gezira Tenants Representative Body would continue to advice on issues like timing and amounts of payments to tenants of the shares in the net proceeds, the Tenants Reserve Fund, the Tenants Welfare Fund, and "any matter affecting the tenants as partners in the Gezira Scheme."[57]

As a showcase for the new British policy in the Sudan—which included the intention to make copies of the note available to the press—considerable attention was given to the Sudanese responsibilities in the future Scheme. The Gezira Board would be responsible to the Executive Council. As half of its members were Sudanese, those members would share "the ultimate responsibility for the direction of the Scheme." Two of the six directors of the Board and half of the members of the Gezira Local Committee would be Sudanese. Recruitment of Sudanese for appointment as inspectors in the scheme would begin in 1949, with the first five vacancies to be filled before the rains of 1950 and fifteen more one year later—as many British staff would retire. A firm statement was made that the government wanted "to devolve full responsibility" for Gezira to Sudanese, with the usual disclaimer that this transition should not endanger "the efficiency of the Scheme on which so much of the country's revenue depends."[58]

After the October 1949 meeting, the Legislative Assembly formed a Select Committee to prepare a final proposal for future Gezira management. The Committee met 24 times, conducted interviews, studied reports and documents, and "toured the Gezira" (again). Its final report was very similar to the note from the Financial Secretary from July 1949. A major change suggested was that a senior member of the Financial Department should replace the Financial Secretary in the Gezira Board—to avoid confusion about the (many) responsibilities of the Financial Secretary. Furthermore, four directors should be appointed by the governor-general, of whom at least three should be Sudanese.[59] Sudanese should be trained for taking up positions with highest responsibility, but any promotion of Sudanese should "continue to be strictly on merit" given the standard disclaimer that "the special contribution of the Gezira Scheme towards the prosperity of the country" should not be damaged. Finally, the Select Committee had computed that a "sound" development orientation would require at least £60,000 to start and some £250,000 each year to develop projects.[60]

At the end of March 1950, the complete set of recommendations was approved by the Legislative Assembly. The Gezira Scheme Bill confirmed the establishment of the Sudan Gezira Board, responsible to the Executive Council through the Financial Secretary, with as its duties: "(a) The management of the Scheme; (b) The promotion of the social development of the tenants and other persons living within the Scheme area and, at the Board's discretion, in areas adjacent thereto; (c) The promotion of research to further the productivity and stability of the Scheme."[61] The Explanatory Memorandum made it perfectly clear that the Bill would not change the distribution of costs[62] and profits, land and water ownership and management, position of tenants, and supervision by field staff. The Board would have the additional duty of promoting "the social development of the tenants and other persons living within the Scheme." As the link to the *Sudan* Plantations Syndicate was important to keep the confidence of the cotton trade, the word "Sudan" was included in the Board's name "to ensure continuity."[63]

After the debate, Deputy Speaker S. Abdel Kerim Mohamed spoke about the "historic day when this great scheme, with all its complicated organisation, patiently built up through a quarter of a century, becomes the absolute property of the Sudanese people." Luckily, the revolutionary changes in Gezira life had taken place "without serious dislocation and trouble even in bad years." Without the companies there would not be a Gezira with its "high level of efficiency" resulting from "the sympathetic understanding and tireless and patient devotion of the men who have built this organisation." To maintain the high standard, and to improve on it, might "tax the capacity of the Sudanese to the utmost," but it would be the future of Gezira and its inhabitants. The gradual Sudanisation of Gezira would be "an advance towards practical self-government of the most far reaching importance."[64]

No Desire for Secrecy as Such

On July 1, 1950, Gaitskell, former manager of the SPS, who had "generously agreed to carry on for a year or two," became Managing Director of the Sudan Gezira Board.[65] Support for Gaitskell seems to have gone far, as apparently "[n]obody attempted to harness Gaitskell." During "the Gaitskell era" of the Sudan Gezira Board, Board members received the information that "Gaitskell saw fit."[66] There was criticism on what Gaitskell did; a continuing area of discourse between SGB and government (obviously) concerned the sales of the cotton. Already in November 1951, there were complaints about the "confusing" way Gezira cotton was sold on the market.[67] At a press conference earlier that same year, Gaitskell (we have to assume) defended a policy of relative secrecy. The SGB had "no desire for secrecy as such," but there was need for "certain discretion in its commercial affairs" like any other firm would do. Informing the public about SGB issues was a good thing, but too much openness would harm the market position of SGB cotton. On the same press conference, Gaitskell defended allowing only little tenants' influence on marketing decisions, as representative "in whom all tenants would have confidence" were not around yet. Involving more people would also work against secrecy during sales negotiations.[68]

Apparently, the successor of Gaitskell, George Raby, was more open to the Board when it came to cotton marketing. With Raby, Board members "for the first time" had the idea that they knew "what is going on." Raby seemed to have related "far more to the Board than Gaitskell ever did." The main problem with Raby seems to have been that he was too open and that "his methods and manner" were "disturbing"—examples of which we will discover below![69] But we must take a step back first to the end of 1951, when it was clear that Gaitskell would leave Gezira.[70] On November 14, 1951, a candidate Managing Director and Chairman of the SGB was selected. Some of the British field staff suggested that Social Development Officer Bill Beers would become Managing Director, which "would be dam funny if it was!."[71] On November 29, 1951, however, Gaitskell himself announced that George Raby would visit Gezira to "to see if he will take on job of M.D. in his, J.G.'s place."[72]

After the arrival of Raby on December 2 in Khartoum,[73] Gaitskell explained the reasons for his own retirement. He did not leave because of disagreements, but for family reasons—his wife did not like Gezira and Sudan at all. He also argued that it would be good to have a new person in charge. Raby was "an outstanding engineer and administrator." He had worked in the steel industry in South Africa, but had resigned when the government changed.[74] After that, he had been appointed as Manager of the Groundnut Scheme.[75] Raby was "remarkably gifted in human relations,"[76] which was confirmed by Ms. Johnson after she met him on December 14, 1951. She found Raby very "pleasant & a business man," and had "no difficulty in getting on with him."[77] On January 4, 1952, rumor spread that Raby had accepted the position;[78] on January 7, 1952, the official announcement was made that Raby "had taken job."[79]

The official announcement also said that Gaitskell was "to stay or rather be an advisor to the Board."[80] In the fall of 1951, Governor of the Blue Nile Province William Henry Tucker Luce had written to Financial Secretary Chick about the "great pressure" from "Sudanese of all sorts [...] to do something about John Gaitskell's departure." Luce felt that given the "tremendous trust" put in Gaitskell "personally," the government should "without hesitation take advantage" of Gaitskell's "willingness" to stay involved, at least for an extra year. Gaitskell's continuing input would prevent a "complete break with the Gaitskell 'touch'." Gaitskell was "impressed and touched by the strength" of this trust, but did not want to stay attached to Gezira beyond April 1953, and certainly not for long periods in Sudan itself. In

March 1952, it was arranged that Gaitskell was to become "Consultant" for another year. Part of the deal was that Gaitskell would write "a full length account of the Gezira Scheme dealing with its social, agricultural and scientific achievements." That was to become the 1959 book.[81]

Raby seemed to have accepted Gaitskell as Adviser or Consultant "or whatever he may be called" for a year;"[82] it seems to be evident that once appointed, Raby made it very clear that he was the one in charge.[83] For some doing exactly that was promising, as Raby "certainly should clear a lot of things."[84] There was a difference, however, between the "Gaitskell touch" and Raby's "remarkable gift" in human relations. One time, Raby met the heads of departments "telling them he expected them to know their jobs & be able to tell him what they were doing not come to him & say what are we to do (as in J.G.'s time). That he expected them not to nuffle."[85] Sometimes Raby's directness had unfortunate effects. The letter he wrote in March 1953 about the integrity of some Barakat staff members was "astonishing," especially as it did not "say anything about being certain of the integrity of the whole of his staff."[86] Soon, Raby's "'integrity' letter" became a "cause célebre."[87] When it came to the decision room of the Managing Director in relation to the powers of the Board, there seems to have been a clash between Raby and the Board as well. In an exchange of letters between Raby and John Carmichael, Deputy Financial Secretary and director of the SGB—probably the most prominent Board member—about a reshuffle of staff that had not been discussed in the Board, the two men simply disagreed on the question whether such decisions had to go through the Board or not.[88]

But there were bigger issues to disagree on, despite the willingness on paper to have a "much closer co-operation" between SGB and the Sudan Government" and the need for the SGB to "restrict its activities to these defined in the Gezira Scheme Ordinance." After all, everyone wished for the "development and advancement of the Gezira in tune with the requirements of the Sudanese Government."[89] One area of disagreement concerned the direction of work of the SGB, the other area concerned who was supposed to do the work of the SGB. Both themes were closely related (again) to the way Raby expressed his—and as such SGB—policy on the matters. In an undated speech, Raby saw big business in cotton, especially its additional products, like stalks for fiber, seeds for oil, and husks for plastics. There were simply no alternative crops "within shooting distance of cotton as a cash crop." He did not hide his frustration about lack of progress made on this front, when he claimed that all this had already been known for years: a "goddam book" on the issue had been in Gezira for 25 years. It was "nearly mildewed by now," but someone "could have spotted that before."[90] It was exactly this rather direct way of making a statement—and acting—that was "the potential danger of the man" giving "cause for anxiety" with several government officials—next to many things Raby said being "often nonsense factually."[91] Several government officials were worried about "the apparent lavish way" of funding new projects by the SGB; introducing Hibiscus Cannabinus was apparently studied. Another complaint was that Raby operated without consulting "the Research folk or Departments, before leaping to expenditure."[92] Raby's appointment turned out to be "very regrettable."[93]

The Sudan Government must have had huge problems with the more political statements of Raby as well, for example, on the issue of who should manage the large Managil Extension—something not officially decided yet at the time (see chapter 7). For Raby, it was clear that "whatever" the government would do, the people doing the management had "got to come from this Board." On development projects in the South of Sudan—a hot topic at that time, as we will see in chapter 7 as well—Raby claimed that the people promoting development were to come from "the North," as it was their job "to be so educated, to be so equipped mentally and physically that they can go down and develop their own country in conjunction with their own countrymen down there."[94]

I would think, however, that the Sudan Government especially disliked the way Raby discussed the role of British staff and experts in Sudan in a period when this subject was one of the most tricky ones the government had to deal with. For Raby, there was so much development work in Gezira and the Sudan that no British staff—or any staff for that matter—should worry "that there won't be a job for them." One just had to be "damn good" irrespective where one came from. In a comment on these ideas, Carmichael remarked that it was such language that might result in the Sudanese getting "bitten and contract Rabies."[95]

With all these differences of opinion between Raby and the Sudan Government, it comes as no surprise that Raby stepped down as Chairman and Managing Director of the SGB in 1955.[96] He was succeeded by Mekki Abbas as Managing Director and Abel Hafiz Abdel Moneim as Chairman. Whatever Raby might have been for the British people in Sudan, the Sudanese Minister of Finance and Economics thanked Raby for leaving behind "a record of achievements and of splendid work which the Sudan will always appreciate and remember." Raby would be remembered for his "faith in the Sudanese as a nation" and his "confidence in their ability to run their own country." In addition, the "courage and trouble with which you publicised your views to the world" were also reason to thank Raby.[97] One person seemed less positive about the Raby years: Raby is not mentioned once in Gaitskell's book on Gezira.[98]

Roll on, Sudanization!

An article in the Sudan Star of March 24, 1954, during Raby's chairmanship, quoted the SGB representative in the United Kingdom—a Mr. Michael Egan—as having said that Gezira cotton was grown "under the supervision of British experts." Carmichael, then Permanent Under Secretary to the Ministry of Finance, could imagine that this might be "good publicity in the U.K.," but for a Sudanese audience it was "hardly tactful in the present situation." The Sudanese might perhaps think that the United Kingdom was "determined not to let the Sudan go free." The least one could have done was adding something on the training of Sudanese to become involved in government work—including SGB duties. There was a definite need for the SGB "to exercise a little more care in the briefing of their publicity agents."[99]

Sudanization was a flagship policy in postwar colonial Sudan, even though it had to be executed in a carefully planned way—similar to devolution discussed in chapter 4. Compared to the SID Sudanization process discussed in chapter 3, SGB Sudanization attracted much more public attention. This was arguably because replacing the British Syndicate cotton inspector with a Sudan Gezira Board agricultural inspector supervising Sudanese farmers was a much stronger symbol of change than replacing a British with a Sudanese engineer to do computations for a new canal in an office. Field inspectors were much more visible. The government desired to devolve full responsibility for Gezira management to Sudanese as soon as possible, but as always the disclaimer was made that this should be done without endangering the efficiency of the scheme as this would lower Sudanese revenue.

In its 1947 report on the future Gezira, the Special Committee (of the Advisory Council for the Northern Sudan) had been clear about the need to involve Sudanese in Gezira affairs. Again, there was no real doubt about the need for experts to run Gezira and instruct its tenants and develop its inhabitants; the main discussion was what nationality these experts should have. The Committee recommended that "a sufficient number of the Syndicate's present capable staff" should be retained by the future Gezira administration. After 1950, Gezira would still benefit from their experience, and the scheme should not "suffer" because "people of no previous experience in the work" would enter. Obviously, the Committee looked forward to the moment that Sudanese would have "a tangible part" in Gezira management. Young Sudanese men (indeed, no women) "with general social culture and sufficient experience to get on well with the people and lead them wisely" and "sound technical knowledge" on agriculture should be trained to "understudy the Syndicate Inspectors."[100] A similar plea to allow Sudanese to "gain experience from their non-Sudanese colleagues" was made in 1949.[101]

On June 30, 1950, at the moment of taking over Gezira, there were 140 British and 1,800 non-British staff in the service of the Syndicate. The British staff occupied the management positions, either in the field as inspector or at HQ in Barakat. At the same 1951 press conference already mentioned above, Gaitskell defended the SPS policy of only recruiting British. The Syndicate had a contractual obligation to "provide always an adequate and efficient Staff to supervise the cultivation" and a demand from its shareholders and "the country to produce the most efficient results they could." Apparently efficient results could not have been achieved with Sudanese.[102] From July 1950 onward, efficiency had to be realized with Sudanese, or at least it should be done as soon as possible without British.

Next to continuing agricultural production, social development brought along new tasks, and new extensions required additional personnel.

The retirement rate of British staff was a continuous source of fear for the SGB. A 1951 forecast indicated that 50 percent of the British staff might have retired by 1953—instead of 1956. Indeed, after 1950 the change in numbers of staff is considerable, but the Syndicate always had to deal with larger changes in staff numbers, for example in the 1930s. The main difference with earlier periods was that this time vacancies would have to be filled with Sudanese in a different political reality, instead of selecting from the many British recruits that wanted to work in a colony overseas. When selecting Sudanese candidates, the SGB seemed to have tried recruiting nothing but sheep with five legs. Candidates should have the "personality, drive, initiative, education, desire or adaptability for an open air life and good physique" to become a good inspector. In addition, "the applicant's character must be beyond reproach." Not surprisingly, the SGB did not easily recruit Sudanese personnel. Additional British recruitment—even though it was against government policy—was considered to be "essential to the Scheme's efficiency." Not that finding enough British inspectors was straightforward; it became "increasingly hard to persuade British to put their lives into the Gezira Scheme," especially without a clear perspective on a career in Sudan "whatever the political future of the country."[103]

In general, there was strong competition with other Government institutions when recruiting Sudanese staff. More or less similar to the SID, the SGB saw itself in a difficult position in competing for Sudanese staff. Life in Gezira "is apt to be dull, and the men [always men] may miss the amenities of town life." Apparently, the British were not that happy with the majority of the Sudanese applicants that were interested. Ms. Johnson—our constant source of inside information—was told that Gaitskell was "very disappointed" by the "Sudanese who came for interview."[104] In 1951, when a new set of Sudanese inspectors was to be recruited, it seems that some of the Sudanese inspectors agreed with their British colleagues that the Sudanese applicants could not "possibly be up to much."[105] Apparently, candidates who were interviewed included a tram driver or ticket collector from the Khartoum Tram Service.[106] Sixteen new Sudanese inspectors were finally recruited, "6 or 8" were "ex schoolmasters."[107]

The results of Sudanization were not always liked by everyone. Even something like car failure was blamed on it. Culwick noticed that the "helpful workshop mechanics" had received instructions "to reduce the outlet in the Bedfords" but had "brightly stopped up the outlet altogether. Roll on, Sudanisation!"[108] About a year later, she made a similar call to Sudanization when discussing the railways, which seemed "to be running true to form" with long delays. "Roll on, Sudanisation! Every day and in every way we get muddlier and muddlier."[109] Although I do understand how frustrating it is that cars do not work and trains do not run, blaming failure on Sudanese doing the work would have reflected standard British views on the Sudanese as backward, a theme we will return to in the epilogue.

Picking Up Ground

As a result of its Herculian task to change its staff, the SGB had managed to appoint five Sudanese inspectors in June 1950 and sixteen in 1951. In the beginning of 1952, the SGB recruited a total of 45 Sudanese inspectors.[110] The new Sudanese inspectors were placed under the guidance of experienced British staff; inspector Johnson—husband of Ms. Johnson—was one of them. He concluded that the Sudanese inspectors were "tactful and firm" with the tenants and "combined well with their British colleagues." His "own" Sudanese inspector had "a lot of ideals for his country,"[111] but if we are to believe Ms. Johnson, the ideals changed quickly. One Sudanese inspector "hated the tenants," and others "thought them an idle lot & that the job was very hard."[112] Later that year, one Sudanese inspector complained to inspector Johnson that he "got really worked up about tenants who didn't work." Johnson replied that he "might as well get used to" it, as they had to do the job in the future without the British.[113]

Those exciting days of huge development plans, uncertainty of British control and increasing Sudanization of posts, made it "extremely difficult" to ensure that the Sudanese would be offered "a really smoothly running organisation" when the British would have left.[114] As late as 1955, just before Sudanese independence in 1956, Carmichael envisaged "a very rapid deterioration in the efficiency of the Ministry" after handing over. He feared that Sudanese "entering into posts of higher responsibilities will be placed in a most unfair position" in terms of results they could achieve.[115]

I will elaborate on the issue in the Epilogue, but one could question whether such doubt ever went away in expatriates' brains whose jobs were to be Sudanized. In a recent blog, someone working for an NGO in South Sudan expressed the belief "that it's good for INGOs to be staffed by nationals from the countries they're functioning in," but also that it might just be "a bit too early to be kicking out the expat human resources."[116] In the next chapter, we will discuss the results that were to be achieved through development policies in post–World War II Sudan and elsewhere, including the position of international, expatriate experts.

Chapter 7

The Everlasting Rectangles: Gezira and International Development

In the early 1950s, it was clear—even to the British—that in a near future Sudan would become much more, if not completely, independent.[1] At the same time, the United Kingdom wanted to keep some kind of influence in the country. One of the possible ways was to clearly offer support to Sudan's development efforts, with irrigation high on that agenda. A completed Managil area required a second dam on the Blue Nile—planned at Roseires—for its water, and it was clear that Sudan did not have all the financial resources itself to build that dam. In 1953, a certain Luce from the Sudan Government Agency in London shared some ideas—after a discussion with colleague Allen from the Africa Department in London—with John Carmichael in Khartoum. Basically, Luce argued, if "maintaining British influence in the Sudan really means anything," supporting Sudan's economic development was "the most effective way of showing it." Luce had also spoken with William Nimmo Allan and Humphrey Alan W. Morrice, both from the SID. The three of them thought that financial support for Roseires Dam, "in much the same way" as the loan for Gezira in the 1920s, would be one of the best options.[2] After all, large-scale irrigation was "always in the forefront of Sudanese minds." A UK loan to the Sudan would have great political value.[3]

In 1929, the UK budget for colonial development had been £1 million per year; in 1940 it had risen to £5 million, and in 1945 a sum of £120 million for ten years was reserved.[4] As part of the larger changes in British colonial policy after World War II associated with these budget changes, the Sudan Government tried to change its policies toward Sudanese development. Irrigation was clearly important, as the Managil project—basically a copy of the Gezira irrigation model—and policies on agriculture in Gezira itself show, but other large-scale projects like the Zande Scheme in the south of the country and mechanized durra cultivation projects on dry lands were started too. The Gezira organizational model served as basic blueprint for these efforts, and people with working experience in Gezira became consultants for them. British colonial officers in several other countries approached their colleagues in London and Sudan asking for information on the wonder scheme in Gezira.

Development as a concept is often described as a post–World War II concept.[5] This chapter does suggest that there are at least strong links between earlier efforts within colonial states and later international development, an issue I will return to in the epilogue. What was certainly different in the post–World War II years compared to colonial times was the Cold War. Development efforts were part of the new international political reality of independent states, spheres of influence and trying to keep close contact with the former colonies. How to achieve such contact (or influence) was not easily agreed upon. The Gezira Scheme became a model for irrigation and development projects elsewhere, both in Sudan and in other African countries. The British who had worked in Gezira—be it for the SPS, the SID, or the civil service—found employment elsewhere, with many of them becoming international development experts in the Truman-induced era of international development.

Little Encouragement

In post–World War II Sudan, Carmichael was working on keeping British interest strong in Sudan, for example, in helping the SGB find contacts with British firms.[6] In general, Carmichael had "different views" on the matter compared to Luce. There would indeed be a need for foreign capital in the Sudan in a very near future, but "for reasons rather different from those" Luce considered.[7] What those different reasons of Carmichael were remains a little unclear, but Luce argued that keeping influence in its former dominion meant that the United Kingdom had to "give" to Sudan "without much expectation of something in return." In general, Luce considered that the United Kingdom could offer defensive support to the new country, financial and economic assistance, continued service of personnel, and education to the Sudanese. Such support could perhaps preserve "all that British effort has been built up over the past 55 years" in Sudan, avoid that Egypt would "move in as the British move out," and assure that the new Sudanese nation would be based "on the foundations already laid."[8]

UK goodwill to assist a "young nation," after having "nursed [it] from infancy to make its way in the world," should be the way to retain Sudanese goodwill. Such a policy in the Sudan would also have a positive effect on similar processes in British African colonial territories.[9] The British Treasury was somewhat less enthusiastic. Sudan "was in rather a special category"; it was not a member of the Commonwealth—it had not been a UK colony— and did not have its own currency.[10] It was UK policy to restrict overseas investment as far as possible to Commonwealth countries.[11] Given this "little encouragement" from UK Treasury and government, Sudan had to turn to others for financial support.[12] Despite these little hiccups in strategies to provide financial support for Sudanese irrigation development, the British continued to develop ideas and strategies on the subject. After all, Sudan had "much greater potential" for agricultural development "than any other country in the Middle-East." In addition, the cotton of Sudan was still important to the Lancashire cotton industry. The British remained optimistic that they would continue to be welcome in Sudan, as "British technicians generally give them the best service." In an optimistic mood, the British were even sure that the contract for the second main canal from Sennar to bring water to the Managil Extension was theirs, as British firms had been "told in so many words" by Sudanese Ministers.[13]

On May 13, 1957, however, it was the German company Messrs. Julius Berg & Philips Holsman that started digging the new Managil Branch Canal within "the first international tender in the history of the Sudan."[14] With UK support less welcome than the British had hoped, other options to buy influence had to be found. Perhaps agricultural research could be one of the "cheapest yet most effective ways" to give assistance to Sudan, as it had become "the key to economic development."[15] The British kept a close eye on what other countries did in Sudan as well. Luckily the Russians "had refused to put up money" for Managil, and Sudan "did not want the Russians" involved in the Roseires Dam Project.[16] Sudan turned to the International Bank for Reconstruction and Development to find support for projects like Managil and Roseires.

Smiling Green Fields

Managil was a major element in the Sudanese policy for so-called productive development along the Nile (simply referred to as "the river"). Other policy directions were productive development "away from the river," expansion of social services, and expansion of security forces. Given the budgetary restrictions in Sudan, social services and security could not "have priority over" any of the two productive development strategies. The plans for social services and security forces needed "no boosting!"; it was difficult enough "to keep them within the limits of reason and finance." Financial support for these projects had to rely on results from the productive ones. As development "away from the river" would require activities in sparsely populated areas, it was not prioritized either. Therefore, the most popular options for productive development included a few "expensive projects," like Managil and Roseires, but also a private pump scheme (Khasim el Girba on the Atbara), and schemes for sugar, rice, tea, coffee, and tobacco in the South. Given the political unrest in South Sudan, northern projects had priority, but once conditions improved, southern schemes "should run concurrently" with northern projects.[17]

In line with earlier colonial policy, Sudan's main development policy was increasing irrigated production along the Nile.[18] The 1957–1962 Development Program focused on Managil and Roseires, although "due consideration" (or lip service) was given to "development of social and administrative services."[19] Within this general policy, the Managil Extension was the "major productive project" with the "quickest substantial results" to provide revenue for the government.[20] Even without additional water from Roseires, extending the irrigated area in the Gezira plains was possible. Managil was also the project that would require a massive investment of government money plus foreign support, as not all the money for Managil and Roseires could come from Sudan's own budget. From the 42 million pounds that would be required in ten years, half had to come from the Sudanese government's budget. This equaled about £2½ million per year, an impressive sum given that net government income had been £1.8 million in budgetary year 1955/56.[21] As such, funds for any other project were limited, especially with high cotton prices something of the past. There was also lack of experienced staff, which was partially blamed on Sudanization. This created a circular problem: Sudanese staff was needed to attract international experts and banks, but to have a sound plan, international support was needed.[22]

An organization to run Managil was needed as well. The SGB claimed Managil in the early 1950s, arguing the SGB was the only existing institution with the experience and the staff needed. The SGB was interested in managing Managil for full control of its cotton—including pest control.[23] In 1955, management by the Ministry of Agriculture was decided against, as the Ministry itself did not want to. A cooperative tenants' society was briefly a possibility, but tenants could (obviously given what we found in chapter 4) "not be expected to develop the essentialities of cooperation" in the time available. Perhaps a limited company similar to the Sudan Light and Power Company Ltd. could be established, but this might cause "some misunderstanding of the role of such a company." A separate, second government board would be a slight overkill. Perhaps the While Nile Schemes Board could do it, but being so close to it taking water from Sennar, Managil would have "to be run on the same terms as Gezira"—including tenants requiring "much the same handling." Management staff was "a scarce factor in and outside the country," which made the SGB the most obvious choice for operating Managil.[24]

In November 1955, the SGB accepted the gift of managing Managil. It was prepared to undertake the task, as long as it was to be permanent—one of the earlier options had been involving the SGB only temporarily. Furthermore, the SGB wanted some "appropriate financial arrangements." In its second meeting, the Sudan Government committee dealing with the issue recommended that the SGB should manage Managil "permanently."[25] This recommendation was accepted in March 1956.[26] When the SGB accepted the management of Managil, it did not exactly know what it was going to manage. Even though it was clear that Managil would be a rather close copy of the Gezira Scheme,[27] some major issues had to be decided upon.

The first issue was along which timeline and in how many phases Managil should be realized. Should one wait for Roseires to be ready before realizing the full Managil area? The SID could either construct the full canal and all the large regulators or start with a smaller regulator at the end of the existing Main Canal for the first 200,000 feddans in Managil—which was the best option according to the Director of Irrigation, who wanted staff available to start working on the Kenana Irrigation Project close to Roseires. However, the government committee agreed with the SID Irrigation Adviser that the Kenana plans were rather vague to build on. It advised to go for the full 800,000 feddan Managil Extension "to enable economies to be made in planning"—the SGB fully agreed that postponing development of their new asset was to be avoided.[28] These recommendations were accepted in March 1956.[29]

Another issue was how to make best use of available Nile water. The existing Nile agreements did not provide the water needed to irrigate all areas in winter. Should Managil or Gezira or both stop irrigating after January 31 each year? How big should Managil tenancies be? Would they be 10 feddan of cotton in a four-year rotation, resulting in 40-feddan tenancies, as in Gezira? Would tenants farm 5 feddan of cotton in a three-year rotation, resulting in 15-feddan tenancies? This last number came close to the size of the many half-tenancies already existing in Gezira[30] and would allow many more Sudanese taking up a tenancy in Managil. However, assuming Managil would have to quit irrigation in February, would such a small tenancy give sufficient financial returns on the cotton?

The SGB favored a system of five cotton-feddan tenancies and proposed that irrigation in Managil—not in Gezira—would be stopped after January 31.[31] Smaller tenancies would give "a better distribution of wealth" and durra and would be "acceptable to most tenants" for its "economic size"— probably a reference to the labor needs. In a meeting on the issue, SGB-adviser Gibbons argued that on average it was the five-feddan tenancies that gave higher cotton yields in Gezira. The existing irrigation infrastructure could irrigate these smaller tenancies as easy as the regular ones, which meant that a smaller standard tenancy did not require changes in the existing SID designs. The committee decided in favor of tenancies with five feddans in cotton.[32]

In the meantime, design and construction of Managil infrastructure within the SID had continued. In the 1952/53 season, the main canal capacity was doubled to allow for Managil flows.[33] The work did not progress as smoothly as one had hoped. In the season 1954/55 only 15,000 feddans could be surveyed, partially because of staff being ill, possibly caused "by the reluctance" of some engineers "to undertake such tedious but essential work."[34] Construction of the first Managil phase started in 1956.[35] The earth work of the Germans was finalized in June 1958, when progress on the main canal was "very satisfactory."[36] On July 20, 1959, the first water was admitted into the Managil Canal "and the happy event celebrated." Managil was ready to "spread civilization in those desolate regions" through "smiling green fields."[37]

An Equally Profitable Substitute

The purest green the fields would be, but what would be the source of the green was not clear yet—and another subject of negotiation, this time with a more pronounced role for the international experts from the International Bank for Reconstruction and Development (IBRD) and the World Bank (WB).[38] The decision to go for the full Managil Extension before Roseires was ready did still mean that Sudan could not realize the two projects with its own financial resources. International support was needed, but before providing a loan, the international banks wanted to ensure that the proper crops were grown on the fields they were asked to finance. The February 1958 report of the IBRD Mission to Sudan—first in a series—complimented the Sudanese development program with its "commendable emphasis on the development of productive resources" and "determination to resist pressures for greatly expanding outlays on social services" without the necessary financial sources. Irrigation was essential to develop the productive resources, but three major problems had to be solved first: the cropping pattern on the new areas, the exact planning (or not) of the Roseires dam, and the interests of countries "other than the Sudan" in Nile water—which in practice meant Egypt.[39]

On the issue of crops and rotations, a three-year cotton-based rotation in Managil would bring more intensified agriculture compared with the Gezira rotation—"although still far from being intensive itself"—which was feasible both from agricultural and economic points of view. The Mission of 1958 did not really see an "equally profitable substitute for cotton," although the importance of cotton should perhaps be "de-emphasized somewhat in the interest of greater income stability." For financial reasons, the mission advised to introduce American short-staple cotton as an alternative for long-staple cotton.[40] The Agricultural Advisory Committee of the Sudan had just completed its own Managil report, advising against American variants as these would not mix well with the long-staple cotton. If it had to be short-staple, it had to be Egyptian. However, the committee wanted to treat Gezira and Managil on equal footing for water, which meant equal cotton too. In an attempt to solve the water issue, stopping irrigation of all cotton areas, including private pump schemes, on February 28 each year, was suggested.[41]

This discussion only postponed the decision on what to do with cotton in relation to water availability in both Managil and Gezira. Without additional Roseires water, Gezira and half of Managil could be fully irrigated until 28 February each season. Early-maturing cotton varieties would be a good option, as these needed less water than the late-maturing Gezira varieties and still be economically feasible.[42] This depended on what economic feasibility actually meant—as ideas about feasibility would be different for different people. However, "comparative rates" of the cotton profits might "appear to" favor the Sudanese government and SGB, not the tenants. Luckily, the Mission emphasized, the tenants had "the full benefits of the dura crop."[43] Apparently, the mission recognized that cotton was not a profitable crop for the tenants, at least not within Gezira arrangements. Possibly with this economic tension between SGB and tenants in mind, the Mission emphasized that "further diversification" of cropping "would undoubtedly be desirable." This might actually be possible on 15-feddan tenancies with only 5 feddans of cotton, which would encourage tenants to do more with their own labor and hire less from outside—where did we hear that one before? As even with those smaller tenancies, outside labor would be needed during the picking season, short-staple cotton would be a good option, as it could be harvested before long-staple cotton, which would spread out the demand for labor over the season.[44]

The IBRD report was followed by a discussion within the Sudanese government itself, assuming the many documents produced by Carmichael on the economic aspects of irrigation development in 1958 are indeed proof of such discussion. It was still not decided what to do with the extra water that a completed Roseires reservoir would provide. In the discussions throughout the years, the assumption had been that the order of preference would be providing Managil with any additional water it required first, followed by developing new pump schemes, and finally starting a major irrigation scheme in the Kenana region. To Carmichael, the 1958 discussions on crop intensification, available labor, and the continuing budget issues in Sudan had brought other options to the foreground—options with the added benefit that "capital overlay" was "comparatively small."[45]

The "other options" were all based on growing more crops within the existing infrastructure, with minor additions of new cotton extensions from Gezira and Managil main canals, and irrigating winter crops after durra or cotton irrigation had stopped. Carmichael claimed that all these options yielded the same profit per unit of water but saved on expensive new projects. In the period before Roseires was to be completed, at least until the 1966/67 season, "further expansion of productive development" was not easy. The rapid expansion of Managil was a good reason to give Sudan "a rest for a while until that expansion has been fully digested." A new scheme like Kenana was simply less profitable.[46]

However, doing nothing outside the larger Gezira plains would not be a feasible option either. What would happen if Roseires would take longer to complete? Carmichael saw options for new pump schemes that would stop irrigation after December plus an increase in cropping intensities in Gezira and Managil. The only thing the development planners had to avoid was Managil becoming more profitable than Gezira, as this would upset Gezira tenants.[47] In the worst-case scenario, should international money not become available for Roseires, the Sudanese government itself had to build the dam. In that case, Managil had to be finished first and Roseires building could only start somewhere in 1962 or 1963. After the earlier shift within development discussions from purely expanding infrastructure toward other uses of irrigation canals and creating better—yeomen—farmers through social development, at the end of the 1950s, the irrigation expansion focus to make money for the Sudan was firmly back on the agenda. Gezira and Managil could be extended with a third canal from Sennar, or schemes in the Kenana or Khasim El Girba areas could be realized.[48] The irrigation possibilities seemed endless!

King or Prince

Once the next international IRBD report had become available in October 1959, endlessness knew its boundaries again. After repeating the generally agreed wisdom that Sudan's options to increase "commercial agricultural production lie in the development of irrigation," the report went on to emphasize the need for more diversification of irrigated production through introducing additional crops into the "cotton rotations." King Cotton was to become Prince Cotton. The mission was extremely positive on possibilities to achieve this new diversity. Managil was well planned "as an extension of the Gezira Scheme"; there was "a strong demand for tenancies," and the mission was "most favorably impressed with the highly competent operation." The infrastructure was "of excellent standard" and the first Managil season reflected "excellent planning, organization and administration" by both SGB and SID.[49]

The mission concluded that the good old partnership model in Gezira and Managil still worked well, at least for two-thirds in terms of partners, with the Ministry of Irrigation operating the irrigation system and the SGB managing production "efficiently." All this optimism aside, the IBRD did complain again on the "lack of flexibility" when it came to cotton—there should have been much more emphasis on short-staple cotton. The mission concluded that future Managil Stages should grow short-staple cotton, which could mature on flood water only. Furthermore, the tenants were a slight problem. They might have "grasped the main requirements of cotton culture"—even in new areas like Managil—but how they dealt with "supplementary crops, which is not supervised, is much less efficient." The mission thought the Sudan approach of irrigated farming had the advantage that farmers "unskilled in modern techniques" could still farm efficiently "under supervision." Good yields required "careful management."

The IBRD brought the focus of centralized planning back on the tenants, when it suggested giving "more technical advice, supervision and some degree of control" to tenants who were growing other crops than cotton, as this could double their yields.[50] With the smaller tenancies in Managil (and in Gezira with the half-tenancies), intensification of the cropping pattern was more possible than ever before for tenants and their labor force. Shifting from night storage to night irrigation was desirable, however, despite the advantages night storage would bring—more efficient irrigation and less inconvenience to farmers. The disadvantages of more expensive canals and silt removal were considered too high, though, by the IBRD. As farmers "in other countries" could "irrigate successfully at night," continuous water distribution had to be considered in Gezira and Managil as well.[51]

All these considerations brought the mission to prioritize development projects in Managil, including restricting seasonal water requirements. Furthermore, allowing existing pump schemes, which were still operating under restricted water licenses, to expand and intensification of cropping patterns in Gezira and Managil without remodeling canals were good policies. Only once these options were fully explored should new pump schemes be considered, followed by expanding the Gezira/Managil area with another 200,000 feddans. The mission was clear that the Kenana Scheme "should be deferred for the time being."[52] It was also very clear that the IBRD would only provide financial support for anything related to Roseires when a Nile agreement was in place between Egypt and Sudan.[53] In 1959, with Egypt eager to please Sudan, this agreement was reached.[54] Internal discussion in Sudan concerning whether to actually build the last Managil phases continued well into that same year. Perhaps Managil water could be used more economically elsewhere? In a typical reasoning of selected—or selective—path dependency, further realization of Managil was decided upon. The machines were there, the organization was there, the canal was there, so why not use them? Even long-staple cotton was not a problem; despite increased competition with short-staple cotton, Sudan could still sell its long-staple cotton. Managil was to be completed.[55]

Changing Traditional Enjoyments

Creating a Gezira Satellite in Managil was not the only thing on the Sudan development agenda in the 1950s and 1960s. What to do with the Gezira Mothership was another main issue. Should the four-year cotton rotation be kept? How should policies of increased Sudanese influence on Gezira business and developing yeomen farmers evolve over time? What did these policies actually have to do with crop growing and selection in Gezira? Should the SGB limit itself to cotton? As with Managil, discussions on these issues were both held within international circles, Sudanese government circles, and the SGB itself. Within the policies and general ideas about Sudanese tenants (who should become farmers) discussed in earlier chapters, the SGB started with experiments on mixed farming in the early 1950s. Several inspectors considered such experiments not in line with "their job of commercial cotton production," but SGB experiments continued, including applying fertilizers, not applying pesticides, growing specific cotton varieties in certain blocks, early water closure, weed control, and weeding on fallow.[56]

A main SGB concern was that their fellow partner in actually growing the crops—the tenant—was not as progressive and enlightened as the SGB considered itself to be. The system of cotton production was in order, but for all other crops, the "traditional enjoyment" tenants had received—free land, free water, free choice of cultural methods, plus seed supply with 100 percent ownership of crops—would have "fostered a conservation," which had become "a stumbling block" when it came to improving production of other crops than cotton in Gezira. The "sorry fact" was that the tenant was the one setting "the pace, exhort him how we will" for development. The SGB reply was considering additional collective measures, for example joint spraying campaigns for all crops. However, financial arrangements for such work within the existing Gezira regulations were tricky. Using a tenant's cotton account was easy enough, but not for activities unrelated to cotton. Such "a complete change of policy" would possibly jeopardize the "great social experiment of the Gezira Scheme system of partnership." Using "force" to ensure tenants did comply was "obviously out of the question," but educating tenants would (still) "take some time."[57]

A feasible short-term option could be providing improved durra seeds to tenants "who will willingly take it." Once those tenants would have higher yields, "even the most conservative of tenants" would change their mind, especially if it was "debited to his account."[58] The unique Gezira partnership worked best when one partner decided what to do and how much the other had to pay—collectively if possible. Therefore, discussions on alternative crops in Gezira continued with a strong focus how these crops could bring financial revenue for the SGB.[59] Even an early policy like Schedule X was enrolled to argue that "dependence on a single cash crop" was not in the tenant's interests.[60] Nevertheless, it was clear throughout the whole debate that the cash crop cotton would remain "the major item" in any cropping policy or program for economic development.[61] Other crops could be part of the cultivation pattern, other types of cotton might be possible, but Gezira was really "ecologically suited to Egyptian cotton" in a way that "not many other parts of the world" were.[62] Cotton was to stay.

International experts gave their ideas on the matter as well. The IRBD stressed that Gezira had started primarily for the cultivation of cotton and that only because tenants liked growing durra, this crop was included in the original three-course rotation. Fitting durra "was not easy," as cotton and durra were competitors. Cotton harvests were actually slightly higher without durra. Nevertheless, Gezira rotations with durra had become "taken for granted." With the lower cotton prices, however, other crops might become more important in terms of revenue. Groundnuts and fodder crops could yield very satisfactory returns. Especially groundnuts might be a better crop in Gezira. Durra could be grown outside Gezira and Gezira groundnut farmers could simply buy the grains they needed. In order to realize any of the changes, however, tenants had to improve their cultivation and "some measure of control" of their practices had to be organized, preferably by the SGB.[63] Giving the lead to the SGB limited the scope for new crops, as the SGB would only consider including crops that it could market itself. This basically meant that only crops already known in Gezira were serious candidates for more intensive production: cotton, durra, wheat, groundnuts, and the forage crops lubia and phillipesara.

Water from Roseires could possibly irrigate 120,000 feddans of wheat and 120,000 feddans of groundnuts in Gezira. These two crops would occupy the first of two fallow years in the four-year rotation before the new cotton crop. In this respect, the Gezira rotation provided "more scope for intensification" than the already dense three-year rotation in Managil and the pump schemes. One could question, however, whether uniform rotation schedules would be the best strategy in an area as diverse in terms of soils, rainfall, and marketing options as Gezira. Tenants close to Khartoum could grow vegetables for a considerable urban market, something not really possible in the south of Gezira.

In the early 1960s, with increasing amounts of half tenancies of 20 feddans and alternative crops, a more diverse pattern in agricultural strategies had already arisen. Within the Gezira model, such developments could obviously not remain spontaneous—the SGB had to have a clear, collective policy for each crop. As smaller tenancies seemed to be better at creating the type of farmers the planners looked for, the smaller farms "should be encouraged" to stimulate and create the "enterprising tenants" that were needed. The strict supervision that still existed for cotton—at least on paper—was "unlikely to be accepted" on any other crop. Supervision had to be replaced with "a new emphasis on extension teaching." Guiding tenants was a task SGB saw for itself. Furthermore, marketing of wheat and groundnut—partially through stimulating cooperative marketing of groundnuts—was a goal as well.[64] Forcing "the indigenous tribes" into cotton might be outdated, but transforming those tribes into cash crop growers on a voluntary basis was "the real problem" that planned economic development in Gezira faced.[65]

In the Ten Year Plan for social and economic development for the 1960s, the cultivable area in the Gezira Scheme was planned to be enlarged with 290,000 feddans. Throughout Gezira, more land was to be used for cropping, and fallow would be reduced. The additional fields would be sown with wheat, groundnuts, vegetables, and fodder crops. What such a change exactly would entail, was unknown. In SGB style—"when in doubt, experiment!"—four Gezira Blocks in different groups were selected to start a trial with a complete, intensive rotation on 720 feddans. Block inspectors had to produce a report every fortnight and were expected to put "particular emphasis" on the issues the tenants would face, including weed growth, pests and diseases, crop yields, cost of production, marketing, and of course, "tenants attitude to change."[66]

Establishing cooperative societies for the production and marketing of groundnuts was stimulated by the SGB. A first one was established in Darwish Block in the Centre Group in the 1961/62 season. In the next season there were 10 such cooperations; in the season 1963/64, 12; and in 1964/65, 24, varying in size from 23 members with a total of 110 feddans with groundnuts to 160 members with 455 feddans.[67] Tenants were also encouraged to develop joint efforts concerning mechanization, for example, to buy combine-harvesters for the wheat crop with a group of farmers. One such group started in 1964.[68] Tenants' cooperatives remained under SGB "supervision"; the SGB also provided logistical support and general guidance, as the "rural mind" remained "resistant to change."[69]

A Working Party

The future of Gezira, especially all the water that would become available through Roseires, remained high on the Sudanese economic agenda. In 1963, a so-called Working Party (WP) was established to discuss the economic future of Gezira and ways to respond to, but very much improve and plan, that future. Its time frame would be the moment when Roseires water would become available, thus the season 1966/67. Its main question was to study options for alternative cropping strategies "for existing and future areas of production" from that season onward. In its 1965 interim report, after the well-known praise for Gezira as successful "large scale farming" through strong cooperation between "tenants and management," the WP stressed that success was under threat. The "high degree of organisation" had been able to combine "the exacting requirements" of cotton with durra and livestock, but a growing population together with increasing expectations for a higher standard of living required a new approach. As "no nation" could accept an average national income "of LS.30,- per head," increased agricultural production was highly needed. Increased production required updating the canal system, as several branch canals could not provide the water supplies needed. Once updated in 1967, proper-sized Gezira canals would bring Roseires water to areas with a cropping intensity of 50 percent throughout the scheme.[70]

The former approach of rapidly expanding the irrigated area had to be replaced with an approach that ensured the original Gezira area would reach "a new and higher stage of production and development." The WP argued that Gezira had to be managed for the benefit of Sudan as a whole, which did not necessarily relate directly to the "specific requirements" of people in the Gezira area. Tenants would (have to) be transferred "out of subsistence agriculture" into commercial production; the productivity of those tenants who had already made the move had to be raised. Obviously, the welfare of the farmers dependent on the scheme could not and was not to be neglected. Their "full co-operation" would not materialize anyway as long as they would not receive "a reasonable return for their efforts."[71]

The WP expected that "willing co-operation" of tenants would require more than "simple monetary incentives." Tenant autonomy had to be checked, though, as it was not necessarily leading to good results for Gezira as a whole. Once again, the employment of outside labor was brought up as an example of wrong tenants' decisions, as it resulted in "under utilisation" of labor available in Gezira. People living in Gezira should find useful employment "within rather than outside" the area.[72] In a comment on the interim report—probably by Hugh Ferguson, chief of the Agricultural Research Division—this phrasing was criticized. The order of priorities itself was fine, but it would have been "psychologically better" to stress "the tenant's specific interests earlier." After all, tenants' representatives reading the report might not agree with the order of tenants' and national interests in the report.[73]

Somehow, the WP did not conclude that the SGB would have to change much. The management model "developed since the days of the companies" was a "major asset" of Gezira. Central planning along lines already set remained the basic strategy to follow. A shift to American cotton was not recommended as the existing ginneries did not work for that type of cotton. The tenancy size was to stay the same and agriculture in Gezira had to remain on a rotation of four years or multiples of four years. The Gezira layout of 90-feddan numbers subdivided into 10-feddan plots was "fixed" as changing it required "a major recasting of the irrigation system." Something like "the present rapid expansion of the wheat area" would be much harder to achieve when Gezira would have "lacked the central organisation" it had—including the Research Farm. Extension services in the Gezira needed to be expanded, with the SGB, its field staff, and its contact with tenants through the village councils and the Tenant Union being the perfect organization to do so.[74] The WP was simply proposing that the SGB should do more of the same rather than doing something different.

Support May Be Infeasible

Within the planning context of the SGB and its WP, the bottleneck for improving Gezira remained the behavior of the tenants. The "marked contrast" between cotton production and the tenants' "own crops" continued to be too large to accept. Given Sudan's scarcity of "capital for development," one could simply not permit Gezira tenants to produce "half crops," when with some extra effort they could grow the "full crops" that benefited the country—and themselves. The basic dilemma for Gezira in the 1960s was that tenant "support may be infeasible" but that it had to be there within Gezira logic. A feasible strategy might be to develop indirect measures. The SGB could start working with contracts whereby tenants accepted obligations regarding production methods in return for price guarantees. Purchasing fertilizer could be made a condition for growing durra. Refusing water to crops that were sown late or were neglected was studied. Water rates were an option to ensure proper irrigation, especially "to prevent the wasteful use of water."[75]

The WP did provide a little bit of a new approach to Gezira farmers in terms of the profitability of the Gezira Scheme for the different partners. Eleven basic scenarios of different cropping patterns were defined with their yields; for each scenario the profits were computed with linear programming—quite a new approach at the time. In fact, the approach was so new that not all scenarios the WP wanted to study could be computed as the time needed to run the scenarios was too long. The scenarios that could be computed showed four scenarios yielding the highest revenue on system level. All four included cotton. Interestingly enough, the WP also developed scenarios maximizing the net profit for tenants. The scenarios most profitable for tenants did include only a small amount of cotton or no cotton at all, depending on the yields assumed. The scenarios showed "a divergence of interest" between the Sudanese government and SGB, on one side, and tenants on the other, caused by the procedure of sharing the cotton proceeds. Only because Gezira had to have cotton, it was a factor for tenants to consider. If not obligatory, cotton would not be a feasible economic option for tenants.[76]

This unattractiveness of cotton for tenants might have been known for a longer time—as discussions in chapter 2 and the IBRD report already suggested—but it was the first time that the basic dilemma of Gezira was spelled out so clearly by the Gezira management itself. Ferguson welcomed the Working Party report for finally clarifying "what many of us felt we knew already": cotton was the revenue owner of Gezira, at the expense of tenants' profits. Writing this down in an official report had "broken the ice" on the position of cotton.[77] "Sharing" net proceeds from cotton might have had many advantages "when the Gezira was primarily a one crop scheme"—when it was run like a plantation—but it was not at all appropriate in a scheme in which a variety of crops were to be produced. Suddenly, the "major handicap in the development of the Gezira," namely the fact that "the tenant's interests diverge markedly from those of the nation," was out in the open.[78]

Nevertheless and not surprising, the WP kept the proper Gezira spirit. It did not doubt that the government and Board had to guide the Gezira Scheme and that cotton had to stay,[79] "strictly enforced as now," as long as it was impossible to show the "advantage of this crop in its proper perspective to the tenant." A "certain degree of flexibility" could be allowed for other crops, although growing (unirrigated) phillipesara—a fodder that was introduced to stimulate growth of the succeeding cotton crop—would require "the same degree of supervision" as cotton. In addition to an adapted model of continued supervision, the WP hoped that financial measures might do the trick of providing the proper perspective to tenants. SGB finances were completely based on cotton, but it might be better to change the financial returns of different crops. Financial measures might decrease "the need for supervision" too. The low turnover of tenancies was a (hopeful) sign that tenants were still better off farming within Gezira than outside it. This suggested at least a "theoretical" option to increase payments by tenants "to the point where they were only marginally better off than they would be farming elsewhere."[80]

In true Gezira spirit, "large scale trials" of intensive cropping were recommended and expected to show that anticipated yield levels could be reached within two or three years—thus in time to profit from Roseires water—assuming a "vigorous" extension program through the SGB was developed. Hopefully, the results would also show that the SGB could relax "detailed control of its operations" somewhat. Institutional changes would have to be "evolutionary rather than abrupt," but despite dreams of less control, the SGB should still recruit additional staff.[81]

The Gezira Study Mission of 1966—another IBRD Mission—came to Sudan to study how to intensify Gezira cropping patterns and evaluate the functioning of the SGB. The mission confirmed pretty much 100 percent of the findings of the Interim Report. The yield problems in Gezira were caused by tenants "reluctant to accept [...] criticism." The necessary intensification could only be reached through "wholehearted cooperation of the farmer," which required clear and sensible advice, extension, and benefits for tenants. The management had to look for "incentives" to stimulate tenants' abilities to grow crops with the required "meticulous care." SGB inspectors had to become field advisers and use demonstration centers to show the best techniques to tenants. A tenant's work would improve as soon as new financial incentives were "made clear to him"—still a him.[82]

The 1966 IBRD mission did provide one new—rather important—perspective. The mission suggested that tenants should have more security of tenure, simply by making the practice that tenancies were automatically but unofficially renewed each year the official system. In a similar way, the government lease of Gezira land—which was also unofficially continued—would need to be officially renewed for another 40 years. Nevertheless, at the end of the 1960s, Gezira planners—whether from Sudan or IBRD—had not come very far in solving the central dilemma of Gezira. How could one reasonably expect that Gezira tenants would "contribute to the nation" when it was not clear how much tenants could expect the nation "to contribute to them"?[83]

Frozen

There might have been many post–World War II discussions within Sudan on how to develop Gezira further—discussions that can be traced back to the 1920s as we have seen in earlier chapters—the Gezira Scheme and its management kept being referred to as the "model for all agricultural development in the Sudan." This was certainly true for the many pumping schemes along the Nile, both public and private, which copied the Gezira cropping pattern, tenancy structure and focus on cotton.[84] The frozen idealized Gezira model brought economic development—the immense revenue made from cotton—and provided the "example to those working for social development elsewhere in the country."[85] This image of Gezira as the model to follow was actively stimulated and enhanced by the Sudanese government and the SGB. Material was made for others to learn about Gezira.[86] The SID decided to write down its own experience with and ideas about Gezira, with the hope to ensure that "others" would not make the "elementary mistakes" once made by the SID that were "now forgotten." SID experience was obviously relevant for Sudan, but as irrigation was a major development policy "all over the world," other countries should profit from SID knowledge as well.[87]

Official Gezira documentation should be "authoritative," and to be based on "official and other specialist sources," which could be referred to, as long the resulting material could "not be attributed specifically to the Central Office of Information." I cannot but conclude that the authoritative documents resulting from this policy tried to hide the fact that Gezira had started as a joint initiative of government and SPS—let alone on the initiative of the SPS, as I would argue. To ensure that the correct message on Gezira would reach the broader public, the role of "the Sudanese people themselves in formulating the new proposals" had to become a very important part of the official history of the Scheme.[88] In post-1950 public material, the Syndicate is hardly mentioned at all. Building on a narrative that was actually already used in 1943, the Sudan authorities were the ones that had decided that Gezira management was "beyond the resources of the Sudan Government"—who had paid Sennar Dam and the canals—and therefore had invited "a third partner"—the SPS—to "manage the undertaking."[89] The assessment of Gezira management being beyond the government might be correct, but the chronology of the narrative was not.

The Gezira model provided "valuable experience" on "how links might be established" between government services and "the cultivator." The Gezira Scheme would have shown what was (not) "technically, economically and socially feasible" for the conversion of the "traditional cultivator into a cash crop producer."[90] It was the combination of economic profit, social goals, and contact with the local people that made Gezira the example for plans for the general development of the southern Sudan in the 1940s. Not much happened during World War II, but after the war, Zande District would set the scene for social and economic development based on cultivating cotton. Resettling the Zande people was key, as this would "comply with the fundamental principle" that cotton growing would be impossible "without adequate agricultural supervision."[91]

Breaching the general Sudanese policy to restrict commercialization of the southern provinces, the Zande Scheme started in 1946. The scheme originated from several proposals for regional and large self-sufficient economic developments, including programs for cotton growing and resettlement. Between 1946 and 1950, over 50,000 families were resettled on geometrically organized holdings; cotton cultivation was made compulsory.[92] If this does not sound familiar, please consider that the management was to become the responsibility of a board, as the Gash Board—the entity that had taken over the Gash Delta from the Kassala Cotton Company (KCC). Obviously, in 1946 the SGB did not exist yet, and a private firm managing the new treasure of Sudan's southern policy was out of the question.[93] The Zande Scheme should be built on a partnership with the Azande.[94] Unfortunately the scheme was highly unpopular with the Azande, as their financial returns were extremely low.[95] Both Zande and Gezira are about forced production of cotton, (more or less open) farmer resistance, disagreements between British colonial officials about balancing economic benefits and social goals, and a small British community of development workers.[96] The main difference with Gezira is that the Zande Scheme has gone into history as a failure.

If the Gezira model was not followed directly, then at least its approach of large-scale operations and mechanical efficiency was taken as source of inspiration. Within the national discussions on agricultural development in Sudan, "development away from the river" was largely shaped as large-scale production of durra by mechanical means. In the 1947–1951 period in particular, government efforts aimed at increasing durra areas "under mechanized cropping."[97] At the end of the 1950s, about 600,000 feddans were under such a system in large-scale commercial operations, financially supported by the International Cooperative Alliance. The 1959 IBRD Mission saw "considerable scope for expansion," although there were some problems because of the "primitive status of the population in these areas."[98]

A Particular Model

Some 15 years earlier, the Sudan Government was only too happy when Gezira attracted attention as early as 1944 from British Guiana, where the British government was thinking about irrigation (in Bonasika) and sugar cane (in Corentyne). The "particular model" to apply was the Gezira scheme, as "there might be a great deal to be learnt from what is happening there now."[99] The news that the concession would not be renewed had reached Guiana, which made the Gezira even more powerful as a model. In Guiana, it "would be quite out of the question to have any private profit-making syndicate" involved—it had to be a government program.[100] What type of government institution it would be was unclear, however. A separate board was unlikely given the small size of the projects, but it was clear that "the only method of efficient cultivation in Guiana was estate production." The option to study the model of the Kolchoz (Russian collective farm) was mentioned in passing, but appeared not to be a serious option.[101] Particularly information on "the relationships between the company and the cultivators and the exact method of organization involved" was of interest. In addition to Gezira and its SPS management, Guiana (and the Colonial Office in London) were also very interested in Kassala, as there the work "discharged by the company" in Gezira was done by the government. Perhaps Kassala would be the model for Guiana, and indeed for Gezira as well.[102]

The booklet that was sent to Guiana was "Notes on the Gezira Irrigation Project," published in January 1926,[103] as the documentation celebrating Gezira as a heaven of government wisdom had not been made yet. Indeed, as the Circulation Notes going with the letters of the Colonial Office indicate, much of the 1926 booklet was not applicable to British Guiana conditions at all—and probably not even to Gezira in the 1940s. How the land acquisition issue had been solved in Sudan was not the model to follow in Guiana. The booklet usefully illustrated, however, that a "central authority" should "have full control of cultivation" in order to be able to remove "an unsuitable cultivator." The experience from the West Indian land settlement schemes confirmed this idea,[104] suggesting that Gezira was an important reference project, but not necessarily the single, overarching model available.

The different responsibilities and profit shares of the partners in Gezira were also of interest. Even without something like a Syndicate in sight, the

Gezira profit division was "interesting," especially the way years of bad harvests could be compensated.[105] Gezira would also show that a scheme with "a single crop"—especially a "cash crop in rotation or on general arable farming"—would require that certain activities that determined "the success of the individual farmer" were to be performed on behalf of that same farmer by an organization, for example, through a Public Utility Corporation. Land ownership did not need to be collective, and not everything needed to be managed centrally, but a scheme "must be a unit."[106]

With the Gezira model, Gezira people spread out over the world as well—actually even more so than Gezira as a model, which pretty much stayed within Africa, as discussed in the epilogue. Although Mekki Abbas, the first Sudanese to chair the SGB, became the first secretary (1958–1963) of the United Nations Economic Commission for Africa in Addis Ababa, which was established in 1958 as part of a UN plan for regional economic commissions on each continent,[107] the "people" spreading out mainly were the British working in Gezira. With employment options in Sudan becoming less, but the world of international development opening up, (former) SPS and SGB employees kept each other informed about possibilities in that emerging international job market. Gaitskell himself is a main example of a former Gezira person moving into several other functions after leaving Gezira. Fairly soon after his leaving Gezira, Gaitskell became member of the Mission to East Africa.[108]

In chapter 5, we have seen the example of Gaitskell informing Gibbons on a project in Pakistan. Another former colleague of Gibbons, who had just had "a most interesting 4 months in the W. Indies, part for Bookers Sugar Estate, part for the Colonial Office," asked in 1959 whether Gibbons was interested in a cotton project in Aden. A shared colleague who was working in the West Indies as "a junior inspector on a sugar estate" among a "tough bunch of all nationalities" had applied for that same job, but he might not be suited as it was "a fairly big scheme with all sorts of problems."[109] From these and other letters from former colleagues to Gibbons, we learn that former SPS people came to work in countries like Canada, Kenya, Rhodesia, and Uganda.[110]

Other British with working experience in Sudan and Gezira came back to that same country as foreign experts. C. H. Smith, SPS inspector since 1929, later an SGB "estate manager," and finally the SGB Agricultural Manager,[111] retired from the SGB in 1955, but a few years later he traveled in Sudan as cotton adviser for Barclays Bank, writing reports about the state of Sudanese cotton on private schemes and writing advice for managers of those private schemes. His advice was based on "experimentation and experience over many years in both the Gezira Research Farm and the Gezira System."[112] Carmichael, the civil servant we met before, retired from the Sudan Civil Service in March 1959; he returned to Sudan in 1967 for a company inspection tour.[113]

Someone who moved away from Sudan in the 1950s to return to the country for a more regular position was Ferguson from the Gezira

Research Farm. His curriculum vitae—undated but most likely written in the early 1950s—described his main strength as a "very intimate knowledge of native agricultural systems and native crops throughout the Sudan." It added that his experience would be useful for "similar environments in other parts of the world," as the Sudanese systems and crops were "typical" for such regions.[114] Ferguson applied in October 1954 for a position at the India Tea Association.[115] Around the same time, he was also considering jobs of Director of Agricultural Research in Jordan[116] and at Fisons Pest Control Limited in England, a chemical company producing, among others, herbicides.[117] As neither of these last two jobs seemed to have worked out—the Jordan Ministry apparently never requested the Foreign Office assistance for the post[118]—Ferguson went to India.

In 1961, when he was about to retire from his post in India, he approached Gaitskell about a project he had read about in the *Economist* of April 1, 1961. Apparently the company Mitchell Cotts and the Ethiopian government were planning to start a cotton scheme, and Gaitskell had some relation to the project. Ferguson "wondered" if he could "be of some use" for the project.[119] In May, Ferguson was invited to apply for the position of Chief of the Research Division in Sudan. Ferguson was not particularly interested in the post, but Bedford from Fisons Pest Control Limited assured him that if he accepted the post "it would be very greatly appreciated not only in the Sudan but also in the UK." Bedford even mentioned that Carmichael had told him that he might be able to "persuade the Foreign Office to subsidise the post if a good man can be found."[120]

Just a few days later, Gaitskell informed Ferguson that the Mitchell Cotts project in Ethiopia was really just in a very early stage, "rather like the Gezira in 1910! In fact there is nothing there but a river and a plain."[121] With Ethiopia not being an option, Ferguson accepted the post in the Sudan in August 1961, after negotiating extra opportunities for leave and the highest possible salary.[122] Gaitskell was delighthed that Ferguson had done so.[123] In his position, Ferguson was member of the Working Party until his final leave from the Sudan in October 1964. After that—second—retirement from Sudan he kept in touch with opportunities for projects, among others on the Rahad project with MacDonald Construction Partners, with International Business Consultants on the Khasm el Girba sugar factory, and with the FAO on the Publication Research and Indexing Project in Sudan. Ferguson was even able to share his Gezira experience with the 1967 Rahad Mission of the Food and Agricultural Organization. He send the delegates his copy of the Gezira Inspector's handbook—the handbook we discussed in chapters 2 and 4.[124] As late as 1970, Ferguson was approached by Coode & Partners concerning a project around Sennar. The company was interested in agricultural possibilities in Gezira and had approached Ferguson after he had been suggested by someone from East Grinstead. This person was W. N. Allan, the former Irrigation Adviser of the SID.[125]

"We Know Our Business"

As another sign of the growing internationalization of the development community after World War II, the SID had become member of the International Commission on Large Dams (ICOLD) and the International Commission on Irrigation and Drainage (ICID) in the early 1950s. A SID delegation joined the 1954 ICID conference in Algiers and made a study tour to North Africa and Southern France.[126] One of the tangible results of this trip was that the Sudanese engineers could finally see a type of automatic gate from French North Africa in actual working order.[127] A few years before the study trip, the SID had started trials in Gezira with these Neyrpic gates, which could offer Gezira "semi-automatic" water control with "less man power."[128] The results of the two gates—one at K77 on the Main Canal and one in the Abu Usher Escape[129]—were rather disappointing. The Neyrpic gates did not maintain the constant upstream water levels they were supposed to do. The Neyrpic firm was asked to visit Gezira and check the gates;[130] it looks like the tests remained unsatisfactory, as the SID Annual Reports remain silent on the matter after 1955.

The "first international advisor" came to the SID and Gezira between December 3–14, 1953. Engineer Inglis from India visited Sudan "to give technical advice on various irrigation problems." Inglis signaled that irrigation water was "used very wastefully." Such wasteful use "was linked with the present system of irrigation." In North Indian irrigation systems, mainly in the Punjab, distributaries (or minors in Gezira jargon) carried constant discharges with so-called semi-modular outlets (giving a relatively constant discharge). This meant that only restricted quantities of water were supplied to Punjab tenants, who were allowed to irrigate as large an area as they desired. Within the Gezira model of restricted tenants' choice, Inglis did see options for the Punjab model. Many SID people considered the Punjab example as irrelevant for Gezira; Inglis found "a strong conviction" that actual Gezira practice "was the only possible system" in Sudan.[131] The SID was much happier that Inglis "supported strongly the widely held opinion" that night storage resulted in "silt troubles." In fact, Inglis had presented an additional other reason to quit night storage, as he would have shown night storage "inevitably led to the wasteful use" of water. Inglis might not have offered all the solutions that the SID had in mind, but his views deserved "the most careful consideration" as they were from "a man with an international reputation as an expert in hydraulics."[132]

Expert he may have been, but Inglis did not share the SID claim that a dedicated Hydraulic Research Station to conduct model experiments on silt and water in Gezira was needed.[133] The SID developed some water control experiments in cooperation with Professor Hendry of Khartoum University, but luckily the SID received a boost with the visit of French irrigation engineer M. Jacques Arrighi de Casanova. This French colleague considered Gezira as "one of the most perfected irrigation schemes from a technical, economical, and social point of view." Obviously, the SID had "never doubted" it was on the right track with irrigation development in Sudan, but hopefully that would "now be accepted without question." Accepting the expertise of SID should make it "unnecessary to call in foreign advisers in order to convince our critics that we know our business."[134] Unfortunately for the SID, the foreign experts kept coming to Sudan, especially in technical missions for Managil and Roseires.[135]

As colonial rule gave way to independence, late-colonial European and/or independent new governments in Africa and Asia requested assistance from international development institutions and consultants in rural development efforts. Occasionally engineers from one country tried out technologies from another area, but more often, foreign irrigation engineers visited other countries that continued to work based on their own, well-known design practices from their respective colonial irrigation approaches.[136] The new spirit of development aid, in which technical and managerial assistance was stripped from its colonial roots, made it possible to reevaluate Gezira from a neutral economic perspective. A combination of control and assistance was to guide African farmers in their productive activities. With fewer options for strict control, however, postcolonial schemes had to focus on building support structures for farmers. A certain translation of colonial coercion and force toward extension and training filled the gap. Forced production schemes were translated into extension programs to support farmers who remained to be "considered ignorant, uneducated and in much need of 'modernisation.'"[137] More often than not, the ignorant farmers were African farmers.

The SGB—especially Gaitskell—did its piece in linking itself to world development. The Gezira model of treating tenants both as a group and as individuals was presented as the early version of the "Third Way" between pure capitalism and communism. "We are not the only people in the world who are up against this sort of problem and have come to this sort of conclusion, and the story of the most famous contemporary parallel, the Tennessee Valley Authority in America, where stupendous regional development has resulted from a National Board being enjoined with such a duty as a charge on its profits, has had a good deal of effect on our minds."[138] By showing its own source of inspiration—the TVA—the SGB (and Gaitskell) claimed that the model of a National Board with a development project under its wings could create revenue and happiness at the same time. In the epilogue, we discuss these and other ideologies of the magical concept of development in more detail.

Epilogue

A Typical Battlefield: Understanding Negotiated Development

Despite the changes in Gezira over the years, Sudanese farmers were still irrigating fields and Sudanese staff was still doing supervisory activities in the Gezira Scheme. Whatever one's ideas about the correct direction of development, something still happened in the Gezira Scheme of the 1970s and 1980s, even though it would not have been what many people had planned or hoped to happen.[1] Gezira became a popular destination for researchers and consultants. Much of their shared interest was on what had gone wrong in Gezira. For engineering consultants, returning to a good functioning irrigation system was key.[2] For academic researchers, understanding the failures in development of Gezira—or at least the differences from what was (supposedly) planned—was the goal.

Based on field work in the early 1970s, Tony Barnett claimed—in a direct response to Gaitskell's story of development—that Gezira was an illusion of development. This idea was as strongly framed within dependency theory as it was contested by many reviewers of the book.[3] Despite the contested theoretical framework, Barnett clearly showed that farmers' strategies in Gezira—actually Managil—were diverse and not at all in line with official Gezira rhetoric. A similar conclusion was reached by researchers entering Gezira within consultancy projects on rehabilitation of the irrigation infrastructure in the 1980s. Their reports suggested that night storage was not practiced anymore. Handwritten comments in a 1987 report suggest that the person making these notes had known this for a much longer time. The commentator welcomed the new evidence, which would have shown "for the first time" that continuous flow irrigation was standard. The official Gezira documents—"the powers-that-be"—kept stating, however, that night-storage was still the standard.[4] Several consultants advised to rehabilitate the night-storage system, by removing silts and weeds, reconstructing canals, and rebuilding the telephone system needed to communicate the indents. The World Bank made restoring the night-storage system a requirement for funding rehabilitation efforts in Gezira.[5]

Whatever one thinks of capitalistic failures or night-time irrigation, Gezira in the 1980s seemed in desperate need of change—or so the many consultants argued. Paraphrasing Culwick, who had contrasted the certainty of the 1950s with the uncertainty of the 1910s, the Gezira "sense of uncertainty and impermanence" of the 1980s came "as a shock" to those who took "the Scheme with all its elaborate organisation for granted."[6] However, such ideas about certainty or uncertainty are not very useful. After reading my account on Gezira, it should never ever be a surprise again that development planning is confusing to some, plans are not realized as originally planned by many, or that outcomes are blurred for most. As much as the uncertainty on the best route for Gezira of the 1980s was not principally different from the huge discussions immediately after World War II, when the "carefree days"[7] were over, Culwick was mistaken in suggesting that the uncertainty she read in early documents was not present in her own time. In this Epilogue I will reflect upon Gezira as a symbol of colonial and postcolonial irrigation development, but I will symbolize Gezira development as modernization by muddling through.

Never Backward

On November 23, 1921, Eckstein—chairman of the Syndicate—gave his annual speech to the SPS stakeholders at the Ordinary General Meeting in London. The SPS managed 60,000 feddans in the Gezira area, of which 12,000 were fully canalized. Eckstein praised Sudan for its "potentialities of wealth," "efficient and economical administration," and "a field for the production of a high-class cotton." After linking Sudanese wealth and imperial cotton so smoothly, he stressed that it was "the duty of the British Government" to ensure that Sudan would become self-supporting, which—obviously—was to be achieved "by stimulating the production of cotton." All the relevant factors were available: soil and climate were highly suited, and the Sudanese population "quite ready and able to do the work under proper supervision." Obviously, good conditions themselves do not produce cotton—making cotton was hard work. Luckily, Eckstein emphasized that the cooperation between Syndicate and government was "all that can be desired."[8] Indeed, relations between Syndicate and government were pretty good; running a system like Gezira without workable relations would not have worked. However, we have seen in different chapters that on certain topics relations between (parts of) the Sudan Government and the Syndicate had been tense.

In 1921, the relationship with the tenants "could not be better" either; Eckstein stressed that tenants knew that they were treated "absolutely fair, straightforward and sympathetic." In chapter 4, we discovered this rosy image was not necessarily accurate either and that—as much as creating happy governments was—ensuring that tenants did what they were supposed to do was not smooth at all. Eckstein acknowledged the importance of the SPS inspectors, on whom the Syndicate depended "a good deal" to shape the correct tenants. Luckily, the field staff was "perfectly loyal and devoted to its duties."[9] In chapter 5, we discovered that perfect loyalty and devotion was not given either. If inspectors showed their loyalty by staying more than five years, loyalty was at least constructed in clouds of alcoholic vapor. In 1950, at the termination of the SPS Concession—discussed in chapter 6—the British Cotton Growing Association praised "the real pioneers" like MacIntyre and Archdale, who had achieved "so much during most trying conditions"— including living in tents "in the desert" in hot weather without other water available than Nile water.[10]

The real pioneers learned the cotton work in the SPS schemes at Tayiba and Barakat—the "nuclei of trained cultivators and of trained British staff" as Syndicate director Asquith put it in 1921. He forgot to mention that the SPS made good money from these schemes as well, but was quick to emphasize the importance of industries, which were "complementary to and not competitive with those of Great Britain." Irrigation would bring the Sudanese people "prosperity and insurance against starvation," but the Sudan cotton would supply the Lancashire mills, which could sell the finished products back to Sudan. The machines needed in Sudan would bring employment in Britain as well.[11] When the Sudanese were subject of debate, it was more on protecting them from "too rapid devolutionary schemes" or "a minority of the big men of the Khut Councils becoming their masters." Although Gezira would "free the Tenants and make them stand on their own feet,"[12] careful planning was also needed to protect the Sudanese from too harsh development.

Particularly before World War II, development equaled economic growth and construction of infrastructure.[13] The profits would "finance a further stage in development" and lead to "progress" and to "the welfare and prosperity of the Sudanese." In the post-1945 spirit discussed in chapter 7, the Sudanese "[f]ive year plan for post-war development" firmly stated that development was "not new to the Sudan." How could it be? It had been "going on continuously for close on half a century."[14] Despite these apparent rhetorical aspects, something had been going on indeed, but exactly what and how? We have learned that Gezira development involved different parties—SPS, field staff, Sudan Government, tenants, cotton industry—and their views on development. Despite the British rhetoric of progress and happiness, we have been warned that such progress was not automatic and that (Sudanese and British) people were needed to do it. Developing Gezira was not a smooth process automatically leading to happy people. It was hard work for all parties involved to negotiate their desired way forward, even when the desired outcome was presented as some kind of indisputable and automatic, glorious future.

Cloning Gezira

An automatic future was certainly a common ideal in the post–World War II world, which promised to be a mechanical age—not just in Sudan. Given the general problem of attracting labor in the late 1940s, "the use of machines wherever possible" was a major aim in Gezira and elsewhere.[15] If machines have even ever been the measure of men,[16] it was in postwar African development schemes. A general "extraordinarily naïve faith" in the major benefits of mechanical cultivation of large tracts of land created schemes like the mechanized durra schemes in Sudan already touched upon in chapter 7, the Groundnut Scheme in Tanzania, and the Niger Agricultural Project.[17] These and other projects were supported by ever-increasing budgets for post–World War II (colonial or postcolonial) development.

The Groundnut Scheme in Tanzania was a set of state farms in different parts of the country rather than one integrated development project. Cultivating groundnuts on a large scale seemed an attractive option for post–World War II Britain with its lack of food and other agricultural products. In April 1946, a mission visiting suitable sites concluded that large-scale cultivation was both possible and desirable. Less than five years later, in January 1951, the Groundnut Scheme was cancelled, after spending £49 million on producing some thousands tons of nuts.[18] Another groundnut wonder-in-the-making was the Niger Agricultural Project (NAP) near Mokwa, Nigeria, which started in 1948. "The whole system was based on that evolved in the Gezira area of the Sudan." Two officials from the Nigerian government had travelled to Gezira in 1948. They were particularly interested in larger-scale mechanized agriculture, which was reason to visit the Mechanized Crop Production Scheme in Ghadambaliya as well.[19]

The NAP was a settlement scheme; the NAP Company—a limited liability company with shares between the company and government—would clear the land and control agricultural operations of settlers from densely populated parts of Nigeria. The first 78 settlers in the Mokwa area were actually farmers from within the project area to be developed, which "was in accordance with arrangements at Gezira", there were just not as many people in Mokwa compared to the original Gezira area.[20] As their later fellows in the Kano region (see below) the early farmers "resented an authoritarian management which told them where to live, where to farm, what to grow and when to perform various agricultural operations."[21] The NAP was closed down in 1954, having established only 163 settlers by 1953.

Projects like the NAP may have been less successful than planned, but they did show that the attractiveness of Gezira did not stop at the Sudan border. The same year that the NAP was closed down, the colonial government in Kenya was considering inviting an expert on Gezira—F. A. Brown,[22] who had worked in Gezira between 1923 and 1951 for both SPS and SGB. The Kenyan government was thinking about rice development in the Mwea-Tebere districts of the Central Province. The project would settle a large number of people who had been involved in the Mau Mau protests—several detainees were digging canals already. The idea was to have between 40,000 to 60,000 acres available for irrigated rice cultivation in a near future. As

Kenya had little experience with such projects, the government was looking for "expert advice at an early stage in the scheme," both on irrigated rice cultivation and "on the organisational aspects" that were expected when resettling the population.[23] Brown had done similar advisory work in British Guiana.[24] He was interested in the Kenya job and suggested that it "would be advantageous" to visit Gezira when en route to Kenya "for three or four days" to study its latest developments.[25] As Brown's knowledge of rice was limited, it was arranged that he would cooperate with rice expert Rhind, who would be in East Africa at the same time anyway for the East African Agriculture and Forestry Research Organisation.[26]

The idea to develop irrigated agriculture on the Mwea plain dated at least back as early as the 1920s. In 1925, the colonial government supported a private initiative to irrigate sugarcane. Four years later, the Embu District Administration of Nyeri Province proposed employing instructors from neighboring Tanganyika to provide training on irrigation. In 1933, the Provincial Agricultural Office suggested establishing irrigated rice farms. A earlier failed attempt to grow cotton made that crop disappear from the discussions.[27] The development of the Mwea system became highly politicized, when anti-colonial protests in Kenya grew in the early 1950s. Land surveys on the plain in 1951 were "a match tossed into a powderkeg,"[28] as Kenyan settlers thought that their land was taken by Europeans. In 1952, when the Mau Mau Revolt started, Mwea was among the regions that witnessed some of the earliest and heaviest violence because of its conflicts over land.[29] As an act of irony, the Mwea Scheme became one of the measures to solve the reasons for the revolt. British control over the Mwea plain was to bring the stability that had been threatened by Kenyan fears that Europeans were taking the same land in the first place.

In 1953, Mwea's main purpose was providing employment to Kikuyu who were detained under the Mau Mau Emergency Regulations.[30] Mwea was planned to be ready in 1960, which would require at least 745,000 pounds.[31] The government of Kenya had set aside 50,000 pounds for the first phase. One thousand Embu families would cultivate plots with rice and maize and live in villages near their plots. Their agricultural work would "be supervised by agricultural officers living on the site"—how much that sounds like Gezira![32] The swampland "which only for a few years ago provided an ideal hideout for Mau Mau terrorists" was gradually transformed into an area producing crops on "highly-productive holdings run by African tenant farmers."[33] By the end of 1950, the 22-mile Mwea furrow was nearly complete. The experimental rice plots on the plains must have had satisfactory yields, as in May 1951 expansion of the area with 3,000 acres was included under the category of "Development Schemes Contemplated."[34] In September 1952, some 6,000 acres had already been surveyed. These plans were clear enough, but starting the whole scheme in 1953 would be "putting the cart before the horse."[35]

There was the "thorny problem" of land tenure; a regulation should be found to ensure each person with cultivation rights in the area would be compensated for loss of land—an issue in Gezira as well. Receiving preferential treatment in leasing an irrigated holding was one option. The land issue was studied by Simpson, who had worked in the Sudan Civil Service on land tenure arrangements in Gezira. A second major issue was water use and distribution from the rivers to be exploited for Mwea-Tebere. The black cotton soils in the area were to be studied before applying irrigation; irrigation would start on the easier red soils. As the only colonial officers with some irrigation experience were hydraulic engineers, options to send an employee of the Agricultural Department "to the Sudan and/or India for training" were considered,[36] or sending an officer for 9 to 12 months to Ceylon or Malaya.[37] Finally, Brown was asked. He advised Mwea to appoint a manager "able to exercise strict disciplinary powers."[38] Other British experts with Sudan experience joined Mwea as well. A British expatriate with long experience in Gezira was appointed General Manager of Irrigation Development Projects for Kenya in 1956. Being an outsider, his work was not easy, as he was seen as "an interloper." He retired in 1960.[39] The District Officer in Mwea came from retirement in Sudan. At least one officer responsible for settlement had experience in Gezira.[40] A manager with extensive experience with mechanical rice farming (E. G. Giglioli) was recruited from British Guiana.[41]

The Mwea system of management was "semi-military," with "close supervision" protecting tenants "from failure."[42] Both management and tenants worked under "a fairly rigid schedule of operations." Agricultural operations were not supposed to differ from the official schedule, as changing production rhythms would upset land preparation and irrigation scheduling. How familiar this all sounds. "Persuasion" was to ensure that farmers accepted the discipline. Field assistants and head cultivators (tenant leaders in each block) were responsible for "inducing farmers to time and synchronize their operations properly."[43] A 1957 note specified that tenants should "look to their Block Inspector for guidance in all matters both personal and cultivational through the medium of the Village Committee."[44] The dominant view in Mwea was that "what was good for production was good for the Scheme." As such, it was also "good for the tenants."[45]

From 1960 onward, tenants were required to subscribe to the Trust Land (Irrigation) Rules. The eight-page document set standards for crop practices, and absenteeism, and allowed the management to take disciplinary action against tenants failing to adhere to instructions.[46] In 1967, rice was grown on approximately 5,000 acres—with head works and main canals able to irrigate 15,000 acres. At that time, the layout of the system units and fields was considered wrong, as it did not take into account mechanized land preparation. Manual work "could not meet the demands of sophisticated agricultural production required by the government's income policy."[47] To allow for mechanized preparation, the layout of plots was made much more regular. Plots of 100 meters by 40 meters were supplied with water through a network of canals.

A Gezira of Wheat?

The example Mwea had set for "a modern, transformed agriculture" inspired the newly independent Nigerian government in the late 1960s and early 1970s to realize three irrigation projects in Northern Nigeria to modernize existing agriculture.[48] In addition to facilitating this transformation, the Nigerian projects were to increase wheat production to reduce imports and save foreign exchange. Through increased productivity, rural living standards would improve as well.[49] Resettling farmers into larger units would allow providing services such as schools, markets, and hospitals, and thus improve rural welfare.[50] In a 1968 report, the FAO had actually suggested to copy the Mwea structure completely, including the tenancy model.[51] The Dutch engineers from NEDECO that took the lead in planning the irrigation system of the Kano plain—the Kano River Project (KRP)—actually argued that private land ownership would stimulate farmers to invest in agricultural production and land protection measures. To ensure proper use and thus profit for the scheme as a whole, NEDECO suggested building an organization "to have the farmers advised and guided in all phases of production."[52] Project management in Northern Nigeria was to have full control over the crop production system, including marketing and input supply. Tenant farmers should be selected for their working ability and their "calculated capacity to adapt the severe discipline of two crops per year irrigation farming."[53]

In 1974 NEDECO wrote an additional short report on the management needs of the Kano system. All decisions on water control, water distribution, and most on agricultural aspects were to be the responsibility of the project management, at least in the initial years. Assuming that the canal system was properly operated and maintained, farmers should be made conversant with irrigation technologies and competent to cultivate irrigated crops. Inputs such as fertilizers, insecticides, agricultural machinery, and credit facilities should be made available. Marketing of agricultural production should be organized. Some nuance was brought in, however, as some participation of the farmers was anticipated upon, for example on crop choice, cropping pattern, water management, mechanization, and marketing of crops.[54] These nuances were needed, as the Kano farmers were extremely eager to participate, though not always officially. The Kano area was densely populated, it was not an "empty" area like Mwea—which had not really been empty either.

Imposing any new arrangement would be difficult, even though "it was assumed that the projects would be given adequate powers to deal with" farmers' resistance.[55] In 1983, an expert concluded almost with regret that experience in the Kano River project—and close to it in Bakolori— suggested that the necessity of "re-allocating the land back to the farmers" had unfortunately resulted in a "lack of control over the land." Therefore, it was impossible to operate "the schemes at their optimum level."[56] This "disquieting evidence" went beyond the Kano project in terms of relevance: it was "Nigerian society" as a whole that had as "yet not become attuned to the discipline imposed by an irrigation-assisted system of farming."[57]

Colonial dreams continued to shape policies for postcolonial irrigation development in Africa. The cases of Mwea and Kano demonstrate that irrigation development activities did not end after the colonial territories attained independence, but that such activities did not stay the same throughout. As late as 1967, Gezira could be used as an example of a project that had managed to provide settlers with an income superior to any of the neighboring areas. Similarly, Mwea would have given settlers incomes high enough to make them accept an unfamiliar village pattern of settlement.[58] With forced reorganization out of the question, however, postcolonial approaches for centralized management in rural development programs had to offer an alternative to direct coercion. Modern management in the 1970s had to move beyond Gezira drive and aimed for providing a package of inputs and welfare services to stimulate increased participation in agricultural production.[59]

At the same time, the many international irrigation specialists that took the Mwea system as a model to follow in African rural development based their preference on the premise that the population had to be reorganized "to meet the requirements of efficient irrigated agriculture."[60] Most African irrigation projects would have failed because they lacked effective centralized management.[61] With such management absent, farmers could not be provided with the services needed to secure their social well-being.[62] The original Gezira model applied control of land and production to ensure (force) cotton production by tenants. Mwea applied a similar system of encouragement. The Kano management, however, might have been based on similar production assumptions as Mwea, but could never develop its underlying management strength. The farmers were expected to adapt to this production system, but they were not to be forced. Instead, "full level extension workers *taught* farmers how to make the most of irrigation technology"; farmers were also *"instructed* in the use" of new seeds, fertilizers, insecticides and pesticides, and to proper irrigation.[63]

The Kano system was never as discipline-oriented as the Mwea system. The inability to "discipline non-cooperative farmers" frustrated management.[64] In 1977, an official complained that farmers' cooperation was "not satisfactory because we have to *ask* farmers, not tell them what to do."[65] One critical account of the optimistic planning of the Kano irrigation system refers to an important reason for such frustration: the unrealistic prospects to realize "a Gezira scheme of irrigated wheat competing with imports" was simply "a very remote prospect."[66] Despite disappointments and counterexamples, the Mwea model could still feature prominently in international development discourse up to the 1980s, for example, in a 1986 Food and Agriculture Organization report—although it is fair to say though that the validity of the Mwea model is only partially acknowledged. After claiming that in the first 10–15 years of its operation, the system would have been very successful in terms of increased production and higher farmer incomes, the authors of the report feared that the strong emphasis on centralized control in the initial stages might have proven not capable to adapt to changing circumstances, with potential social and economic stagnation as a result.[67]

Call Me God[68]

With the new independent African states and the policies and projects they wanted to realize, international consultants arrived to do much of the work—as we have also seen for Gezira. For European engineers and development consultants, their (originally colonial) knowledge became exportable through development aid.[69] Changing from colonial employees to foreign experts did not mean they changed much. Actually, the experts embodied the colonial approaches which they brought with them to the new development efforts.[70] Gaitskell was one of them. After his 30 years' career in Sudan, he was board member of the Tanganyika Agricultural Corporation, member of the Committee of Experts on the Development of Africa, member of the East African Royal Commission on Land, board member of the Commonwealth Development Corporation, consultant to the UN Special Fund for Sudan,[71] to name just a few of his many other positions. Gaitskell became one "of these colonial wanderers" who followed the wave of independence on the African continent "with a southward drift through the continent."[72]

As we have seen, Gaitskell started as a junior SPS inspector in 1923, sent his first letter—at least the first one we can find in the archives—on what went wrong in Gezira to SPS management in 1928 and has kept writing ever since. In 1949, in his last year as Gezira manager for the Syndicate, Gaitskell was awarded a Companion in the Order of St. Michael and St. George (CMG) "in recognition of his services to the Country."[73] Just one year later, he became the first Managing Director of the SGB. In his career, Gaitskell moved from writing about general development to shaping it in Gezira. After his resignation from the SGB, he stayed active in the field of international development. In the literature, for example reviews about his 1959 Gezira volume, Gaitskell is portrayed as the successful manager with huge and relevant international experience.[74] He was used as one example "that behind every successful settlement in Africa there is an exceptionally able administrator"[75]—as if he had been in charge of Gezira all the time and had done everything himself. Despite this flattering attention, the person Gaitskell is hardly discussed in the existing Gezira literature. When scholars do discuss Gaitskell, he is not analyzed beyond just being an important person in the Gezira story. Nowhere Gaitskell appears as a major force in shaping that same story—his own story basically.[76]

Instead of the individual player affected by the larger Gezira developments,[77] I would argue that Gaitskell shaped those developments for a large part—and definitely tried to do so. Gaitskell had shown his major interest in social development in the Gezira on many occasions, with as expressions of this interest quite a number of letters and notes to SPS management—Gaitskell's first note in the archives dates from 1928, only five (!) years after he came to Gezira. Gaitskell had much influence on the ideas of Robertson, from the other SPS, the Sudan Political Service, who was influential when the post–World War II British policies in Sudan were being shaped. This influence, one expects, would have been enhanced by the two being friends "since about 1924."[78]

In 1938, the same year that the SPS started negotiations on renewing the concession, Gaitskell sent his thoughts on Gezira to the SPS management—first as a draft to the management in the Sudan, later as an official document to the directors in London. Anticipating on the possibility that on June 30, 1938, the Sudan Government could announce the termination of the concession in 1939, he suggested that it was "a good time to take stock of the Gezira Scheme and of our position in it."[79] For Gaitskell, it was clear that the changing sociopolitical context—including increasing political unrest in Sudan and Egypt—should be a major concern for the SPS. The "absence of policy" had caused "embarrassing differences of opinion between the Syndicate and the local province staff." The government, "acting on the influence of the present Governor," was thinking about changing the Gezira Agreement, "the one uniform basis which we do possess." He warned against a SPS policy of ignoring the changing context. A strategy of "letting sleeping dogs lie" would be a disaster, as the dogs were "not sleeping and unless someone takes an initiative the dogs will do so."[80]

Gaitskell continued with suggesting that the SPS actually had the duty of ensuring that the Gezira Scheme would successfully pass on to others after "it has passed out of our hands"—which under normal conditions would still be only after 12 years when he wrote these words.[81] The Syndicate should take a leading part in "infusing the spirit" of the policy of devolution in the Gezira Scheme. This would be good business, because it would prevent the SPS being forced to do the same on terms set by the government; it would also prevent the impression that the SPS had made a move under Sudanese political pressure.[82] The Gezira Scheme existed for 20 years, the SPS had to start "teaching people to run their own show instead of waiting for [British Inspectors] to do it all."[83] In a 1943 note to the SPS management in Barakat, Gaitskell went further than he did before, when he claimed that it was not in "the interest of our Company" to strive for a concession beyond 1950. There would not be much financial gain anymore, but foremost it would be difficult "to avoid constant political attack." The management required after 1950 would not be "easily assessable in terms of financial costs and profits." But the SPS should certainly help the government, "if they want our assistance." If the Government did not want SPS assistance, the Syndicate would "be free of public responsibility for the future here."[84] In his writings for a more general public, Gaitskell liked to present the case of broader development in Gezira emerging over time as a topic much agreed upon in Gezira circles, but also something that was simply inevitable and well received in general.

In remarks made by SPS inspectors, their wives, and SGB workers—encountered in the archive—we get a more divided approval of his ideas.

Gaitskell did not seem to have hidden this idea to the British field staff. Inspector Hunt remembers a meeting with Gaitskell in November 1938—so quite some years before 1943—when they went together to the cinema and had dinner afterwards. "Chief among what he said was that he thought that the Govt. would take over the Synd, even in 1950 when the concession ended" as Gaitskell thought that the Government was "a trifle disillusioned about the progress the nas [natives] were making towards getting things done themselves." The main problem was that "schemes aimed at improving the self-running potentialities of the nas" were not taken very seriously, "if in fact there are any such schemes."[85] After Gaitskell visited Hunt in 1945 to check whether he would stay on as inspector, Hunt thought that Gaitskell "talked well" but "gave his attention only to 1950 and the years after that." Hunt thought "it rather idealistic talk about a group of Englishmen slowly pushing the natives towards a higher standard of living and a higher culture by means of this great scheme," but noted Gaitskell was "extremely enthusiastic" about that idea. In an attempt to bring "him down to earth somewhat" Hunt remarked "what a ghastly country it was," very "unsuitable [...] for wives who have nothing to do." Gaitskell actually agreed with that.[86] In the diaries of Ms. Johnson, Gaitskell appears more interested in what was further away in the future than in day-to-day realities. Johnson refers to Gaitskell's ideas on decolonization as "optimistic," as he seemed to have allowed the British many more years to guide the Sudanese than many other British thought realistic in the turbulent years after 1945.[87] Gaitskell is described as "an impossible person to really discuss things with as he lives in the clouds,"[88] but also as a man who "had not the control of the situation he would like to have."[89]

A Sense of Mission

Impossible, brilliant or in clouds, Gaitskell was the person to defend Gezira as an experimental place for development.[90] After discussing Gezira as model to follow in the 1950s, Gaitskell presents his overarching history of Gezira in his 1959 volume.[91] Perhaps his version is not as different from the history I present in this book as one would perhaps expect. Gaitskell does discuss the differences of opinion between SPS and the government, and he does not hide the differences of opinion within the Syndicate either. He does tend to focus on discussions with the Civil Service, however, and ignores the differences between, for example, the SID and the Syndicate. Night storage simply featured as "an ingenious solution" from Butcher for a clearly identified problem. Especially when discussing the 1940s and 1950s, the years he was partially on the steering wheel, his own ideas come forward and his version starts to differ more from the one in this book. For example, Gaitskell keeps using the phrase "we in Gezira," which after reading my account of Gezira should not be a useful concept anymore.[92]

The 1959 book was basically the extended version of his earlier writings, in which he already claimed that the Gezira success was based on "partnership" and a "limit to foreign capital." The combination of Britain having the capital, Sudan having the land and the labor, and the Syndicate wanting a profit was the secret of their success. Obviously, the SPS had the objective to make a profit, but had been "prepared to withdraw" from Gezira "after it had served its purpose." The SPS would have handed over "a new productive asset" to Sudan, a result from the partnership, but "belonging in the end to the country itself."[93] The main two problems with Gaitskell's writings becomes apparent in this passage: first his tendency to claim that goals in the 1940s were already steering Gezira development in the preceding decades and second that the goals for Gezira were clear-cut in the first place.

A major role in Gaitskell's version of Gezira history—and an observation shared by many who write about Gezira later—is reserved for the so-called experimental setup of the many arrangements and actions in Gezira. The period between 1911 and 1925—between the start of Tayiba and the opening of Sennar Dam—would have been—and was praised for—a deliberate period of testing the right procedures. The period would be part of the "scientific approach" that was characteristic for "the project from its inception."[94] Even though Gezira has had its fair share of experiments, like the Hosh scheme, tests on rotations, etcetera, it goes a bit far to explain all the changes in Gezira history as experiments. Thinking of Gezira as a result of scientific experiment only works when Gezira is allowed to exist only after Sennar Dam was delivering its water in 1925. Gezira existed long before. We should not forget that Kitchener, Wingate, Eckstein, MacGillivray, and others wanted to build Gezira as soon as possible after the 1913 agreement. They had simply not planned experiments, they wanted to realize the real thing quickly. It was World War I and the rising prices after the war that spoiled the fun. Obviously, Tayiba was indeed started as an experiment to test whether winter cotton could grow in Gezira soil. The first pump schemes could become the training grounds for tenants and staff—as often mentioned by Syndicate management—but training is not the same as experimenting. Once the cotton option in Gezira was confirmed by the Tayiba success, the pump schemes were installed to start making money as soon as possible. They did serve to practice, but not to experiment: the Zeidab tenancy system was to become the model, the cotton was to be grown, and the tenants were supposed to be silent and obedient.

Gaitskell's view on the collective elements in Gezira is another case in point of his way of reasoning. The collective arrangements—like the Reserve Fund, ploughing and labor management—not only turned into the key feature of Gezira, but were also the British (even Western) answer to communism in the Cold War. In contrast to the Soviet Union who "made the mistake of moving from the nursery to prison," the West had "a better answer in moving from the nursery to partnership." The relevance of collective arrangements became even higher—or so Gaitskell claimed—once one realized that these "were planned by extremely conservative people," who had "a sense of mission."[95] It were not the socialists—nor the UK Labor Party—that designed the brilliant collective Gezira, but shrewd businessmen. This might actually make the point Gaitskell wanted to make, in the sense that collectivism is not necessarily a hobby of the left. To suggest that collective arrangements in Gezira were designed for the better of the Sudanese tenants, however, within a shared view of a bright future goes a little far. The only sense of mission detectable in the position of the Syndicate was financial: collective arrangements were cheap or necessary to cover its own financial risks after the economic crisis.

In his 1959 Gezira volume, Gaitskell presented Gezira as an attempt to bring "civilization"—in the definition of "ordered liberty."[96] This combination of order and "direction" leading to modernization[97] is the final element typically used in discussions on the relevance and the inevitability of the Gezira model. Direction was needed in "any large agricultural enterprise such as the Gezira"; Gezira's "certain inevitable peculiarities" would affect too many people at the same time. One simply could not but "keep social changes and trends under constant observation." Especially in the 1950s, when "a highly-disciplined agricultural community" had to become "a stable agricultural society" with initiative, social institutions, and "a fuller life," constant guidance was the only way.[98] Perhaps the Gezira focus had been too much on cotton—still "without doubt the best money earner" to Gaitskell in 1959—and perhaps mechanization and mixed farming had not advanced enough yet, but it was just a matter of time, and proper planning. Even though the SGB had been too slow to ensure Sudanese participation in management and had not yet reached the "initiative and independence among the farmers" that was aimed for, the situation would improve.[99] The Gezira motto was to solve problems possibly created by planning with more planning—as the future was known anyway.

When I started considering this book project many years ago, I imagined a main story of how ideas and ideals of development and planning as phrased by Eckstein, Gaitskell, MacIntyre, and many others that were realized in Gezira did influence later African irrigation schemes, like Mwea and Kano. I thought I could argue that central planning of the empty African landscape was a specific characteristic of these systems.[100] With this in mind, I simply started with the archives most accessible to me, those on Sudan in Durham, United Kingdom. The amount of material on Gezira proved to be so huge and its contents so rich that I decided quite soon to focus the book on Gezira only. I saw great options to discuss topics less well represented in the literature on colonialism—the inconsistencies of colonial empires, the internal struggles, the continuous improvisation, the experiences of British colonial officers, etcetera. In working with the material, I discovered that my earlier view was not necessarily wrong—even though actually I did make several mistakes—but that it was rather incomplete. Now, with this book written, I do think that centralized planning of perceived empty space is certainly relevant, but also less specific for colonial irrigation in Africa. Central planning was important in countries as diverse as Russia, the United States, Spain, and the Netherlands.[101] Specific is how centralized planning and modernization are realized and shaped—in Gezira mainly through the tenant model.

Considering the emptiness of African landscapes is not the most relevant theme either, as the dominant justification of many British efforts was the undesirability—and occasional irrelevance—of a recognized existing African reality.[102] A new reality had to be rolled out over an existing wrong one. In Gezira, the new reality was wrought in rectangular tenancies within an overwhelming grid of canals with an overload of bored British inspectors.

This new reality was certainly disciplining, if not completely on the ground at least in the minds of quite a few colonial British actors. I do not think, however, that the one-sided approach suggested by Bernal on the disciplining nature is very helpful.[103] First and foremost, "the" British did not exist, as for example ideas and interests of Government and Syndicate were different. The SPS was powerful, but certainly not a "colonial authority" in the typical sense either. Within the Sudan Government, we encounter different ideas about what to do. As I have shown, the Sudan Government did not ignore preexisting patterns of land use and social arrangements, but the government did decide that these patterns were less relevant for their agenda. By stressing the ideological aspects of disciplining, including suggesting that an image of Gezira positioned on a map of the United Kingdom is part of disciplining Sudan, Bernal misses a key aspect of ideology and discipline: these are not abstract forces of British colonialism, but concrete activities on the ground by British people.[104] There is no need to deny any skewed power relations in colonialism, but what matters more is the—perhaps surprising—mundane character of colonial rule. In their daily actions, British colonialists did take many things for granted, including the discipline, the force, the connection between Sudan and the United Kingdom, and many more. As such, they did confirm the colonial power relations for sure, but they did so as many actors in other times and realities would have confirmed reality through their daily actions.

Some British were in Gezira to help Sudan and its inhabitants. For others Gezira was an interesting business proposition. Gezira might not have been "very successful in terms of socio-economic performance"—although the exact meaning of this claim needs closer scrutiny—but for one of the three partners it was a project for profit. The fact that the Syndicate was in it for the money was all the more reason for the Sudan Government to distrust many of the Syndicate actions and to build in checks and balances in Gezira regulations. The whole point of Gezira is that it represented both "capitalist value" and "moral and political value." Being the result of continuous negotiations between representatives of (versions of) these different values, it is hard to imagine Gezira reality being anything else but "somewhat arbitrary."[105]

An Anomalous Giant

Ten years after Gaitskell's Gezira book, Robert Chambers published his claim-to-fame, a book on settlement schemes in Africa, with a focus on Mwea. Chambers had been a District Officer in the Kenyan government between

1958 and 1961, and lectured at the Kenya Institute of Administration between 1962 and 1964. He then moved into research, wrote about the Mwea system and made an impressive career in the Institute of Development Studies in the United Kingdom.[106] In a comparative study of settlement schemes in Africa, Chambers could not neglect Gezira nor Gaitskell. Indeed, he does discuss both and praises Gaitskell, who had "contributed his classic work on Gezira with its wise precepts about development," including warnings against over-ambitious goals, the need for experimentation, and a plea for "gradualism." Chambers admired the book as an exception within the larger set of literature on settlement for its breath. Gezira was to be admired for its "deliberation which makes many modern schemes appear irresponsibly rushed" with which Gezira had been developed. As I have already discussed, the idea that "plans worked out in the first decade of the century led to a pilot pump scheme in 1910 and the completion of the Sennar dam in 1925"[107]—the perception of deliberately planned experimentation before Sennar—is problematic.

Chambers pointed out as well, however, that within the larger context of settlement schemes Gezira was an exception too. It was "an anomalous giant" not replicated elsewhere. Within the larger debate created by the Labor government in the United Kingdom—similarly as in Sudan—group or cooperative farming became popular after World War II, in which individual small farms would profit from a system of centralized services.[108] The Israel model of the moshav became highly popular, as the Gezira model with its centrally planned production was somehow less relevant in the new political setting. Chambers actually welcomed this change in ideas, which "included many sound insights about agricultural development projects." The scale of Gezira "had been misleading" and replication had proven very difficult as elsewhere in Africa "physical conditions" were not "as favourable." So, the model was possibly good, but simply too large. Large-scale mechanical cultivation was problematic, partially because of maintenance issues, but mainly because it created problematic labor peaks in the non-mechanized farming activities.[109]

Both Gaitskell and Chambers started their career within the colonial system, but where Gaitskell explicitly took up the move from the colonial hierarchy to the new world order of independence to discuss what "the West" should do—first and foremost against the Soviets in the Cold War—Chambers did use the colonial setting as a scenic backdrop at most. His analysis of settlement schemes did conclude that several aspects had been rather utopian; his solution was to work "with greater care and understanding." Chambers was a clear proponent of looking at development projects as a whole, to understand what happens, with a particular focus on the administration of the schemes. Throughout his work, Chambers proved to be a pragmatic person. He showed no signs of strong ideological roots, he rather described development as something that had to happen. His two books on Mwea within the context of planned development in Africa both received favorable reviews.[110]

In the 1980s and 1990s, Chambers was a main exponent of promoting stakeholders' involvement in developing efforts; he published extensively on involving users in irrigation design and stakeholders in development in general. In 2005, he published a book about his own development when writing about development, as main example of continuity and change. Concerning the lessons he drew from his work on African settlement schemes, he concluded that much of his work was "now of mainly historical interest." However, the "three neglected angles" he had defined in 1969 were still important: "commitment, continuity and irreversibility." Another need he identified in his self-reflection was the need to "plan planning and plan management, both in headquarters and in the field."[111] The new approach to development should include much more of the recipients' voice and should be as non-bureaucratic as possible, but planned nonetheless. The—fascinating—move from the Chambers promoting well-planned colonial settlement schemes toward the Chambers promoting planned participatory development is illustrative for the move of development discussions and practice in a broader sense.[112] Also typical is that his approach seemed to relate strongly to a need for European input—or American for that matter—in development elsewhere. The need for external experts is not questioned; Chambers wanted experts with commitment to stay and bring development. He was looking for the type of commitment "to 'their' projects and to the settlers" he found in the late colonial settlement projects in Africa. Commitment and continuity was vital for gaining the knowledge and understanding of a country a "foreigner" needed. The simple question why a foreigner could actually go to a country to develop it is not posed by Chambers. He focused on the how.[113]

Africans Are Doing It for Themselves

It would be unfair to accuse Chambers of neocolonialism and European-centric political ideas, but reading his material does not take away a feeling of discomfort—well, at least not my feeling. Focusing on what needs to be done in development without even taking into account the colonial roots of development ignores the question why the model works as it does: why do the foreign experts go to Africa, and not the other way around? It also ignores the realities of colonial inheritance, in terms of infrastructures, bureaucracies, and ideologies.[114] Finally, such a pragmatic focus tends to hide that underneath the foreign experts' involvement the idea is still that Africans are apparently not able to do development for themselves—especially those in the rural areas.[115] In colonial times, the African territories were simply perceived as regions without recognized histories. "European racism placed Africans at the bottom of the colonised world."[116] In 1965, English historian Hugh Trevor-Roper considered that African history had nothing more to offer than "the unrewarding gyrations of barbarous tribes in picturesque but irrelevant quarters of the globe."[117] Africa was "backward, primitive, barbarian."[118]

Already in the 1970s, the Kano farmers in Northern Nigeria—in one of the planned clones of Mwea—defeated the typical image of African farmers being lazy, not interested, or barbarian. Their actions showed that the tenancy model and strong production control was not universally accepted, and certainly not by these Kano farmers.[119] According to several experts, the desired market incentives for agricultural production were not well developed in African rural societies. Even when African farmers were acknowledged with the ability to respond to prices of products, they were accused of having "a keen ability to choose the least burdensome (in terms of labor) way of attaining the income they want, or, to put it in other words, the ones likely to give them the highest return for the amount of work they are willing to do" when confronted with alternative opportunities.[120] One could actually argue that such an approach to labor versus income is rather economical. However, in the African context, development agencies associated it with ignorance. With farmers being unaware of their need to be modernized and unable to raise production through such modernization, raising total production still required rather strong central management with "a considerable measure of autonomy." However, the institutions that were to assist farmers through strict control should not be too strictly controlled by higher authorities. The message was clear: Everybody needed to be controlled, but some needed more control than others.[121]

In such debates within Sudan—we have encountered enough examples of Sudanese being blamed by British—it was not always clear what the Sudanese were. Were they Africans? Were they Egyptians? Were they something in between? The general "scaling" of the colonial British was to rank the Sudanese from the north closer to the (higher developed) Egyptians than to the (less developed) Southern Sudanese. An important reason for the British to defend their actions in Sudan—both related to colonial and postcolonial efforts—was their "patient and disinterested consideration" of the issues at stake when developing a country.[122] As late as in 1982, former SPS inspector Smith could explain the role of the British field staff in terms that would not have stood out in the concession days of Gezira. British inspectors were there "to ensure that the tenant should cultivate the land in a proper manner according to rotations laid down."[123] For this purpose, the field inspectors lived in the Gezira area in their Block, among the tenants. Smith would be one of the persons to stress this aspect of Gezira life, as we learn about him that when he was working for the SPS—as a Group Inspector in the South of Gezira[124]—he was quite keen on "paying surprise visits to the tenants and staff in the field in the very early hours of the morning. This kept them all alert and on their toes."[125] Smith and his fellow foreigners brought "development."

Embodiments of Utopia

The colonial and neocolonial hierarchy of development could only be maintained with an independent measure for development, including a ranking.

Even though development projects like Gezira and the African settlement schemes "could be all things to all men, embodiments of diverse Utopian aspirations,"[126] a measure of success was the only way to defend somehow that development was a useful concept. Even when distinguishing between development as an immanent and objective process, and development as a subjective course of action or intentional practice,[127] the whole point of the claim for "development" as measurable is that the intentional practice of developing is backed by the objective process of development. A recent attempt to measure development (actually "civilization" as the title suggests) shows how problematic it is to discuss social change in terms of regions being ahead or behind in the development of sociocultural complexity.[128] Apart from the methodological problem of the dominance of the category of energy capture, which basically overshadows the other categories, why would using more energy mean a region is developed—let alone civilized?

Measurability of development, however, goes across ranks, political preferences, and scientific disciplines. Neo-Marxist studies can claim that Gezira was not developed or underdeveloped, because they knew exactly how to measure development in terms of capitalistic properties of production. The anti-Marxist answer in terms of "stages of economic growth" is as much a version of measurable development as the Marxist original.[129] As Adas has forcefully argued, however, it is actual European dominance in the last couple of centuries that has framed the development discourse. Being able to overpower other regions—this is an overstatement, as the Gezira story shows that overpowering is always in the making—the European standard of societal prosperity and its shape—through settled urban populations in an industrialized economy—has become the norm of development.[130] European perceptions of scientific and technological superiority were reason for and shaped interactions with colonized people overseas in need of similar development. Scientific and technological measures of human worth became the frame to argue for and find evidence of Europe's racial supremacy and its logical "civilizing mission."[131] Accepting this reconstruction of European standardization of what "development" or "the good life" should look like, is accepting that "development" as a measurable effort does not exist. This means that development projects cannot fail. The logical implication is that it is impossible that colonialism would have "distorted" development either.[132]

Modern society is obviously as much a construct as developed society. No society has ever really been modern; any distinction between Western and other societies on the modernity front is not backed by evidence.[133] All societies know spirituality, liberty, rationality, the past, just in different ways and shapes. Development as a process—actions that are taken to become "developed"—would be different from development as a product or goal. The idea that "development" as result would be realized through "development" as a process, or a verb needs to be refused.[134] The contrast between the object of development and the subject developing itself or others is nonexistent. Development is always a verb, is always political and is never an objective—let alone—measurable entity. Continuous negotiations shape results as entities and as contested symbol at the same time.[135]

The theoretical and methodological underpinning of my Gezira history is that development is local and constructed within networks of actors.[136] These networks are continuously created and recreated by human actors engaging with other human actors and nonhuman intermediaries, simultaneously at different localities. In such a process, micro and macro are irrelevant concepts, as the micro shapes the macro while being created by the macro. The actor "reveals the narrow space in which all of the grandiose ingredients of the world begin to be hatched"; the network explains "through which vehicles, which traces, which trails, which types of information, the world is brought *inside* those places," how these vehicles are transformed, and then "pumped back *out* of its narrow walls."[137]

A methodological consequence may be—has been for me—to bring what Gezira was because of and for the many actors as closely as possible to the agency of these actors. Cars being driven into canals by drunken drivers are as relevant to a story of development as grant government policies. History as a discipline should be good at dealing with the sources in such a way; as such it is the field of history where Latourian approaches promise to be of immense value. Missing one or two documents, however, can be a problem. A mundane issue of not finding a letter can become highly important. Take one of the two cradles of Gezira, Tayiba. As far as I have been able to detect, the Tayiba pump scheme started in 1911 as a joint decision of the Sudan Government and the Syndicate. My sources suggested that the government would have started the scheme anyway: the pumps had already been ordered. Missing this start in 1911, however, has rather a huge effect on the interpretation of what happened in 1913, when the Sudan Government was to take over management.[138] Suddenly, 1913 Tayiba appears as a government scheme taken over by the Syndicate instead of a threat of the Sudan Government taking over a Syndicate-managed project not being fully executed. Details matter.

Captain Latour

I think I have shown throughout this book that Gezira does provide a well-suited starting point for moving into the details of the networks of development which matter. Recognizing that there is no "inside" or "outside," or "local" or "context," but that networks are continuously (re)created, is important, as it warns those studying networks that any presuggested division in terms of levels, contexts, or relations needs to be avoided. This is not an argument that hierarchies, arenas, and institutions do not exist. They do, Gezira is real, skewed development efforts happen. The resulting hierarchical constructions themselves, however, can never be used as explanatory forces for human agency.

Gezira was both product of development as a verb and development as a concept. Continuous negotiations shaped the scheme's canals, cotton, and tenants' resistance, as well as the contested symbol of development itself. Gezira officials knew that items like engineering, social development, cotton growing, and many other political debates were closely linked. Throughout Gezira history, in colonial and postcolonial times, Gezira officials knew that they were changing society through their practices. In fact, they wanted to change society, improve the colony or the country, and improve the lives of the tenants.[139] That does not mean, however, that daily Gezira practice was heavily changed by these larger-scale development debates. Despite the changing role in development discourse, daily practices by Gezira management and inspectors were rather stable—in practice and clearly on paper.

We may need to conclude that development discourse does not necessarily influence developmental practices, but rather that those practices are claimed as symbols of desired development over time. Practices are used to (re)construct desired images of development, to claim that the glorious future is around the corner if only we were prepared to act in the correct way. This hijacking of practices for a political agenda happened as much in Gezira as in the many other development projects that continued their work rituals despite changing development policies of donors. [140]

Gezira has been a symbol for many different development visions and efforts. In the twentieth century, Gezira was the main road to economic development for Sudan. After World War II, its cooperative model was supposed to be the Western answer to Soviet communism and a major model showing how to include farmers in economic and social development. In the 1970s and 1980s, Gezira was the symbol of failed development, either because it did not behave according to Marxist theory or because its infrastructure needed to be rehabilitated. The immense plain between the Blue and White Niles south of Khartoum provided the dramatic stage for a pas de deux of Captain Jack Sparrow and Bruno Latour. Gezira was planned out and made up as agents went along.

Archival Sources

Sudan Archive, Durham University

Entries

Aglen P.
Allan W. N.
Carmichael J.
Clayton G. F.
Crompton C. W. L.
Culwick G. M.
Daly M. W.
Ferguson H.
Gaitskell A.
Gibbons E. P.
Hunt F. B.
Hunt L. S. J.
Johnson W.
Morgan C. E. F.
Robertson J. W.
Simpson S. R.
Smith C. H.
Smith R. J.
Sudan Plantations Syndicate
Taylor E. A. P.
Tottenham P. M.
Wesson A. D.
Williams W. M.
Wingate General Sir Reginald
Wolfe J. M. B.

Secondary Material

Sudan Government List
    The Sudan Gezira Board British staff register: Incorporating the defunct Sudan Plantations Syndicate & Kassala Cotton Company; Sudan Gezira Board; Sudan Plantations Syndicate; Kassala Cotton Company. 1953.

National Archives, Kew, London

Entries

CO111/784/10
CO822/1105
CO822/1106
CO822/1554
CO822/156
CO822/1730
CO822/193
CO822/213
CO822/550
CO822/962
FO141/451/6
FO141/497/4
FO141/578/1
FO141/579
FO141/692/14
FO141/758/9
FO141/795/1
FO371/102759
FO371/1427
FO371/41423
FO371/97000
IWC5
LAB8/1890
T236/6261
TI/12095

Notes

Introduction   Settling Certain Details Coming to a Deal

1. Sudan Archive, 880/12/28: Letter from Pearson to Miss Crompton, dated 18/2/1907.
2. SA, 880/12/28: Letter from Crompton to Sister-in-law, dated 17/7/1905.
3. SA, 880/12/42: H. D. Pearson, W. L. Crompton and S. M. Vines 1907 Notes on the marking out of the Gezira in the Anglo-Egyptian Sudan into minutes of latitude and longitude, London...Born in England in 1860, Crompton migrated to Australia in 1877, where he worked as a surveyor from 1885. Prior to moving to the Sudan in 1902, he surveyed in the Gold Coast (modern Ghana).
4. SA, 626/1/29–39: W. N. Allan, The development of irrigation in the Gezira, dated 21/1/1939.
5. J. C. Scott (1999) *Seeing like a state: How certain schemes to improve the human condition have failed* (New Haven, CT: Yale University Press).
6. SA, 880/12/28: Letter from Crompton, undated.
7. SA, 415/8/152: "Cotton from a Wilderness," *Daily Mail*, dated 14/1/1925.
8. SA, 415/8/152: "Cotton from a Wilderness," *Daily Mail*, dated 14/1/1925.
9. SA, 415/8/152: "Cotton from a Wilderness," Sudan was officially part of the Anglo-Egyptian condominium (1899–1956) and administered by a governor-general appointed by Egypt. Obviously, even though the British were officially only in Egypt in an advisory role, they controlled Egypt and as such Sudanese appointments. Sudan itself was administered as a British imperial possession, but under the Foreign Office, as Sudan was not a colonial territory. See R. O. Collins (2008) *A history of modern Sudan* (Cambridge: Cambridge University Press); M. W. Daly (1991) *Imperial Sudan: The Anglo-Egyptian Condominium, 1934–1956* (Cambridge: Cambridge University Press); D. W. Daly (1986) *Empire on the Nile: The Anglo-Egyptian Sudan, 1898–1934* (Cambridge: Cambridge University Press); M. W. Daly (1985) *Modernization in the Sudan: Essays in honor of Richard Hill* (New York: Lillian Barber Press); J. Ryle, J. Willis, S. Baldo, and J. M. Jok (2011) *The Sudan handbook* (Woodbridge: James Curry); R. L. Tignore (1966) *Modernization and British colonial rule in Egypt, 1882–1914* (Princeton, NJ: Princeton University Press).

10. Sayed Mohammed Afzal, Director of Research, Pakistan Central Committee, after his visit to the Gezira in 1946, quoted in Press & Information Officer (ed.) (1959) *The Gezira scheme from within: A collection of articles by heads of departments* (Khartoum: Middle East Press), 3.
11. Actually, the system and its dam at Sennar were officially opened in 1926, but water from Sennar reached the cotton fields for the first time in 1925.
12. Walt Disney's "Pirates of the Caribbean: At World's End" (2007).
13. See, for example, A. Gaitskell (1959) *Gezira: A story of development in the Sudan* (London: Faber and Faber); V. Bernal (1997) "Colonial Moral Economy and the Discipline of Development. The Gezira Scheme and 'Modern' Sudan," *Cultural Anthropology*, 12, 447–479. A good counterweight of inevitability is offered by A. I. Clarkson (2005) *Courts, councils and citizenship: Political culture in the Gezira scheme in condominium Sudan* (PhD Thesis, Durham University). See for a small story of the type I tell C. Katz (1991) "An Agricultural Project Comes to Town: Consequences of an Encounter in Sudan," *Social Text*, 28, 31–38.
14. It might be good to think about development as "counterfactual history" or "virtual history," not only because it might be healthy to discover that events are never as inevitable as people think but also because developers try to create virtual worlds of wonder. See N. Ferguson (ed.) (1997) *Virtual history: Alternatives and counterfactuals* (New York: Basic Books).
15. T. Barnett (1977) *The Gezira Scheme: An illusion of development* (London: Frank Cass), 1.
16. The history of the percentages is more complex, as they were changed later, as will be explained in the book.
17. M. W. Ertsen (2006) "Colonial Irrigation: Myths of Emptiness," *Landscape Research*, 31, 147–167.
18. Gaitskell (1959).
19. Ertsen (2006); M. W. Ertsen (2008) "Controlling the Farmer: Irrigation Encounters in Kano, Nigeria," *TD: The Journal for Transdisciplinary Research in Southern Africa*, 4, 209–236.
20. L. B. Rand (1989) *High stakes: The life and times of Leigh S. J. Hunt* (New York: Peter Lang), 177 and 179.
21. SA, 802/1/52: *Sudan Times*, around 1904.
22. SA, 802/1/52: *Sudan Times*. Much later, looking back at his Sudan adventures, Hunt referred to Sudanese tenants as "the wild men of the desert" (SA, 802/1/41–47: Sworn Statement by Hunt, dated 19/1/1931).
23. SA, 802/1/3–5: undated Letter from Leigh Hunt to Lord Cromer, probably early 1904. The *Sudan Times* put rather bluntly that Hunt proposed "an exodus of the American niggers to their Canaan." (SA, 802/1/51: *Sudan Times*). From Hunt's letter and other written material, it is not immediately clear what he means with "the Negro problem." He might refer to the ideas of William Edward Burghardt Du Bois, who had just published a paper with that term in 1898 (W. E. B. Du Bois (1898) "The Study of the Negro Problems," *The Annals of the American Academy of Political and Social Science*, 11, 1–23). Du Bois and others around him outlined the difficulties that the black community in the United States faced and wanted to suggest solutions to these problems. Perhaps Hunt saw emigration to the Sudan as a solution.

24. SA, 802/1/41–47: Sworn statement. See Rand (1989) for much more detail on these and other actions of Hunt. See H. Dickson (2012) *Old reliable in Africa*, Forgotten Books, original publication 1920 (New York: Frederick A. Stokes Company).
25. SA, 802/1/7–8: Letter from Wernher, Beit & Co to Hunt, dated 28/4/1904.
26. SA, 415/9/1–2: Minutes SPS, dated 2/9/1904. SA, 415/8/2–3: Report of the Directors SEPS, dated 16/11/1904. Furthermore, a Mr. Seymour Fort and Mr. Luigi de Chastillon became directors, a Mr. Lionel Phillips became the chair. The financial relations that Hunt mobilized shows that cotton development in Sudan was done by large firms with a huge international portfolio. See also S. M. Mollan (2008) *Economic imperialism and the political economy of Sudan: The case of the Sudan Plantations Syndicate, 1899–1956* (PhD Thesis, Durham University).
27. SA, 415/9/3–5: Minutes SPS, dated 9/10/1904. Yes, really! Quentin Tarantino would love it!
28. SA, 415/9/11: Minutes SPS, dated 11/7/1905.
29. SA, 802/1/15–18: Note of Arrangement with Mr. Leigh Hunt made by Director of Agriculture & Lands at Interviews on February 18–19, 1906, dated 20/2/1906.
30. SA, 802/1/41–47: Sworn Statement.
31. SA, 418/5/99–100: The Start of the SPS Gezira Scheme, "Above received from MacI in 1945" (Note from Gaitskell).
32. SA, 415/9/13–15: Minutes SPS, dated 19/10/1905.
33. SA, 802/1/19: Notice of a Proposed Transfer of Shares into £41,000 Debentures.
34. SA, 802/1/41–47: Sworn Statement. With Hunt, MacGillivray had started the Bank of Abyssinia, and Hunt "induced [him] to come into the Show." SA, 418/5/99–100; The Start of the SPS Gezira Scheme.
35. SA, 802/1/41–47: Sworn Statement. This purchase was reason for Hunt to defend his few years in the Sudan. After all, he had made from the sale "a modest profit, which would seem to indicate that the Company was not then on the rocks [...] as Mr. MacGillivray was no tenderfoot, but one of the shrewdest of businessmen."
36. SA, 802/1/34: Special Resolution.
37. SA, 415/9/21: Minutes SPS, dated 2/5/1907.
38. SA, 415/8/7: Balance Sheet SPS, dated 30/4/1908. Another major player in the future, Lord Lovat joined the SPS Board as well. Lovat was a Scottish landowner (Gaitskell [1959]).
39. See, for example, J. M. Hodge, G. Hödl, and M. Kopf (eds) (2014) *Developing Africa: Concepts and practices in twentieth-century colonialism* (Manchester: Manchester University Press).
40. In 1922 the SPS started the Kassala Cotton Company (KCC). The Sudan Government and the SPS signed an agreement on November 7, 1922. On completion of the railway extension to Kassala in East Sudan, the KCC would be awarded a 40-year concession over the Kassala cotton-growing area. In May 1923, it was agreed that the KCC would take over the running of the cotton growing area in Kassala at some point in 1924 (Mollan, 2008). A few years later, in 1927, when both the results were less satisfactory and the government ideas about Kassala changed, the KCC was asked

to leave. To compensate the KCC loss of options to make profit in Kassala, it was offered cotton areas in Gezira (see chapter 2). T. Niblock (1987) *Class and power in Sudan: The dynamics of Sudanese politics, 1898–1985* (Albany: State University of New York Press); S. J. Von (1978) "Cotton in Kassala: The Other Scheme," *Journal of African Studies*, 5, 205–243). The KCC—basically part of the syndicate, although officially separate— had negotiated a 40-year concession for their Kassala cotton area, in return for a 20 percent share in the profits. The SPS directors all agreed that for future negotiations in Gezira, both extensions of area and concession, these Kassala arrangements were very interesting. (SA, 415/10/68–70: Minutes SPS, dated 12/10/1922: SA, 415/10/73–75; Ordinary General Meeting SPS, dated 2/11/1922). However, in the first half of 1923, the Sudan Government "were not inclined" to change anything in the agreement. In the light of changing financial arrangements—notably the part of the loan reserved for the SPS—the government was actually "prepared to consider the matter," but it had to wait. Eckstein was actually in favor of sacrificing 5 percent of the Syndicate's share of profits and repay the loan in return for extending the concession period. (SA, 416/2/19–22: Minutes SPS, dated 9/5/1923).

41. SA, 417/1/56: 21st Ordinary General Meeting SPS, dated 14/11/1928.
42. Gaitskell (1959). See also Gaitskell A. (1964) "Resource development among African countries" in M. Clawson (ed.) *Natural resources and international development* (Baltimore, MD: Johns Hopkins Press); A. Gaitskell (1955) *What have they to defend?* (The Africa Bureau Annual Anniversary Address); A. Gaitskell (1952) "The Sudan Gezira Scheme," *African Affairs*, 51, 306–313. Barnett (1977). See also T. Barnett (1979) "Why are bureaucrats slow adopters? The Case of Water Management in the Gezira Scheme," *Sociologia Ruralis*, 19, 60–70; T. Barnett and A. Abdelkarim (1991) *Sudan: The Gezira Scheme and agricultural transition* (London: Frank Cass); Bernal (1997). See also V. Bernal (1991) *Cultivating workers: Peasants and capitalism in a Sudanese village* (New York: Columbia University Press); V. Bernal (1990) "The Politics of Research on Agricultural Development: An Instructive Example from the Sudan," *American Anthropologist*, 92, 732–739; V. Bernal (1988) "Coercion and Incentives in African Agriculture: Insights from the Sudanese Experience," *African Studies Review*, 31, 89–108.
43. Although being large does not necessarily imply central management, I would argue. See the United States for that matter. The European Union comes to mind as well.
44. Both Barnett and Bernal use Gaitskell's 1959 book as historical source, which is problematic as Gaitskell's work is not an academic treatment of Gezira history but a political publication on how to develop nations. Barnett and Bernal may suggest they write a history of Gezira, but at closer look, they both include several less accurate statements and interpretations in their work. Having said that, so did I in my own paper from 2006 (Ertsen [2006]).
45. The Gezira was, as much as the colonial project as a whole, continuously in the making and as such "unfinished." See J. Darwin (2012) *Unfinished Empire: The global expansion of Britain* (London: Allen Lane/Penguin). See M. M. Van Beusekom (2002) *Negotiating development: African farmers*

and *colonial experts at the Office du Niger, 1920–1960* (Portsmouth, NH: Heinemann); M. M. Van Beusekom (1989) *Colonial rural development: French policy and African response at the Office du Niger, Soudan Français (Mali), 1920–1960* (PhD Thesis, Johns Hopkins University) for an account of interactions between colonial officials and farmers in the Office du Niger in French West Africa. Van Beusekom tends to put less focus on the disagreements between colonial officials than I would do.

46. H. J Sharkey. (2003) *Living with colonialism: Nationalism and culture in the Anglo-Egyptian Sudan* (Berkeley: University of California Press), 1.
47. P. Barron, R. Diprose, and M. Woolcock (2011) *Contesting development: Participatory projects and local conflict dynamics in Indonesia* (New Haven, CT: Yale University Press).
48. Barron et al. (2011; 250 and 261).
49. Paraphrases of Scott's titles of 1999, 2009, and 1985, respectively. J. M. Scott (2009) *The art of not being governed: An anarchist history of upland Southeast Asia* (New Haven, CT: Yale University Press); Scott (1999); J. M. Scott (1985) *Weapons of the weak: Everyday forms of peasant resistance* (New Haven, CT: Yale University Press). See also J. M. Scott (1976) *The moral economy of the peasant: Rebellion and subsistence in Southeast Asia* (New Haven, CT: Yale University Press).
50. As a process of structuration, see W. H. Sewell (2005) *Logics of History: Social theory and social transformation* (Chicago: University of Chicago Press); A. Giddens (1984) *The constitution of society: Outline of the Theory of Structuration* (Oakland: University of California Press); A. Giddens (1979) *Central problems in social theory: Action, structure and contradiction in social analysis, Contemporary social theory* (London: Macmillan). See also A. Giddens (1990) *The consequences of modernity* (Cambridge: Polity Press).
51. If only because comparing such vastly unequal categories is methodologically problematic and I would argue that the Sudanese tenants were not controlling anything as a united entity anyway. See A. Mikhail (2011) *Nature and empire in Ottoman Egypt: An environmental history* (Cambridge: Cambridge University Press), 291.
52. M. W. Ertsen (2010) *Locales of Happiness: Colonial irrigation in the Netherlands East Indies and its remains, 1830–1980* (Delft: VSSD Press). Such modernizing efforts are not typical for colonial powers either, but I skip that discourse for the moment. See M. E. Aubet (2013) *Commerce and colonization in the ancient Near East* (Cambridge: Cambridge University Press); G. E. Areshian (ed.) (2013) *Empires and diversity: On the crossroads of archaeology, anthropology, and history* (Los Angeles: Cotson Institute of Archaeology Press); E. H. Cline and M. W. Graham (2011) *Ancient empires: From Mesopotamia to the rise of Islam* (New York: Cambridge University Press); I. Morris and W. Scheidel (eds) (2009) *The dynamics of ancient empires: State power from Assyria to Byzantium* (Oxford: Oxford University Press).
53. Ertsen (2008).
54. Barron et al. (2011; 2). Modern Gezira is still such an environment, but that is another book with other sources.
55. SA, 889/7/39–40: Letter from Bill to Alison, dated 24/7/1981.

56. SA, 889/7/41–42: Letter from Bill to Barbara, dated 6/8/1981 (probably). Bill, the former inspector, also thought that "anyway John Gaitskell has already done it, extremely well and from the inside."
57. The poem, published in 1899, was written for Queen Victoria's Diamond Jubilee, but exchanged for "Recessional." Wikipedia, accessed May 2013. See M. Adas (1989) *Machines as the measure of men: Science, technology, and ideologies of Western dominance* (Ithaca, NY: Cornell University Press), on ways in which Europeans shaped their scientific and technological superiority in interactions with people overseas (Sub-Saharan Africa, India, and China). See also M. Adas (2006) *Dominance by design: Technological imperatives and America's civilizing mission* (Cambridge: Harvard University Press). See also D. R. Headrick (1988) *The tentacles of progress* (New York: Oxford University Press); D. R. Headrick (1981) *The tools of empire: Technology and European imperialism in the nineteenth century* (New York: Oxford University Press).
58. SA, 408/1/41: Letter from the Governor of the Blue Nile Province, dated 26/06/1946. He quoted the verses from "savage wars of peace" to "have done with its childish days." See J. M. Hodge (2007) *Triumph of the expert: Agrarian doctrines of development and the legacies of British colonialism* (Athens: Ohio University Press); Hodge et al. (2014) for more material on the change in British colonial policy after World War II.
59. SA, 408/1/41: Letter.
60. I might even agree with Barnett that Gezira development could be well described as an illusion, but not for his reason that capitalistic development did not take place.
61. P. Ennis, M. Healey, and F. Purdue (2004) *Negotiating development: Rationales and practice for development obligations and planning gain* (London: E&FN Spon); G. Lachenmann and P. Dannecker (eds) (2008) *Negotiating development in Muslims societies: Gendered spaces and translocal connections* (Plymouth: Lexington Books); Van Beusekom (2002). See also E. F. Arnold (2013) *Negotiating the landscape: Environment and monastic identity in the medieval Ardennes* (Philadelphia: University of Pennsylvania Press).
62. The diaries of several of these women—available in the SA—do make a fascinating read and will be used in later chapters.
63. SA 418/2: G. M. Culwick (1955) *A Study of the human factor in the Gezira Scheme*. Also available as publication G. M. Culwick (1955) *A study of the human factor in the Gezira Scheme* (Barakat: Sudan Gezira Board).
64. The majority of archival material for the book was collected at the Sudan Archive of Durham University in the United Kingdom, through several short visits in 2009, 2010, 2011, and 2013, and mostly during a four months' fellowship at the Institute for Advanced Study in 2012. Additional material was collected from the collections at the National Archives in Kew, London, during short visits in 2011 and 2013.
65. Some claim linking Gezira and Sudan policies, but do not really succeed, like A. Q. Cheeseboro (1993) *Administration and change in the Gezira scheme and the Sudan, 1938–1970* (PhD thesis, Michigan State University).
66. I fully understand that attention should be paid to the colonized. And yet, I think that studying those that were supposed to be in power—and were—will yield interesting perspectives as well. Compare with J. Oorthuizen

(2003) *Water works and wages: The everyday politics of irrigation management reform in the Philippines* (Hyderabad: Orient Longman), one of the rare modern irrigation ethnographies focusing on irrigation managers. See also E. T. Jennings (2006) *Curing the colonizers: Hydrotherapy, climatology, and French colonial spas* (Durham, NC: Duke University Press).
67. Barron et al. (2011; 2).
68. Barnett (1977) does pay attention to such realities, but in terms of what they should be or what did not happen. I do not necessarily see predictable outcomes from human agency, let alone that I would know what would be desirable.
69. Sudan Gezira Board (1967) *The Sudan Gezira Board: What it is and how it works.*
70. L. Bader (1927) "British Colonial Competition for the American Cotton Belt," *Economic Geography*, 3, 210–231; T. J. Bassett (1988) "The Development of Cotton in Northern Ivory Coast, 1910–1965," *The Journal of African History*, 29, 267–284; W. H. Himbury (1918) "Empire Cotton," *Journal of the Royal African Society*, 17, 262–275; A. Isaacman and R. Roberts (eds) (1995) *Cotton, colonialism and social history in Sub-Saharan Africa* (Portsmouth, NH: Heinemann); W. Moseley and Gray L. C. (eds) (2008) *Hanging by a thread: Cotton, globalization, and poverty in Africa* (Athens: Ohio University Press); S. Onyeiwu (2000) "Deceived by African Cotton: The British Cotton Growing Association and the Demise of the Lancashire Textile Industry," *African Economic History*, 28, 89–121; B. M. Ratcliffe (1982) "Cotton Imperialism: Manchester Merchants and Cotton Cultivation in West Africa in the Mid-nineteenth Century," *African Economic History*, 11, 87–113; R. L. Roberts (1996) *Two Worlds of Cotton: Colonialism and the Regional Economy in the French Soudan, 1800–1946* (Stanford: Stanford University Press); J. Robins (2013) "Coercion and resistance in the colonial market: Cotton in Britain's African empire" in J. Curry-Machado (ed.) *Global histories, imperial commodities, local interactions* (London: Palgrave Macmillan); T. Sunseri (2001) "The Baumwollfrage: Cotton Colonialism in German East Africa," *Central European History*, 34, 31–51.
71. See Adas (1989); D. Biggs (2010) *Quagmire: Nation-building and nature in the Mekong Delta* (Seattle: University of Washington Press); Daly (1985); B. Bush (2006) *Imperialism and postcolonialism* (Harlow: Pearson Education); D. K. Davis (2007) *Resurrecting the granary of Rome: Environmental history and French colonial expansion in North Africa* (Athens: Ohio University Press); H. J. Hoag (2013) *Developing the rivers of East and West Africa: An environmental history* (London: Bloomsbury); Hodge (2007); Hodge et al. (2014); P. M. Holt and M. W. Daly (eds) (2011) *A history of the Sudan: From the coming of Islam to the present day*, 6th ed. (London: Longman); Mitchell (2002) *Rule of experts: Egypt, techno-politics, modernity* (Oakland: University of California Press); S. Moon (2007) *Technology and ethical idealism: A history of development in the Netherlands East Indies* (Leiden: CNWS Publications); D. Mosse (2003) *The rule of water: Statecraft, ecology and collective action in South India* (Oxford: Oxford University Press); S. B. Pritchard (2012) "From Hydroimperialism to Hydrocapitalism: 'French' Hydraulics in France, North Africa, and beyond," *Social Studies of Science*, 42, 591–615; W. D. Swearingen (1987) *Moroccan mirages: Agrarian dreams*

and deceptions, 1912–1986 (Princeton, NJ: Princeton University Press); W. D. Swearingen (1984) *In search of the granary of Rome: Irrigation and agricultural development in Morocco, 1912–1982* (PhD Thesis, University of Texas); Van Beusekom (2002); Van Beusekom (1989).

72. See Barron et al. (2011); E. Akyeampong, R. H Bates., N. Nunn and J. A. Robinson (2014) *Africa's development in historical perspective* (New York: Cambridge University Press); D. Anderson (2002) *Eroding the commons: The politics of ecology in Baringo, Kenya, 1890–1963* (Oxford: James Currey); A. B. Atkinson (2014) *The colonial legacy: Income inequality in former British African colonies* (Helsinki: UNU-WIDER); D. Aw and G. Diemer (2005) *Making a large irrigation scheme work: A case study from Mali* (Washington: World Bank); A. Byerley (2012) *Africa as laboratory and the complexity of epistemic decolonisation* (Uppsala: Nordic Africa Institute); K. M. Cleaver (1993) *A strategy to develop agriculture in Sub-Saharan Africa and a focus for the World Bank* (Washington: The World Bank); P. Collier (2008) *The bottom billion: Why the poorest countries are failing and what can be done about it* (Oxford: Oxford University Press); D. Constantini (2008) *Mission Civilisatrice: Le role de l'histoire coloniale dans la construction de l'identite politique francaise* (Paris: Editions La Decouverte); G. Diemer and L. Vincent (1992) "Irrigation in Africa: The Failure of Collective Memory and Collective Understanding," *Development Policy Review*, 10, 131–154; F. Dufour (2010) *De l'idéologie coloniale à celle du développement* (Paris: L'Harmattan); Easterly W. (2006) *The white man's burden: Why the West's efforts to aid the rest have done so much ill and so little good* (New York: Penguin); FAO (1987) *Consultation on irrigation in Africa* (Rome: Food and Agricultural Organization); N. Gilman (2009) "Review of 'Modernization as a Global Project'," *Diplomatic History*, 23, H-Diplo Article Reviews at http://www.h-net.org/~diplo/reviews/, No. 238-B (A two-part Forum Review), published on July 29, 2009; J. G. Hariri (2012) "The Autocratic Legacy of Early Statehood," *American Political Science Review*, 106, 471–494; L. Heldring and J. A. Robinson (2012) "Colonialism and Development in Africa," *African Economic History Working Paper Series*, 5; J. Leigh-Smith (2014) *Works in progress: Plans and realities on Soviet farms, 1930–1963* (New Haven, CT: Yale University Press); W. Maathai (2009) *The challenge for Africa* (London: Arrow Books); M. Meredith (2005) *The state of Africa: A history of the continent since independence* (London: Simon and Schuster); P. Mishra (2013) *From the ruins of empire: The revolt against the West and the remaking of Asia* (London: Penguin); J. R. Moris and D. J. Thom (1990) *Irrigation development in Africa: Lessons of experience* (Boulder: Westview Press); D. Moyo (2009) *Dead aid: Why aid is not working and how there is a better way for Africa* (New York: Farrar, Straus and Giroux); S. J. Ndlovu-Gatsheni (2013) *Empire, global coloniality and African subjectivity* (New York: Berghahn Books); J. Poncet (1961) *La colonization et l'agriculture européennes en Tunisie depuis 1881* (Paris: Mouton); P. Préfol (1986) *Prodige de l'irrigation au Maroc: Le développement exemplaire du TADLA 1936–1985* (Paris: Nouvelle Editions Latines); Reader J. (1998) *Africa: A biography of a continent* (London: Penguin); I. Stone (1984) *Canal Irrigation in British India: Perspectives on technological change in a peasant society* (Cambridge: Cambridge University Press); H. Tilley (2011) *Africa as a living laboratory: Empire, development, and the problem of scientific*

knowledge, 1870–1950 (Chicago: University of Chicago Press); H. Tilley and R. J. Gordon (eds) (2007) *Ordering Africa: Anthropology, European imperialism, and the politics of knowledge* (Manchester: Manchester University Press); T. Tvedt (2004) *The river Nile in the age of the British: Political ecology and the quest for economic power* (London: I. B. Taurus); J. Ubels and L. Horst (eds) (1993) *Irrigation design in Africa: Towards an interactive method* (Ede: Technical Center for Agricultural and Rural Cooperation).
73. See R. Rottenburg (2009) *Far-Fetched Facts. A Parable of Development Aid* (Cambridge: MIT Press), for a great book on this theme. This is also a good lesson for many engineers, but the market among them for historical monographs is not huge.
74. See B. J. Andres (2014) *Power and control in the Imperial Valley: Nature, agribusiness, and workers on the California borderland, 1900–1940* (College Station: Texas A & M University Press); Arnold (2013); C. F. Ax, N. Brimnes, N. T. Jensen and K. Oslund (eds) (2011) *Cultivating the colonies: Colonial states and their environmental legacies* (Athens: Ohio University Press); T. J. Bassett and D. Crummey (eds) (2003) *African savannas: Global narratives and local knowledge of environmental change* (Oxford: James Curry); J. Beattie, E. Melillo, and E. O'Gorman (eds) (2015) *Eco-cultural networks and the British empire: New views on environmental history* (London: Bloomsbury); W. Beinart and L. Hughes (2007) *Environment and empire* (Oxford: Oxford University Press); R. A. Butlin (2009) *Geographies of empire: European empires and colonies, c. 1880–1960* (Cambridge: Cambridge University Press); J. H. Casid (2005) *Sowing empire: Landscape and colonization* (Minneapolis: University of Minnesota Press); Davis (2007); D. K. Davis and E. Burke (eds) (2011) *Environmental imaginaries of the Middle East and North Africa* (Athens: Ohio University Press); M. Fiege (1999) *Irrigated Eden: The making of an agricultural landscape in the American West* (Seattle: University of Washington Press); Hoag (2013); D. Joergensen, F. A. Joergensen, and S. B. Pritchard (eds) (2013) *New Natures: Joining environmental history with science and technology studies* (Pittsburgh: University of Pittsburgh Press); E. Kreike (2013) *Environmental infrastructure in African history: Examining the myth of natural resource management in Namibia* (New York: Cambridge University Press); R. MacLeod (ed.) (2000) "Nature and Empire: Science and the Colonial Enterprise," *Osiris*, 15; Mikhail (2011); A. Mikhail (ed.) (2012) *Water on sand: Environmental histories of the Middle East and North Africa* (Oxford: Oxford University Press); S. B. Pritchard (2011) *Confluence: The nature of technology and the remaking of the Rhône* (Cambridge: Harvard University Press); M. Reisner (1993) *Cadillac desert: The American West and its disappearing water* (New York: Penguin); M. Reuss and S. H. Cutcliffe (eds) (2010) *The illusory boundary: Environment and technology in history* (Charlottesville: University of Virginia Press); E. Russell (2011) *Evolutionary history: Uniting history and biology to understand life on earth* (New York: Cambridge University Press); R. A. Sauder (2009) *The Yuma reclamation project: Irrigation, Indian allotment, and settlement along the lower Colorado river* (Reno: University of Nevada Press); A. Sluyter (2002) *Colonialism and landscape: Postcolonial theory and applications* (Lanham, MD: Rowman and Littlefield); S. Sörlin and P. Warde (2011) *Nature's end: History and the environment* (Basingstoke: Palgrave);

S. Stunden Bower (2011) *Wet prairie: People, land, and water in agricultural Manitoba* (Vancouver: University of British Colombia Press); J. A. Tropp (2006) *Natures of colonial change: Environmental relations in the making of the Transkei* (Athens: Ohio University Press); D. Worster (1985) *Rivers of empire: Water, aridity, and the growth of the American West* (Panthenon Books: New York); D. Zeisler-Vralsted (2015) *Rivers, memory and nation-building. A history of the Volga and Mississippi Rivers* (New York: Berghahn Books). See for the relation between human agency and environment, M. W. Ertsen (2014) "Step after step the ladder is ascended: Human agency in irrigated (anti) landscapes" in D. Nye and S. Elkind (eds) *The anti-landscape* (Amsterdam: Rodopi); M. W. Ertsen (2013) "A poor workman blames his tools or how irrigation systems structure human actions," in E. Casella, G. Evans, P. Harvey, H. Knox, C. Mclean, E. Silva, N. Thoburn, and K. Woodward (eds) *Objects and materials. A Routledge companion* (London: Routledge); M. W. Ertsen, J. T. Murphy, L. E. Purdue, and T. Zhu (2014) "A Journey of a Thousand Miles Begins with One Small Step. Human Agency, Hydrological Processes and Time in Socio-hydrology," *Hydrology and Earth Systems Sciences*, 18, 1369–1382.
75. B. Latour (2013) *An inquiry into modes of existence: An anthropology of the moderns* (Cambridge: Harvard University Press); B. Latour (2005) *Reassembling the social: An introduction to actor–network theory* (Oxford: Oxford University Press); B. Latour (1996) *Aramis, or the love of technology* (Cambridge: Harvard University Press); B. Latour (1993) *We have never been modern* (Cambridge: Harvard University Press); B. Latour (1987) *Science in action: How to follow scientists and engineers through society* (Cambridge: Harvard University Press). See also Giddens (1979, 1984, 1990); G. Harman (2014) *Bruno Latour: Reassembling the political* (London: Pluto Press); G. Harman (2011) *The prince and the wolf: Latour and Harman at the LSE* (Winchester: Zero Books); G. Harman (2009) *Prince of networks: Bruno Latour and metaphysics* (Melbourne: re.press); M. Sahlins (2004) *Apologies to Thucydides: Understanding History as Culture and Vice Versa* (Chicago: University of Chicago Press); Sewell (2005). See for work focusing on technology and theory, Bijker W. E. (1995) *Of bicycles, bakelites, and bulbs: Towards a theory of sociotechnical change* (Cambridge: MIT Press); L. Camprubi Bueno (2011) *Political engineering: science, technology and the Francoist landscape (1939–1959)* (PhD thesis, University of California, Los Angeles); T. J. Misa, P. Brey, and A. Feenberg (eds) (2003) *Modernity and technology* (Cambridge: MIT Press); B. Pfaffenberger (1990) "The Harsh Facts of Hydraulics. Technology and Society in Sri Lanka's Colonization Schemes," *Technology and Culture*, 31, 361–397; I. Van de Poel (2003) "The Transformation of Technological Regimes," *Research Policy*, 32, 49–68.

1 Cotton from a Wilderness: The Early Negotiations

1. Sudan Archive, 635/5/6: Tottenham on Sudan irrigation, dated 21/3/1972.
2. Dupuis may have used other suggestions as well, perhaps the German "savant" who traveled in the Sudan in 1845 (probably Lepsius) or otherwise Sir Samuel Baker from the 1860s. SA, 626/1/29–39: W. N. Allan, The development of irrigation in the Gezira, dated 21/1/1939.

3. National Archives, FO141/578/1: C. E. Dupuis, Inspector General S.I.S. Project for irrigation of the Gezira Plain. Preliminary report April 1908.
4. S. M. Mollan (2008) *Economic imperialism and the political economy of Sudan: The case of the Sudan Plantations Syndicate, 1899–1956* (PhD Thesis, Durham University).
5. SA, 112/1/6–22: Irrigation Report by Inspector General of Sudan Irrigation Service, dated 2/3/1910.
6. See also Mollan (2008).
7. SA, 469/1/1–2: Letter from Wingate to Clayton, dated 8/8/1908.
8. Quote from Sir H. MacMichael's "The Sudan." SA, 635/5/6: Text from Tottenham on Sudan irrigation, dated 21/3/1972.
9. SA, 469/2/22–25: Letter from Wingate to Clayton, dated 6/3/1910. Luckily, next to financial considerations, these men had "thoroughly patriotic sentiments" as well, reason to keep stressing that investing in Sudan would keep "the British flag flying when it is now so seriously threatened by the political attitude of both British and Egyptian Governments."
10. SA, 626/1/29–39: W. N. Allan, The development of irrigation in the Gezira, dated 21/1/1939.
11. SA, 469/2/22–25: Letter from Wingate to Clayton, dated 6/3/1910.
12. SA, 469/2/28–31: Letter from Wingate to Clayton, dated 13/3/1910.
13. Mollan (2008).
14. Dupuis mentioned the crops dura, cotton, wheat, and simsim (sesame) as the principal crops, but—as in Garstin's plans—wheat seemed to have been the priority as this crop would not interfere with Egypt's water demands.
15. SA, 469/1/8: Letter from Wingate to Bonham Carter, dated 11/9/1909.
16. SA, 469/1/4–5: Letter from Wingate to Clayton, dated 1/9/1908. Less than two years later, Wingate did consider it "very undesirable [...] to say anything about the matter to the Egyptian Irrigation Department." (SA, 469/2/28–31: Letter from Wingate to Clayton, dated 13/3/1910).
17. SA, 415/10/26–28: Ordinary General Meeting SPS, dated 17/12/1919; SA, 626/1/29–39: W. N. Allan, The development of irrigation in the Gezira, dated 21/1/1939.
18. SA, 418/5/99–100: The start of the SPS Gezira Scheme, "Above received from MacI in 1945" (Note from Gaitskell).
19. NA, FO141/578/1: Letter from Grey to Gorst, dated 11/11/1910.
20. NA, FO141/578/1: Letter from Garstin to Gorst, dated 12/11/1910.
21. NA, FO141/578/1: Letter from SPS, dated 14/11/1910.
22. NA, FO141/578/1: Report of negotiations between SPS and Government Committee.
23. NA, FO141/578/1: Letter from Governor General to Russell (Cairo), dated 11/1/1911.
24. NA, FO141/578/1: Letter from Stark to "Dear Clive," dated 27/1/1911. Pumping for cotton was not uncommon in Sudan, see Central Research Farm (1915) *Pump irrigation in the northern Sudan: With special reference to the cotton crop* (Khartoum: Central Research Farm). We will meet Kennedy later in this chapter in a slightly different role. Serels (2013; 148) mentions that Tayiba was a government station after the negotiations were broken off in 1911 and that the SPS only came back in 1913, but this seems to be based on inadequate availability and reading of the archival sources. This is a pity,

as his book on Sudan grain policies is interesting enough in itself. Serels S. (2013) *Starvation and the State: Famine, slavery, and power in Sudan, 1883–1956* (New York: Palgrave Macmillan). In his great book on British Nile policy, Tvedt (2004; 92–93) mentions the Tayiba agreements as well, but seems to underestimate the undercurrent of tensions between the government and SPS. He describes everything as quite smooth and focuses particularly on the discussions between Sudan and Egypt on Nile water. This is understandable given his focus but, at the same time, runs the risk of overemphasizing the Nile discussions with Egypt for realizing Gezira. Tvedt T. (2004) *The river Nile in the age of the British: Political ecology and the quest for economic power* (London: I. B. Taurus).

25. NA, FO141/578/1: Letter from Secretary of State to Gorst, dated 10/2/1911.
26. NA, FO141/578/1: Suggestions for carrying out Gezira Irrigation Scheme on Co-partnership lines, Cairo, dated 20/1/1911.
27. NA, FO141/578/1: Letter to Wingate, dated 2/2/1911. It would not have helped that Mather, the one bringing the proposals to the FO was a "terrible bore" wasting "two hours of [FO] valuable time."
28. NA, FO141/578/1: Letter from Gorst to Wingate, dated 5/2/1911. NA, FO141/578/1: Letter to Grey, dated 19/2/1911.
29. Or in case such land was not available 30,000 acres at a lower rate in rain fed land (NA, FO141/578/1: Letter from Clayton to Stack, dated 17/2/1911).
30. NA, FO141/578/1: Telegram from Clayton to Stack, dated 3/3/1911.
31. NA, FO141/578/1: Letter to Grey, dated 2/3/1911; Sometimes the spelling "Tayeba" is used.
32. SA, 418/5/99–100: The start of the SPS Gezira Scheme.
33. NA, FO141/578/1: Agreement with the Soudan Plantations Syndicate re Experimental Station at Taiba, dated 31/3/1911.
34. With SPS-employee Wright as manager. SA, 418/5/99–100: The start of the SPS Gezira Scheme.
35. SA, 415/8/27–28: Ordinary General Meeting SPS, dated 11/10/1911.
36. SA, 469/3/17–18: Letter from Wingate to Clayton, dated 23/7/1911.
37. He was clearly also very happy with his own role in bringing the parties back together in January 1911. (NA, FO141/578/1: Letter from Mather to Grey, dated 16/5/1911).
38. BCGA chairman Hutton became one of the Syndicate directors. (SA, 415/8/27–28: Ordinary General Meeting SPS, dated 11/10/1911).
39. SA, 469/3/17–18: Letter from Wingate to Clayton, dated 23/7/1911.
40. NA, FO141/578/1: Telegram van Governor General Council to Sudan Agency, Cairo, dated 12/8/1911.
41. The report also discussed storage reservoirs on the White Nile and on Lake Tana in Ethiopia. (NA, FO141/578/1: Inspector General of Irrigation, Note on 'The possibilities of the Soudan for cotton growing by mean of artificial irrigation, dated 14/12/1911).
42. SA, 635/5/6: Text from Tottenham on Sudan irrigation, dated 21/3/1972.
43. NA, FO141/578/1: Letter from Grey to Kitchener, dated 13/12/1911.
44. NA, FO141/578/1: Letter from Kitchener to Grey, dated 31/12/1911.

45. NA, FO141/578/1: Draft report to the Council of the British Cotton Growing Association on the possibilities of Cotton growing in the Anglo-Egyptian Soudan (undated, probably early 1912).
46. NA, FO141/578/1: Letter from Harvey to Kitchener, dated 13/2/1912.
47. NA, FO141/578/1: Letter from Wingate to Kitchener, dated 16/4/1912.
48. NA, FO141/578/1: Letter from Eckstein to Lovat, dated 27/2/1912.
49. SA, 415/8/35–36: Ordinary General Meeting SPS, dated 7/12/1912.
50. NA, FO141/578/1: Telegram from Governor General to Stack, dated 11/3/1913.
51. NA, FO141/578/1: G. R. Sawar (Central Research Farm) Note on Tayiba, dated 6/3/1913.
52. NA, FO141/578/1: Telegram from Clayton to Russell, dated 5/3/1913.
53. NA, FO141/578/1: Note on Tayiba. Interesting is the remark that "In this connection an enquiry was made as to the reason for limiting irrigation to the hours of day light, and it was found that lack of training in the use of water on the part of the tenants rendered this practice compulsory." Chapter 3 will discuss the topic of day and/or night irrigation in much more detail.
54. NA, FO141/578/1: Telegram from Governor General to Stack, dated 8/4/1913.
55. NA, FO141/578/1: Letter from Ministry of Finance to Kitchener, dated 14/4/1913.
56. NA, FO141/578/1: Letter from Governor General to Clayton, dated 4/3/1913.
57. SA, 112/3/14: Government owned land in 500,000 feddan area, Gezira Canal Scheme.
58. SA, 108/15/24: Letter from Wingate to Stack, dated 10/4/1913.
59. SA, 108/15/24: Letter from Wingate.
60. Given Kitchener's reputation, one could question whether Wingate ever had been driving, but early 1913 was clearly a moment Kitchener forced himself to the front of the driving bus—or train. Kitchener's re-emergence in Sudan in 1911 when appointed Consul-General in Cairo marked a renewed interest—or interference—in the governance of Sudan. "It is clear that he regarded Sudan as his personal territory. Not forgetting that his aristocratic title was styled 'of Khartoum,' his view of empire building was military rather than commercial and his manner authoritarian. Kitchener wanted the Sudan Government to retain a high degree of control over the operations of the business." (Mollan 2008).
61. NA, FO141/578/1: Letter from Cairo to Governor General, not dated.
62. SA, 108/15/28: Letter to Stack, dated 19/4/1913.
63. SA, 108/15/49: Letter to Stack, dated 6/5/1913.
64. SA, 108/15/44: Letter to Stack, not dated, possibly early May 1913.
65. SA, 108/15/41: Letter from Stack to Wingate, dated 30/4/1913.
66. SA, 108/15/41: Letter from Stack.
67. SA, 418/5/99–100: The start of the SPS Gezira Scheme.
68. SA, 108/15/44: Letter to Stack.
69. SA, 108/15/37: Letter to Stack, dated 29/4/1913.
70. SA, 108/15/62: Letter from E. B. Wilkinson to Wingate, dated 3/6/1913.

71. SA, 108/15/28: Letter to Stack. Gezira was a main item on the list, but irrigation works at Tokar and Kassala were also included in the planned loan.
72. NA, FO141/578/1: Letter from Hutton to Wingate, dated 22/5/1913.
73. SA, 108/15/67–69: Letters from Stack to Wingate, dated 8/6/1913 and 11/6/1913.
74. SA, 109/11/23–27: Resume of proceedings at the meeting held at No. 1 London Wall Buildings on Monday, 21/7/1913.
75. SA, 109/11/17: Letter from McGillivray to Colonel Asser, dated 25/7/1913.
76. The issues to be clarified included drainage and health. Kitchener refused to include drainage, but he did not object to it being dealt with in the future. For him, the issue was simple: the Syndicate had to prevent emergence of "any swamps or morasses" as a result of irrigation (SA, 109/11/23–27: Resume of proceedings).
77. SA, 109/11/23–27: Resume of proceedings.
78. Naturally, given the investments and the available time to make a profit, the percentages payable to the government, Syndicate, and tenants were discussed as well. The two partners at the table considered it "essential for success" that "the native"—the third partner, but not at the table—should profit as well. Both parties agreed that many tenants would need financial support, but Eckstein pointed out that security for this support was weak, precisely because tenants would have no land to offer. The government accepted taking over any outstanding debts at the end of the concession, but only those made on a yearly basis (SA, 109/11/23–27: Resume of proceedings).
79. SA, 109/11/23–27: Resume of proceedings. See also SA, 416/1/23–24: Ordinary General Meeting 6 SPS, dated 18/12/1913.
80. SA, 109/11/7: Letter from Wingate to Lovat, dated 22/7/1913.
81. SA, 109/11/6: Letter from Lovat to Wingate, dated 21/7/1913.
82. SA, 109/11/8: Letter from Lovat to Wingate, dated 23/7/1913.
83. SA, 109/11/9: Letter from Wingate to Lovat, dated 24/7/1913.
84. SA, 109/11/15: Letter from McGillivray to Wingate, dated 25/7/1913.
85. SA, 109/11/16: Letter from McGillivray to Poyntz-Wright, dated 25/7/1913.
86. SA, 109/11/13: Letter to Wilkinson, dated 25/7/1913.
87. SA, 109/11/14: Letter from Wingate to Kitchener, dated 26/7/1913.
88. Mollan (2008).
89. SA, 109/11/21: Letter from Lovat to Wingate, dated 29/7/1913.
90. SA, 109/11/28: Letter from Wingate to Kitchener, dated 30/7/1913.
91. SA, 109/11/33: Letter from Kitchener to Wingate, dated 3/8/1913.
92. SA, 109/11/92: Letter from Wingate to Clayton, dated 3/9/1913.
93. SA, 109/11/58: Letter from Wingate to Clayton, dated 26/8/1913.
94. SA, 109/11/111: Letter from Wingate to Clayton, dated 7/9/1913.
95. SA, 109/11/36: Letter from Wingate to Lovat, dated 5/8/1913; SA, 109/11/42: Letter from Wingate to Kitchener, dated 10/8/1913.
96. SA, 109/11/38: Letter from Wingate to Bonham Carter.
97. SA, 109/11/54: Letter from Wingate to Clayton, dated 17/8/1913.
98. SA, 109/11/92: Letter from Wingate to Clayton, dated 3/9/1913.
99. SA, 109/11/92: Letter from Wingate.
100. SA, 109/11/86: Letter from Lovat to Wingate, dated 2/9/1913.

101. SA, 109/11/88: Letter from Wingate to Lovat, dated 3/9/1913; SA, 109/11/87: Telegram from Wingate to MacGillivray, dated 3/9/1913.
102. SA, 109/11/114: Letter from Wingate to Kitchener, dated 11/9/1913; see also SA, 416/1/16: Minutes SPS, dated 10/9/1913.
103. Black Adder III "Amy and Amiability."
104. SA, 108/15/73: Letter from Bernard to Wingate, dated 10/9/1913 (emphasis in original).
105. SA, 109/11/118: Letter from Wingate to Kitchener, dated 11/9/1913.
106. SA, 109/11/119: Letter from Davidson to Wingate, dated 11/9/1913.
107. SA, 109/11/58: Letter from Wingate to Clayton, dated 26/8/1913.
108. SA, 109/11/129: Draft agreement.
109. SA, 109/11/135: Letter from Blackett to Bernard, dated 22/9/1913.
110. SA, 109/11/144: Letter from Davidson to Bernard, dated 29/9/1913; see also SA, 416/1/18–19: Minutes SPS, dated 9/10/1913.
111. SA, 469/5/30–31: Letter from Clayton to Wingate, dated 3/11/1913.
112. SA, 469/5/49–51: Letter from Wingate to Clayton, dated 26/11/1913.
113. SA, 469/5/54–56: Letter from Wingate to Clayton, dated 29/11/1913.
114. SA, 469/5/70–71: Letter from Clayton to Wingate, dated 17/12/1913.
115. SA, 498/8/2–4: Note from Kitchener to Sir Edward Grey, dated 31/1/1914 (text from 24/1/1914, with annex Note from Garstin, Webb, MacClure and MacDonald, dated 25/1/1914). They also approved the dam on the White Nile. SA, 635/5/6: Text from Tottenham on Sudan irrigation, dated 21/3/1972.
116. SA, 498/8/2–4: Note from Kitchener.
117. SA, 498/8/2–4: Note from Kitchener; see also SA, 469/6/4–5: Letter from Wingate to Clayton, dated 17/1/1914.
118. SA, 416/1/28: Minutes SPS, dated 30/1/1914.
119. NA, FO141/578/1: Bonham Carter Note to accompany draft Gezira Agreement, dated 15/2/1914.
120. SA, 469/6/22–23: Letter from Wingate to Clayton, dated 5/2/1914.
121. NA, FO141/578/1: Letter from Wingate to Kitchener, dated 16/2/1914.
122. SA, 469/6/53–55: Letter from Clayton to Wingate, dated 25/2/1914.
123. SA, 469/6/56–60: Letter from Clayton to Wingate, dated 2/3/1914.
124. SA, 469/6/88–90: Letter from Clayton to Wingate, dated 1/4/1914.
125. SA, 112/5/2: Wingate Note as to the effect of the war on raising of the Capital required for the Gezira Irrigation Scheme, dated 14/8/1916.
126. Sudan Government Financial Secretary Bernard seems not to have been the easiest person to work with. Even Kitchener understood "the disabilities of the Sudan Government in so far as its Financial Secretary is concerned." SA, 469/6/125–126: Letter from Wingate to Stack, dated 25/4/1914. At the same time, MacDonald did not want to "be controlled by Bernard which they say means no progress." Furthermore, he distrusted "Kennedy and don't want his finger in the pie." SA, 469/6/122–124: Letter from Stack to Wingate, dated 23/4/1914.
127. SA, 469/6/96–98: Letter from Wingate to Clayton, dated 9/4/1914. MacDonald's plan was to use the Alessandrini firm to execute the works. MacDonald knew Alessandrini from the Aswan Dam and other works in Egypt. Alessandrini would not work under contract, but receive a percentage of the total cost—if costs would be lower the firm would receive a bonus, if costs were higher the percentage would be reduced. This procedure would

128. SA, 469/6/99–102: Letter from Clayton to Wingate, dated 15/4/1914.
129. SA, 469/6/110–112: Letter from Clayton to Wingate, dated 20/4/1914; see also Tvedt (2004).
130. SA, 469/6/122–124: Letter from Stack.
131. Part of the larger Nile strategy, see Tvedt (2004).
132. SA, 469/6/154–155: Letter from Wingate to Clayton, dated 8/5/1914.
133. SA, 469/6/166–167: Letter from Wingate to Clayton, dated 16/5/1914.
134. SA, 416/1/34: Minutes SPS, dated 22/7/1914.
135. SA, 416/1/35: Minutes SPS, dated 23/7/1914.
136. SA, 112/3/53: Letter on behalf of Lloyd George to Wingate, dated 27/7/1914.
137. SA, 469/7/19–21: Letter from Wingate to Clayton, dated 8/10/1914.
138. SA, 469/7/44–45: Letter from Clayton to Wingate, dated 6/11/1914.
139. SA, 416/1/39: Minutes SPS, dated 15/12/1914.
140. SA, 415/8/53–54: Ordinary General Meeting SPS, dated 13/1/1915.
141. NA, FO141/578/1: MacDonald, A note on the Blue Nile Weir & Gezira Scheme, dated 12/2/1915.
142. SA, 109/5/1: MacDonald A note on the Blue Nile Weir & the Gezira Scheme, dated 12/2/1915.
143. SA, 109/5/1: MacDonald. The note mentioned a reservoir at Lake Tsana to increase storage, but did not elaborate on it. An Egyptian report stressed the need for Tsana as well. "It is to be noted also that development in the Gezira of the Sudan will tend still further to reduce the Nile flood and will, in low years, increase the possibility of [fallow] in Upper Egypt. […] Both [Lake Tsana and the White Nile projects] have the same object, to abstract water from the crest of the flood wave & store it until required to augment Egypt's summer supply. […] ultimately Egypt will require all the water that can be stored by both the White Nile & Lake Tsana Dams. Moreover it has to be noted that the Sudan Gezira Scheme will eventually require the assistance of a storage reservoir, even though summer crops, in the Egyptian sense, will not be produced. The volume so required will however amount only to a fraction of that made available by a Lake Tsana Dam, thus allowing much of the greater proportion of the stored water to pass on to Egypt." (SA, 109/4/1: Ministry of Public Works A note on proposed new works [in Egypt]). MacGillivray explained that an SPS representative in the mission to Tsana "might be very useful later on were I able to explain to our City friends what extra water would mean for cotton growing in the Gezira." SA, 112/4/54: Letter from MacGillivray to Wingate, dated 15/7/1915.
144. SA, 112/4/2: Article from The Times, dated 14/1/1915.
145. SA, 112/4/27: Telegram from Kennedy to MacGillivray, dated 13/3/1915.
146. SA, 112/4/28: Telegram from Hakiman to MacDonald; see also SA, 112/4/29: Letter from MacGillivray to Wingate, dated 14/3/1915.
147. SA, 112/4/30: Telegram from MacDonald to Hakiman, dated 14/3/1915.

148. SA, 112/4/35: Memorandum from SPS Managing Director.
149. SA, 416/1/49: Minutes SPS, dated 8/7/1915.
150. SA, 112/4/52: Letter from Lovat to Wingate, dated 9/7/1915.
151. SA, 469/10/3–4: Letter from Wingate to Clayton, dated 10/7/1915.
152. SA, 112/4/54: Letter from MacGillivray to Wingate, dated 15/7/1915.
153. SA, 469/10/5–7: Letter from Wingate to Clayton, dated 17/7/1915.
154. SA, 112/4/64: Letter from Wingate to MacGillivray, dated 5/8/1915.
155. SA, 416/1/57–58: Ordinary General Meeting 8 SPS, dated 21/12/1915.
156. SA, 112/5/8: Letter from Wingate to Lloyd George, dated 14/8/1916.
157. SA, 112/5/2: Wingate Note as to the effect of the war on raising of the Capital required for the Gezira Irrigation Scheme, dated 14/8/1916. Apparently, MacDonald had also produced new material on the Gezira scheme leading to a need to amend the Gezira Agreement. SA, 112/5/1: Letter from Bonham Carter to Wingate, dated 17/5/1916.
158. SA, 416/1/63–64: Minutes SPS, dated 23/8/1916.
159. SA, 416/1/66–67: Minutes SPS, dated 19/10/1916.
160. SA, 112/5/16: Letter from Bonham Carter to MacGillivray, dated 3/12/1916.
161. SA, 112/5/16: Letter from Bonham Carter; SA, 112/5/11: Letter from Wingate to Grey, dated 2/12/1916.
162. SA, 416/1/77+80–81: Minutes SPS, dated 24/1/1917.
163. SA, 416/1/78–79: Ordinary General Meeting SPS, dated 24/1/1917.
164. NA, FO141/578/1: Letter from Hutton to Wingate, dated 9/1/1917; emphasis in original.
165. NA, FO141/578/1: Letter from Ainscough (Financial Department London) to Hutton, dated 12/1/1917.
166. Durham Library: Report of the delegates, dated 1917.
167. SA, 112/7/5: Letter from Wingate to Garstin, dated 3/5/1917; SA, 109/3/1: MacDonald, Nile Control, dated 14/5/1917.
168. SA, 112/7/26: Letter from MacDonald to Wingate, dated 30/6/1917.
169. SA, 112/7/18: Letter from Lindsay to MacDonald, dated 2/6/1917; In addition, the benefits for Egypt should be "secured." See SA, 112/7/21: Letter from Finance Minister to Public Works Minister, dated 6/6/1917.
170. SA, 112/8/1: Letter from Bonham Carter to Wingate, dated 9/7/1917.
171. SA, 416/1/82: Minutes SPS, dated 5/7/1917.
172. SA, 112/8/1: Letter from Bonham Carter.
173. Durham Library: Report of the delegates.
174. SA, 112/8/5: Letter from Hamilton to Bonham Carter, dated 18/7/1917.
175. SA, 112/5/23: Garstin and Stubbs, Egypt and the Sudan. Proposed works on irrigation, drainage and flood protection, dated July 1917.
176. Although in a dry year (once in ten years), the area would have to be reduced to 500,000 feddans. SA, 112/7/13: Note on the sufficiency of water for the Gezira Project as at present contemplated, dated 20/5/1917.
177. SA, 626/1/29–39: The development of irrigation in the Gezira.
178. SA, 112/5/23: Garstin and Stubbs.
179. SA, 112/8/16: Letter from Garstin to Wingate, dated 10/9/1917.
180. Rather less polite ways of saying the latter included that people "flock in from all sides" and "breed like rabbits." SA, 112/10/1: Minutes of Evidence Board of Trade Committee Growth of Cotton in the British Empire, dated 1/8/1917.

181. Where the Empire Cotton Growing Committee was absolutely delirious on the possibilities of growing cotton in Sudan—specifically Gezira was "the greatest hope of the Sudan," but Tokar was also interesting and Zeidab was mentioned to be the likely "model for other similar areas in the Sudan"—and was convinced it would not threaten Egyptian cotton growing, at the same time, the ECC suggested that care was needed given "criticisms in regard to both the schemes above-mentioned." SA, 112/10/1: Minutes of Evidence.
182. SA, 112/8/26: Letter from Bernard to Wingate, dated 14/9/1917; see also SA, 112/9/9: The extension of Cotton-growing in British Dominions, Colonies and Protectorates, with especial reference to Egypt, dated 1/10/1917.
183. SA, 112/9/2: Note of meeting between Sudan Government and SPS, dated 2/8/1917; SA, 112/9/22: Letter from Bernard to Eckstein, dated 3/10/1917, with Minutes dated 11/9/1917.
184. SA, 112/9/22: Letter from Bernard.
185. Durham Library: Report of the delegates, dated 1917; SA, 112/9/22: Letter from Bernard.
186. SA, 112/8/21: Letter from Eckstein to Bernard, dated 12/9/1917.
187. SA, 112/8/26: Letter from Bernard to Wingate, dated 14/9/1917.
188. SA, 112/9/41: Letter from Bernard to Wingate, dated 5/10/1917.
189. SA, 112/9/51: Letter from Bernard to Wingate, dated 30/10/1917.
190. SA, 112/9/48: Letter from MacDonald to Wingate, dated 19/10/1917.
191. SA, 112/9/61: Telegram from MacDonald, dated 5/11/1917.
192. SA, 112/9/62: Foreign Office, dated 8/11/1917.
193. Durham Library: Letter from High Commissioner to Foreign Office, dated 13/11/1917, Report of the delegates.
194. SA, 112/9/66: Letter from Cecil to Wingate, dated 12/11/1917.
195. SA, 112/9/72: Letter from Ainscough to Hutton, dated 12/11/1917; see also SA, 112/9/67: Letter from Hutton to Wingate, dated 14/11/1917.
196. NA, TI/12095: Letter from Chairman British Cotton Growers Association to Bonar Law, dated 16/11/1917.
197. Durham Library: Report of the delegates.
198. SA, 416/1/100–102: Minutes SPS, dated 9/10/1918.
199. SA, 156/4/24: Letter from Stack to Wingate, dated 24/11/1918. Works at Tokar and Kassala added another £400,000 making the total bill almost £4 million. Including everything in one loan necessitating one bill was considered smart.
200. SA, 416/1/106–107: Ordinary General Meeting SPS, dated 20/12/1918.
201. SA, 156/5/1: Telegram from Wingate, dated 2/1/1919.
202. SA, 416/1/108–110: Minutes SPS, dated 5/3/1919; SA, 415/10/18–19: Minutes SPS, dated 28/5/1919.
203. SA, 415/3/18–20: Letter accompanying the Agreement between SPS and Sudan Government, dated 17/10/1919.
204. SA, 415/10/26–28: Ordinary General Meeting SPS, dated 17/12/1919. SA, 415/10/22–23; Minutes SPS, dated 23/10/1919. We will discover in chapter 2 that in the late 1920s the SPS was all too happy to accept lower profits for a longer concession period.
205. SA, 417/4/495–497: Letter from MacIntyre to Bernard, dated 11/5/1920; SA, 415/10/30–31: Minutes SPS, dated 28/1/1920.

206. SA, 417/4/500–504: Letter from Bernard to MacIntyre, dated 7/2/1922; SA, 417/4/500–504: Letter from MacIntyre to Bernard, dated 10/2/1922.
207. SA, 415/10/32–36: Extraordinary General Meetings SPS, dated 24/3 and 8/4, 1920.
208. SA, 415/10/20–21: Minutes SPS. On October 11, 1919, the "Sudan Guaranteed Loan Ordinance 1919" was passed, which approved that £6 million was made available for the "works of the Gezira Scheme including repayment of temporary loans raised under Government of Soudan Acts 1913 and 1915," which totalled £4.9 million with £700,000 additionally allocated for development of the railway system in Sudan and £400,000 allocated for the irrigation of the Tokar region (Mollan 2008).
209. Kennedy had worked himself on Gezira plans, even though none of them were selected as MacDonald's ideas prevailed. His letters start as early as April 1917, but he seems to have written many up to at least 1921.
210. NA, FO141/578/1: Letter from Director of Forestry to Wingate, dated 24/12/1918.
211. SA, 156/1/36: Letter from Wingate to Stack, dated 23/2/1918.
212. SA, 156/1/51: Letter from Willcocks to Wingate, dated 25/2/1918.
213. SA, 156/1/36: Letter from Wingate to Stack, dated 23/2/1918.
214. Committee members included engineers Gebbie, Simpson, and Corey. See Tvedt (2004).
215. SA, 108/1/1: Draft minutes of evidence Nile Projects Commission Part 1, dated 25/5/1920.
216. SA, 592/4: Report of the Nile Projects Commission, dated 25/8/1920. The Commission clearly did not accept "the whole method (employed by Sir William Willcocks)" or terms like "liars" for government officials. It even remarked that unless Willcocks was "prepared to concede that this would have applied to himself when he was a Government official, he has no right to postulate it of others."
217. SA, 108/14/15: Letter from MacDonald to Wingate, started at 14/7/1920, dated 3/8/1920.
218. Despite the continuing debates and the apparent influence Kennedy and Willcocks (two rather different engineers) could have on political decision making. In 1921, Kennedy referred to the Committee report as "a welter of dishonesty, fraud, and calculated misrepresentation." (SA, 156/5/7: Letter from Kennedy to High Commissioner of Egypt, dated 18/8/1921).
219. SA, 415/10/59–62: Ordinary General Meeting SPS, dated 23/11/1921. I will return to this issue in the epilogue, but would like to make the point already that in contrast with most scholars (including A. Gaitskell (1959) *Gezira: A story of development in the Sudan* (London: Faber and Faber) and Tvedt [2004]), these Gezira pumping schemes were no longer experiments once the Tayiba crisis had been arranged in 1913. All efforts after 1913 were to make money and prepare the Syndicate, its inspectors, Gezira, and its population for the large scheme.
220. SA, 415/10/56–58: Minutes SPS, dated 2/11/1921.
221. SA, 415/10/53–55: Minutes SPS, dated 12/5/1921.
222. SA, 415/10/64–65: Minutes SPS, dated 2/3/1922; SA, 626/1/29–39: The development of irrigation in the Gezira.
223. SA, 416/2/29–32: Ordinary General Meeting SPS, dated 8/11/1923.
224. SA, 416/2/42–43: Minutes SPS, dated 1/10/1924.

225. SA, 416/2/56–57: Minutes SPS, dated 19/8/1925.
226. SA, 416/2/61–62: Ordinary General Meeting SPS, dated 28/10/1925.
227. SA, 416/2/61–62: Ordinary General Meeting SPS, dated 28/10/1925.

2 A Task of Some Magnitude: Gezira Management Logic

1. Sudan Archive, 498/8/5–9: Note on Gezira Irrigation Scheme, dated 19/4/1936.
2. National Archives, FO14/579: Minutes meeting on Gezira Extension, dated 23/5/1923. See also R. W. Allen (1924) "Irrigation in the Sudan," *Journal of the Royal African Society*, 23, 257–264; R. W. Allen (1926) "The Gezira Irrigation Scheme, Sudan," *Journal of the Royal African Society*, 25, 229–236.
3. NA, FO141/579: High Commissioner to J. R. MacDonald 22/5/1924. After all, the early 1920s were complicated years for London-Cairo relations, see R. O. Collins (2008) *A history of modern Sudan* (Cambridge: Cambridge University Press); M. W. Daly (1986) *Empire on the Nile: The Anglo-Egyptian Sudan, 1898–1934* (Cambridge: Cambridge University Press); J. Ryle, J. Willis, S. Baldo and J. M. Jok (2011) *The Sudan Handbook* (Woodbridge: James Curry); T. Tvedt (2004) *The river Nile in the age of the British: Political ecology and the quest for economic power* (London: I. B. Taurus).
4. Daly (1986); Tvedt (2004). It is clear that along with discussions on actual water availability of the Nile, internal political concerns of the British in Egypt were at least as important. The British colonial government in the Condominium was afraid "of giving the extremist factions so good a ground for uniting against us." (NA, FO 141/579). For more detail on Egyptian irrigation, see J. Barnes (2014) *Cultivating the Nile: The everyday politics of water in Egypt* (Durham: Duke University Press); C. J. Cookson-Hills (2013) *Engineering the Nile: Irrigation and the British Empire in Egypt, 1882–1914* (PhD thesis, Queen's University); Mitchell (2002) *Rule of experts: Egypt, techno-politics, modernity* (Oakland: University of California Press).
5. A. Gaitskell (1959) *Gezira. A story of development in the Sudan* (London: Faber and Faber). See also chapter 6.
6. NA, FO141/579: Letter of Stack to High Commissioner, dated 31/3/1924.
7. J. Robertson (1974) *Transition in Africa* (London: Hurst & Company).
8. NA, FO141/579: Letter of Stack.
9. NA, FO141/579: Letter of Stack.
10. NA, FO141/579: Letter of Schuster to Murray, dated 23/12/1924.
11. NA, FO141/579: Letter of Schuster.
12. NA, FO141/579: Letter of Schuster.
13. Furthermore, water stored at Sennar in the Blue Nile would be used earlier. Instead of January 18, water would be drawn from the storage on the 10th of that month.
14. NA, FO141/579: Letter of Schuster to Murray, dated 7/3/1925; Note MacGregor, dated 2/2/1925.
15. NA, FO141/579: Letter of Schuster.

16. NA, FO141/579: Letter of Schuster.
17. NA, FO141/579: Minutes Governor-General Council of Sudan, February 1926.
18. SA, 417/1/20–21: Minutes SPS, dated 24/11/1926.
19. Durham Library: Gezira Irrigation Scheme Northern Extension Heads of Agreement, dated 9/11/1926.
20. Nile Commission. 1925. Nile Commission: Report. Members were Cremers (Chairman, deceased before the report was finished), MacGregor (officially "British Delegate," but—as we shall see in chapter 3 as well—the Sudan delegate), and Soliman from the Egyptian Department of Public Works.
21. SA, 415/6/19: Letter from Sudan Government London Office to Eckstein, dated 22/11/1926.
22. SA, 417/1/27: Minutes Adjourned Ordinary General Meeting SPS, dated 8/12/1926.
23. SA, 417/1/13–14: Minutes SPS, dated 17/6/1926.
24. SA, 498/9/32: MacGregor to Schuster, dated 20/9/1925.
25. NA, FO14/579: 23/2/1927.
26. NA, FO14/579: 23/2/1927.
27. NA, FO14/579: 23/2/1927.
28. SA, 990/1/36: Proposed Basis for Rearrangement, Financial Secretary to Mr. Eckstein, dated 24/2/1927.
29. NA, FO14/579: 23/2/1927.
30. NA, FO14/579: 23/3/1927.
31. Sudan Archive, 990/1/50: Agreement between SPS, KCC, and Sudan Government. In 1933 the Kassala Cotton Company (KCC) concession was 45,000 feddans. The KCC was basically a sub-company of the SPS. It entered Gezira after having left the Kassala Delta. See Note 38 in the Introduction.
32. SA, 417/1/30–32: Minutes Meeting 207, dated 7/4/1927.
33. SA, 415/6/28–46: Note of a Discussion Held in the Finance Department on March 8 and 9, 1928 to Consider Various Alterations Proposed by Mr. MacIntyre.
34. SA, 415/6/28–46: Note of a discussion.
35. SA, 415/6/28–46: Note of a discussion.
36. NA, FO141/579: 8/5/1929.
37. A Sudan Irrigation Department memo of December 1928 on further extensions of the Gezira Scheme, suggested that a strategy of risk spreading—extending the irrigated area in several distinct regions—would be beneficial, given the differences in rainfall, soil quality, and vulnerability for diseases, even within Blocks. NA, FO141/579: Note on further extension of the Gezira Irrigation Scheme, dated 6/12/1928.
38. NA, FO141/579: Governor General to High Commissioner, dated 5/4/1929.
39. NA, FO141/579: Huddleston, dated 7/4/1929.
40. NA, FO141/579: 27/5/1929.
41. NA, FO14/579: 11/6/1929.
42. It seems to have been his private view as well, as we will discover in chapter 4.
43. NA, FO14/579: 12/6/1929.

44. SA, 417/1/65: Minutes SPS, dated 20/6/1929; SA 417/1/66–69: Minutes SPS, dated 18/10/1929; also SA, 417/1/73–75: Ordinary General Meeting SPS, dated 6/11/1929.
45. SA, 417/4/419–427: Added in Handwriting: "Government will not drain off any of the Syndicate's staff."
46. NA, FO141/579: Note of Financial Secretary on Gezira Policy, dated 15/11/1929.
47. SA, 495/3/23: Robertson Note on the Western Gezira Division 1935.
48. SA, 594/4/3: Letter MacGregor to Robertson, dated 20/4/1935.
49. SA, 594/4/12: Letter Allan to Director of Irrigation, dated 29/9/1938.
50. SA, 416/2/61: 18th Ordinary General Meeting SPS, dated 28/10/1925.
51. SA, 417/1/56: 21st Ordinary General Meeting SPS, dated 14/11/1928.
52. SA, 416/2/61: 18th Ordinary General Meeting SPS, dated 28/10/1925.
53. As the government has asked to wait with announcing the recent extension agreements, mentioned earlier.
54. SA, 417/1/24: Adjourned 19th Ordinary General Meeting SPS, dated 26/12/1926. Chapter 3 explains the complicated canals and why Eckstein had indeed ample reason to credit the Sudan irrigation engineers.
55. SA, 417/1/24: Adjourned 19th Ordinary General Meeting.
56. SA, 417/1/24: Adjourned 19th Ordinary General Meeting.
57. SA, 415/9/40–42: Ordinary General Meeting SPS, dated 8/10/1908.
58. SA, 418/5/99–100: The start of the SPS Gezira Scheme, "Above received from MacI in 1945" (Note from Gaitskell).
59. SA, 415/8/19–21: Ordinary General Meeting SPS, dated 7/10/1910.
60. D. J. Shaw (1965a) "The development and contribution of irrigated agriculture in the Sudan," in *Agricultural Development in the Sudan* (Khartoum: Philosophical Society of the Sudan, Sudan Agricultural Society). See also SA, 415/4/1–8: Number of model contracts between tenant and SPS for Zeidab, somewhere 1930s. Looking backward in 1965, Shaw stressed that the importance of the Zeidab concession "as a model for land grants and a microcosm of the factors which were to influence the major developments in irrigated agriculture in the Sudan for much of this century can hardly be over-emphasized."
61. SA, Shaw (1965a).
62. SA, 415/8/19–21, Ordinary General Meeting SPS.
63. SA, 415/9/40–42, Ordinary General Meeting SPS.
64. SA, Shaw (1965a).
65. SA, 415/4/1, Tenancy Agreement Zeidab. In 1919 the rent seems to be 50 percent of the cotton crop (SA, 415/4/10: 1919 Tenancy Agreement Zeidab).
66. As part of the agreement between the government and SPS, as discussed in chapter 1.
67. SA, 109/11/16: Letter from MacGillivray to Poyntz-Wright, dated 25/7/1913.
68. SA, 109/11/46: Telegram from MacGillivray to Wingate, dated 11/8/1913. In a text from 1945, MacIntyre suggests that it was the government takeover that caused disturbances. "The first thing that Government did was to alter the tenant rent upwards and the Tenants revolted and went on strike and the Camel Corps had to be sent to Tayiba to keep order" (SA, 418/5/99–100: The start of the SPS Gezira Scheme). However, he may have confused two

moments. In 1913, we read in a letter from the government official in charge at Tayiba for that short time in 1913: "I am very glad it has been decided to charge the tenants the same rate as last year, as if it had been increased I am sure there would have been a certain amount of trouble, as the tenants had got hold of all sorts of stories of what the Government was going to do etc., evidently spread so as to make things difficult for us on taking over" (SA, 108/15/62: Letter from E. B. Wilkinson to Wingate, dated 3/6/1913).
69. SA, 109/11/46: Telegram from MacGillivray to Wingate, dated 11/8/1913. SA, 109/11/49: Letter from MacGillivray to Wingate, dated 11/8/1913. Written in this letter: "12/8/1913. Your telegram received. Have wired Asser. If you concur Syndicates views Tayiba tenants act accordingly."
70. SA, 109/11/54: Letter from Wingate to Clayton, dated 17/8/1913.
71. And they seemed to have been right! The original arrangement gave a net profit of about £E73—based on a gross return of 220, cost of cultivation 71,500, and a rent (water rate etc.) of 250 per feddan totalling 75,000 for 30 feddans. The new arrangement ensured a tenant 40 percent of the same gross produce, with costs still to be deducted but no rent to be paid. This yielded 40 percent of 220 = 88 gross return minus a cost of 71,500 equals 16.5 pounds for a tenant. The rent was thus increased to 132 pounds! (SA, 112/3/16: Statement of results at Tayiba for the two seasons 1911–1912 and 1912–1913).
72. SA, 109/11/63: Memorandum from Dickenson to the Director of Agriculture, dated 9/8/1913.
73. SA, 109/11/63: Memorandum from Dickenson to the Director of Agriculture, dated 9/8/1913.
74. SA, 109/11/58: Letter from Wingate to Clayton, dated 26/8/1913.
75. SA, 109/11/61: Letter from Asser to Wingate, dated 11/8/1913.
76. SA, 109/11/84: Telegram from MacGillivray to Wingate, dated 1/9/1913.
77. SA, 109/11/84: Telegram from Wingate to MacGillivray, dated 1/9/1913.
78. SA, 109/11/66: Telegram from Wingate to Asser, dated 26/8/1913.
79. And not because the Sudan Government "forcibly changed" the Syndicate business model. G. M. Yousif (1997) *The Gezira Scheme: The greatest on earth under one management* (Khartoum: Africa University House for Printing). Along the Nile, north of Khartoum, in the basin areas of Kerma, Letti and Affat, tenancies were seen as "very unsatisfactory," as tenants did not have enough interest in the land. For the basins, the solution would be the arrangement that Sudan farmers would "become owners of the land," a model definitely not in sight for Gezira. SA, 112/1/35: 1912 Annual Report on Basins.
80. SA, Shaw (1965a).
81. SA, 112/10/1: Minutes of Evidence Board of Trade Committee Growth of Cotton in the British Empire, dated 1/8/1917.
82. Gaitskell (1959; 121–122).
83. SA, 408/2/16: Morgan—Some notes on practical agricultural in the Gezira Irrigation Scheme. Morgan is referred to as "member of the staff of the Sudan Gezira Board" suggesting the document is from after 1950. In the document, however, explicit reference is made to the partnership between the Sudan Government, local inhabitants, and "the Companies,"

suggesting the text is from before 1950. Continuity in approaches between SPS and SGB, and Gezira in general, is discussed in chapters 6 and 7.
84. SA, 408/2/16: Morgan.
85. SA, 498/12/42: Griffin—Irrigation Notes, dated February 1941.
86. Which indeed proved to be the case (SA, 418/3/216: SPS—a report on Gezira rotations and Abdel Hakam Experimental Block).
87. Drawing based on information from SA, 408/2/16: Morgan.
88. SA, 408/2/16: Morgan.
89. SA, 850/9/1: The Sudan Gezira Board, Annual Crop Report 1958/50.
90. S. M. Mollan (2008) *Economic imperialism and the political economy of Sudan: The case of the Sudan Plantations Syndicate, 1899–1956* (PhD Thesis, Durham University).
91. SA, 416/3/49: 24th Ordinary General Meeting SPS, dated 28/10/1931.
92. SA, 416/3/68: 25th Ordinary General Meeting SPS, dated 3/11/1932.
93. SA, 416/3/82: 26th Ordinary General Meeting SPS, dated 15/11/1933.
94. "A kantar is the official Egyptian weight unit for measuring cotton. It corresponds to the US hundredweight, and is roughly equal to 99.05 pounds, or 45.02 kilograms. It is equal to either 157 kilograms of seed cotton or 50 kilograms of lint cotton." (http://en.wikipedia.org/wiki/Kantar).
95. SA, 498/10/2: SPS—The Gezira Irrigation Scheme, dated 31/12/1938; SA, 498/10/11: KCC—The Gezira Irrigation Scheme, dated 31/12/1938.
96. SA, 416/3/41: Circular to shareholders SPS, dated 15/4/1931.
97. SA, 416/3/41: Circular to shareholders.
98. SA, 416/3/68: 25th Ordinary General Meeting.
99. SA, 416/3/41: Circular to shareholders.
100. SA, 416/3/68: 25th Ordinary General Meeting.
101. SA, 416/3/68: 25th Ordinary General Meeting.
102. SA, 416/3/82: 26th Ordinary General Meeting SPS, dated 15/11/1933.
103. SA, 416/3/71: Minutes SPS, dated 23/3/1933.
104. SA, 416/3/68: 25th Ordinary General Meeting.
105. SA, 408/2/16: Morgan.
106. SA, 408/2/16: Morgan.
107. SA, 498/3/10: Griffin—Water consumption and peak demand in the Gezira related to the crop area, dated 19/12/1936.
108. SA, 626/11: Sudan Irrigation Department, Sennar Dam Regulation Rules 1933.
109. SA, Griffin—Irrigation Notes, in A Hand Book for new personnel.
110. SA, 626/12: Sudan Irrigation Department, Gezira Canal Regulation Handbook 1934.
111. SA, 408/2/16: Morgan.
112. SA, 626/12: Gezira Canal Regulation Handbook 1934.
113. SA, 408/2/16: Morgan.
114. SA, 408/2/16: Morgan.
115. SA, 626/12: Gezira Canal Regulation Handbook 1934.
116. SA, 626/12: Gezira Canal Regulation Handbook 1934.
117. SA, 408/2/16: Morgan.
118. SA, 416/3/20: 23rd Ordinary General Meeting SPS, 1930.
119. SA, 626/12: Gezira Canal Regulation Handbook 1934.
120. SA, 408/2/16: Morgan.
121. SA: Griffin—Irrigation Notes.

122. SA, 500/19/30: Handing over notes, dated June 1942.
123. SA, 408/2/16: Morgan.
124. SA, 408/2/16: The paragraph 'Chains' is based on Morgan's description.
125. In chapter 4, we will discuss how the British argued that the Sudan tenants were not even doing what they were supposed to do in that respect.
126. See E. M. Crowther and F. Crowther (1935) "Rainfall and cotton yields in the Sudan Gezira," *Proceedings of the Royal Society of London, Series B, Biological Sciences*, 118, 343–370.
127. SA, 408/2/16: Morgan. MacGillivray claimed that tenants were not charged for seed, but "simply" debited with it. Costs were "knocked off" the 40 percent each year. (SA, 112/10/1: Minutes of Evidence).
128. SA: Notes on the Gezira Irrigation Project, January 1926.
129. Sudan Archive, 417/3/65: Specimen of tenants account.
130. SA, 415/6/49: Letter from Bernard to Eckstein, dated 8/11/1922.
131. SA, 112/10/1: Minutes of Evidence.
132. SA, Gezira Scheme. A Handbook for new Personnel, undated, but from Sudan Gezira Board.
133. A. I. Clarkson (2005) *Courts, councils and citizenship: political culture in the Gezira scheme in condominium Sudan* (PhD Thesis, Durham University).
134. SA, 415/5/1: 1929 Tenancy Agreement Gezira.
135. SA, 415/5/14: 1941 Tenancy Agreement Gezira.
136. SA, 416/5/23: 30th Ordinary General Meeting SPS, 1930.
137. SA, 417/2/34: Explanation of the changes involved in the New Ancillary and Tenants' Agreements.
138. SA, 416/3/42: Minutes SPS, dated 10/6/1931.
139. NA, FO141/692/14: Letter from Huddleston to MacIntyre, dated 25/2/1931.
140. NA, FO141/758/9: Letter from Financial secretary to Ryder, dated 15/3/1933.
141. Mollan (2008).
142. SA, 416/3/85: Minutes SPS, dated 25/3/1934.
143. SA, 416/4/51: Minutes SPS, dated 22/10/1936.
144. Mollan (2008).
145. SA, 417/2/34: Explanation.
146. NA, FO141/758/9: Letter from financial secretary to Ryder, dated 15/3/1933.
147. SA, 415/6/13, Note on SPS objections to the incorporation in the New Agreement of a provision whereby SPS must bear a proportion of any increase in the Tenants' share, 1926. Also SA, 416/3/38: Minutes SPS, dated 18/3/1931. Even the way individual tenants sold their cotton to the Syndicate was important and a good example how the arrangements between the government and SPS created problems for the tenants without the tenants being involved in them. The question was whether the tenants were subject to UK income tax! In the original Concession Agreement, the Syndicate had a share in the Cotton Agreement, but at the end of the 1920s the idea was that the SPS would pay the tenants for the cotton a price that included the tenants' and the government's shares of the total profit. The tenants would pay, from what they received "in theory," for the government's share, "although of course, in fact, the Syndicate collects this money from the Tenants by deduction and hands it over to the

Government." (SA, 417/4/27: Letter from Cutforth to Fletcher, dated 14/11/1928). This created the issue of tenants possibly being involved in British taxes; something no one was keen on—"cultivating tenants in the Sudan shall not be subjected by British taxation." (SA, 417/4/215: Letter from Fraser to Fletcher, dated 18/1/1929). Therefore, the change was not made, and the "the complicated machinery" to prevent tenants from being subjected to British income tax was not needed anymore as the original form of the Agreement kept in use (SA, 417/4/267: Letter from Fraser to Fletcher, dated 10/4/1929).
148. SA, 415/6/13, Note on SPS objections.
149. SA, 498/10/2: SPS—The Gezira Irrigation Scheme, dated 31/12/1938; SA, 498/10/11: KCC—The Gezira Irrigation Scheme, dated 31/12/1938.
150. SA, 417/2/38, Agreement SPS and Government, dated 10/6/1945. See Mollan (2008) for details.
151. Clarkson (2005).
152. See Mollan (2008) for details.
153. Mollan (2008). Whether this Gezira dependence was beneficial and/or planned/foreseen/better to be have been avoided by the Sudan Government is a topic of debate in the literature.
154. Mollan (2008).
155. Compare with R. E. Elson (1984) *Javanese peasants and the colonial sugar industry: Impact and change in an East Java residency, 1830–1940* (Singapore: Oxford University Press), on the impact of the Javanese sugar industry.
156. Mollan (2008).
157. Clarkson (2005).

3 No Man Can Serve Two Masters: Designing Gezira Irrigation

1. Sudan Archive 501/2/34: Sudan Irrigation Department Annual Report 1946. I have not found his first names, so R. J. It is!
2. SA 415/3/1: Agreement SPS and Sudan Government, dated 17/10/1919.
3. SA, 635/5/0: Text from Tottenham on Sudan Irrigation, dated 21/3/1972.
4. National Archives, FO141/579: Letter to Kerr 25/5/1923. See T. Tvedt (2004) *The river Nile in the age of the British: Political ecology and the quest for economic power* (London: I. B. Taurus), for more on the position of MacDonald between Egyptian, Sudanese, and his own interests.
5. SA, 469/6/31–35: Letter from Wingate to Clayton, dated 12/2/1914.
6. SA, 495/2/2: Sudan Irrigation Department Memorandum No. 1 23/12/1950.
7. NA, FO 141/579: undated text.
8. SA, 495/2/3: Sudan Irrigation Department Memorandum No. 1 23/12/1950.
9. NA, FO141/579: Stack to Allenby, dated 17/5/1923.
10. NA, FO 141/579: Letter Financial Adviser, dated 22/3/1924.
11. NA, FO 141/579: Gardiner to Wiggin, dated 3/5/1924.
12. As said, these numbers do not include Sudanese field personnel—the ones actually setting the gates. Furthermore, especially in the Irrigation Department after World War II, engineering positions could be held by

Sudanese. More details on the relation between British and Sudanese engineers and government officials in general can be found in chapter 6.
13. SA, 495/2/3: Sudan Irrigation Department Memorandum No. 1 23/12/1950. One could doubt whether colonial Sudan was ever completely independent from Egypt within the Condominium and particularly on Nile/irrigation affairs.
14. SA, 626/9/1–80: Sudan Government Irrigation Department 1927, adapted 1934.
15. SA, 495/2/4: Sudan Irrigation Department Memorandum.
16. SA, 626/9/30: Sudan Government Irrigation Department.
17. SA, 498/6/25: MacGregor, Small Irrigation and Storage Schemes, dated 30/12/1927.
18. SA, 495/2/2: Sudan Irrigation Department Memorandum No. 1, dated 23/12/1950.
19. SA, 415/3/1: Agreement between SPS and Sudan Government, dated 17/10/1919.
20. NA, FO141/692/14: Letter from Huddleston to MacIntyre, dated 25/2/1931.
21. NA, FO141/692/14: Letter from MacIntyre to Huddleston, dated 27/2/1931.
22. NA, FO141/579: Note by Mr. W. D. Roberts and Mr. R. M. MacGregor on certain points connected with the working of the Gezira Scheme 10/2/1924, original date 22/1/1924.
23. NA, FO141/579: Roberts and MacGregor.
24. NA, FO141/579: Roberts and MacGregor.
25. A. Gaitskell (1959) *Gezira: A story of development in the Sudan* (London: Faber and Faber), 123.
26. NA, FO141/579: Note by Roberts and MacGregor.
27. NA, FO141/692/14, Huddleston to MacIntyre, dated 25/2/1931.
28. NA, FO141/579: Note by Roberts and MacGregor.
29. NA, FO141/579: Note by Roberts and MacGregor.
30. SA, 626/1/2: Butcher, Note on Night Watering, dated 4/6/1924.
31. SA, 626/1/17: Butcher.
32. SA, 626/1/19: Butcher.
33. SA, 626/1/12: Butcher.
34. See, for example, the Dutch irrigation engineers and their obsession with measuring water in M. W. Ertsen (2010) *Locales of Happiness: Colonial irrigation in the Netherlands East Indies and its remains, 1830—1980* (Delft: VSSD Press).
35. SA, 626/1/2: Butcher. When stable, irrigation canals have a predictable water depth given a certain flow. When the flow is continuously changing, and water levels too, using that stable relation is not possible anymore.
36. SA, 626/1/21: Butcher.
37. SA, 498/9/22: Letter of MacGregor to Schuster, dated 16/8/1925.
38. SA, 498/9/26: MacGregor to Schuster.
39. SA, 498/9/26: MacGregor to Schuster.
40. SA, 498/9/30: MacGregor to Schuster, dated 20/9/1925.
41. SA, 497/11/17: Note to Chief Engineer, from Griffin, Divisional Engineer Abu Usher, October 1933.
42. SA, 497/11/26: Note to Chief Engineer.

43. SA, 497/11/26: Note to Chief Engineer.
44. SA, 498/15/36: Letter from Griffin to Manager SPS, dated 3/3/1944.
45. SA: Memorandum on Working Arrangements.
46. SA, 498/15/32: Divisional Engineer Western Gezira to Director Sudan Irrigation Department 20/11/1925. Thomas Hobson (1544–1631, Cambridge, England) owned a stable of some 40 horses which he rented out. Although customers might have thought that they had a choice, customers had to choose the horse closest to the door. Hobson wanted to prevent that the best horses were always chosen, to avoid these horses becoming overused.
47. H. Pluesquellec (1990) *The Gezira Scheme in Sudan: Objectives, design and performance* (Washington: World Bank), 21. See also B. Wallach (1988) "Irrigation in the Sudan since Independence," *Geographical Review*, 78, 417–434, who makes a similar remark on page 422. The myth had grown strong!
48. NA FO141/579: Remodelling the Main Canal, dated 10/5/1928.
49. SA, 498/10/10: Griffin, Water consumption and peak demand in the Gezira related to the crop area 19/12/1936.
50. SA, 416/4/36: Minutes SPS, dated 18/4/1935.
51. SA, 497/14/22: Note Griffith, Divisional Engineer Western Gezira, dated 12/1/1936.
52. In chapter 6, we will discuss the debate around the new Managil extension, where a similar decision was finally made but where the principle was debated upon first rather heavily.
53. SA, 496/13/119: Design Sheet No. 54 14/5/1927. Compare with the standardization in the Netherlands East Indies (Ertsen, 2010). For other engineering fields, see A. Picon (2004) "Engineers and Engineering History. Problems and Perspectives," *History and Technology*, 20, 421–436; A. Picon (2000) "Technological traditions and national identities: A comparison between France and Great Britain during the XIXth century," in E. Nicolaidis and K. Chatzis (eds) *Science, technology and the 19th century state* (Athens: Institut de Recherches Neohelleniques). For standardization and its link to control in general, see B. Etemad (2007) *Possessing the world: Taking the measurements of colonisation from the 18th to the 20th century* (New York: Berghahn Books); Alexander J. Karns (2008) *The mantra of efficiency: From waterwheel to social control* (Baltimore, MD: Johns Hopkins University Press).
54. SA, 496/13/120: Design Sheet No. 54, dated 14/5/1927.
55. SA, 496/13/199: Design Sheet 106.
56. SA, 497/12/13: Note Divisional Engineer Maintenance to Sub Divisional Engineer. The handbooks and management rules for Sennar served a similar purpose. Everyone new in the SID had to learn working with the agricultural rhythm—from the SPS—and Sennar rhythm—from negotiations with Egypt—as soon as possible.
57. SA, 496/13/89: Design Sheet No. 32.
58. SA, 496/13/90: Design Sheet 32A, January 1928.
59. SA, 496/13/66: Design Sheet No. 17; see also 497/13/19: Divisional Engineer to Sub Divisional Engineer 8/2/1937.
60. SA, 592/7/33: Design Data Sheet A-21.

61. SA, 500/15/32: Abu Ushar Sub-Division Handing-over Note, dated 7/3/1929.
62. SA, 498/12/2: Smith to Divisional Engineer Maintenance, dated 3/11/1928.
63. SA, 497/13/29: Divisional Engineer to Sub Divisional Engineer, dated 8/2/1937.
64. SA, 497/12/12: Letter to Griffin, dated 17/10/1928.
65. SA, 500/15/31: Abu Ushar Handing-over Note.
66. SA, 500/17/19: Handing-over Note Wad el Nau Sub-Division, October 1934.
67. SA, 500/17/18–19: Handing-over Note Wad el Nau.
68. SA, 500/19/22: Handing-over Note Torr to Smith, June 1942.
69. SA, 497/12/29: Director of Irrigation, Note on future canal layout in the Gezira, dated 1/3/1944.
70. SA, 497/13/21: Divisional Engineer Wad Medani aan Sub Divisional Engineer Messellemia, dated 8/2/1937.
71. SA, 497/13/27: Director of Irrigation to Divisional Engineers Wad Medani and Abu Usher, dated 28/3/1939.
72. SA, 497/13/50: Director of Irrigation to Divisional Engineers and Resident Engineer, dated 23/4/1939.
73. Remember the insistence of MacIntyre in chapter 2 on defining positions of pipes (FOPs).
74. SA, 497/13/33: Divisional Engineer to Assistant Divisional Engineer and Sub-Divisional Engineers, August 1928.
75. SA, 497/13/34: Divisional Engineer.
76. SA, 501/2/19–30: SID Annual Report for 1945, undated.
77. SA, 501/4/4–18: SID Annual Report for 1948, undated.
78. SA 501/4/35: Sudan Irrigation Department Annual Report 1950–1951.
79. SA, 495/2/4: R. J. Smith SID Memorandum No. 1, dated 23/12/1950.
80. SA, 501/4/35: SID Annual Report 1950–1951.
81. J. M. Hodge (2007) *Triumph of the expert: Agrarian doctrines of development and the legacies of British colonialism* (Athens: Ohio University Press); J. M. Hodge, G. Hödl, and M. Kopf (eds) (2014) *Developing Africa: Concepts and practices in twentieth-century colonialism* (Manchester: Manchester University Press).
82. SA, 499/13/46–47: Note from Chick, Appointment of a Development Committee of the Executive Council and of a Commissioner for Development, dated 7/1/1951.
83. SA, 499/14/2–8: Sudan Government, Five Year Plan for Post-War Development, undated.
84. SA, 499/14/31–34: Chick, Finance Circular Letter 30, dated 10/7/1950.
85. SA, 499/14/2–8: Five Year Plan.
86. SA, 501/3/22–38:SID Annual Report for 1947, undated.
87. To many in Gezira, tenants coming with political demands was a surprise, but as we will discover in the next chapter, Gezira tenants were never the quite, obliging characters colonial officials wanted them to be.
88. SA, 501/2/34–45: SID Annual Report for 1946, dated 4/5/1947.
89. SA, 495/2/4: R. J. Smith SID Memorandum No. 1, dated 23/12/1950.
90. SA, 501/2/5–14: SID Annual Report for 1944, dated 6/4/1945.

91. SA, 593/4/1–18: Annual Report of the Ministry of Irrigation & H.E.P. 1955–1956, undated.
92. SA, 593/5/1–21: Annual Report of the Ministry of Irrigation & H.E.P. 1956–1957, undated.
93. SA, 993/7/141–143: Letter from Carmichael to Serpell, dated 24/9/1953.
94. SA, 594/4/9–12: Note from Allan to Director of Irrigation, dated 29/9/1938.
95. SA, 416/5/29: SPS Board Minutes meeting 262, 28/7/1938. These larger extensions in the South-West are the ones that were included in the postwar Development Program.
96. SA, 499/14/10–21: W. N. Allan, Note on the first stages of planning and development of irrigation in the postwar period, not dated, perhaps, July 1944.
97. SA, 499/16/4–30: Note on Managil Branch Development, dated 8/4/1946.
98. SA, 499/14/23–24: R. J. Smith, SID. Note "A" The location of the Gezira Extension, dated 6/3/1947.
99. SA, 719/7/20–31, Extension to Gezira Scheme, Legal Department, dated 29/11/1949. The expropriation of the land was similar to the Abdel Magid area under the Land Acquisition Ordinance of 1930 (first version from 1903). More details on the Gezira lease model is provided in chapter 4.
100. SA, 499/15/2–18: Report on the North-West extension to the Gezira Scheme & the enlargement of the Gezira Main Canal, dated December 1952.
101. SA, 499/14/37: Progress Report 1946/51 Development Programme, Scheme No. 4/4 Extension to Gezira Canalisation, dated 19/1/1952; SA, 499/15/2–18: Report on the North-West extension.
102. SA, 499/15/2–18: Report on the North-West extension.
103. SA, 594/4/25–31: Letter from Director of Irrigation, dated 9/8/1941; SA, 499/15/2–18: Report on the North-West extension.
104. SA, 517/10/18–48: Report on the Future of the Gezira Scheme, dated 27/7/1946, SA, 517/10/1–17: Financial Secretary's Statement on Report Gezira Special Committee of 1944, undated.
105. SA, 499/14/22: Letter of R. J. Smith to Gezira Advisory Board, dated 6/3/1947, accompanying letter with Notes "A" and "B" on Gezira Extensions with same date.
106. SA, 499/14/25–26: R. J. Smith, SID. Note "B" The organisation of the extension area, dated 6/3/1947
107. SA, 499/15/2–18: Report on the North-West Extension.
108. SA, 501/4/22–38: SID Annual Report for 1950–1951, undated.
109. SA, 501/4/22–38: SID Annual Report; SA, 501/5/1–34: Annual Report of the Ministry of Irrigation & H.E.P 1954–55, undated. Cutting water would increase water use efficiency if measured by the crop yield per cubic meter of stored water. If efficiency was measured by the crop yield per feddan, it would not pay to stop watering early. Compare with the distinction between the Dutch and British colonial irrigation strategies in Asia (M. W. Ertsen [2007]. "The Development of Irrigation Design Schools or How History Structures Human Action," *Irrigation and Drainage*, 56, 1–19; Ertsen [2010]).

110. SA, 593/1/1-28: SID Annual Report 1952-1953.
111. SA, 593/1/1-28: SID Annual Report 1953-1954.
112. SA, 501/5/1-34: Annual Report of the Ministry of Irrigation & H.E.P 1954-55, undated.
113. SA, 499/14/42: Letter from R. J. Smith to Financial Secretary on Future of Development, dated 13/4/1950.
114. SA, 499/14/47-49: R. J. Smith to Commissioner for Development, Development of Gezira, dated 5/2/1953.
115. SA, 499/14/47-49: Smith to Commissioner for Development.
116. SA, 594/4/33-37: The Managil Extension, dated 5/9/1955.
117. SA, 495/2/79: A Note of Sudanisation in the Sudan Irrigation Department, dated 7/3/1953.
118. SA, 501/3/22-38: SID Annual Report for 1947, undated; SA, 501/4/22-38: SID Annual Report for 1950-1951, undated.
119. SA, 495/6/48: L. J. Dunn, Implementation of the Mills Scaling System, dated 26/11/1952.
120. SA, 501/3/22-38: SID Annual Report for 1947, undated.
121. SA, 499/14/42: Letter from R. J. Smith to Financial Secretary on Future of Development, dated 13/4/1950.
122. SA, 499/14/47-49: R. J. Smith to Commissioner for Development, Development of Gezira, dated 5/2/1953.
123. SA, 495/2/79: A Note of Sudanisation in the Sudan Irrigation Department, dated 7/3/1953.
124. Sudan Archive, 593/2/1-40: Annual Report of the Irrigation Department 1953-1954.
125. SA, 752/12/21: Diary Ms Johnson, dated 10/1/1952.
126. SA, 495/2/72: R. J. Smith, "Mills" for SID Engineers, dated 8/3/1953.
127. SA, 495/2/78: A Note of Sudanisation in the SID. The importance of the Resident Engineer would only increase once Roseires Dam came into service—a plan discussed in chapter 7.
128. SA, 495/2/38: R. J. Smith Reorganisation of the Sudan Irrigation Department, dated 10/3/1950.
129. SA, 495/2/47: R. J. Smith, Organisation of Sudan Irrigation Department, dated 28/3/1952.
130. SA, 495/2/57: L. J. D., Forthcoming Promotions, dated 10/11/1952.
131. SA, 495/2/70-71: R. J. Smith, Organisation of Sudan Irrigation Department, dated 17/3/1953.
132. SA, 501/5/1-34: Annual Report of the Ministry of Irrigation & H.E.P. 1954-1955, undated.
133. SA, 501/5/1-34: Annual Report Irrigation & H.E.P. 1954-1955, undated. "The field of selection is limited to those who can speak either English or Arabic, and the problem is not made easier by the present worldwide shortage of experiences Irrigation Engineers. The immediate result of the present situation is that work on development has very nearly been brought to a standstill. Even routine maintenance will be something of a problem."
134. SA, 593/4/1-18: Annual Report of the Ministry of Irrigation & H.E.P. 1955-1956, undated.
135. SA, 593/6/1-22: Annual Report of the Ministry of Irrigation & H.E.P. 1957-1958, undated.

136. SA, 593/5/1-21: Annual Report of the Ministry of Irrigation & H.E.P. 1956-1957, undated.
137. SA, 499/15/2-18: Report on the North-West extension.
138. SA, 499/14/31-34: Chick, Finance Circular Letter 30, dated 10/7/1950. This was a continuous worry: "What however, is serious is the fact that the boom following the record cotton crop of 1950/51 and the general advance of civilisation and of technical development have so far outstripped the capacity of the present Government and the administrative organisations generally of the Sudan to deal with them." (NA, FO371/102759: Letter from Riches to Allen, dated 26/8/1953).
139. SA, 593/1/1-28: SID Annual Report 1952-1953.
140. SA, 499/14/42: Smith to Financial Secretary.
141. SA, 495/2/26: R. J. Smith SID Memorandum No. 1, dated 23/12/1950.
142. SA, 501/4/22-38: SID Annual Report for 1950-1951, undated.
143. SA, 993/7/141-143: Letter from Carmichael to Serpell, dated 24/9/1953.
144. SA, 495/2/26: SID Memorandum No. 1.

4 Making the Best of a Rotten Deal: Tenant Realities and Resistance

1. Sudan Archive, 408/2/16: C. E. F. Morgan—Some notes on practical agricultural in the Gezira Irrigation Scheme. "Tenants like to think that their Inspector is interested in them and their work and if he gets to know them by name, their relationship with other Tenants and their circumstances, he will find that they are more responsive than if they are just so many 'folio numbers' to him."
2. SA, 408/2/16: Morgan.
3. SA, 408/2/16: Morgan.
4. W. Beinart and L. Hughes (2007) *Environment and empire* (Oxford: Oxford University Press).
5. J. M. Scott (1985) *Weapons of the weak: Everyday forms of peasant resistance* (New Haven, CT: Yale University Press). See J. A. Briggs (1978) "Farmers' Responses to Planned Agricultural Development in the Sudan," *Transactions of the Institute of British Geographers, New Series*, 3, 464–475; S. H. Lees (1986) "Coping with Bureaucracy. Survival Strategies in Irrigated agriculture," *American Anthropologist*, 88, 610–622; and A. Memmi (1991) *The colonizer and the colonized* (Boston, MA: Beacon Press). See also R. Oldenziel and M. Hard (2013) *Consumers, tinkerers, rebels: The people who shaped Europe* (Basingstoke: Palgrave Macmillan) for a more general discussion on the topic of everyday resistance through engagement with technologies. Even though attention to farmers' ideas increased in the 1970s (like D. W. Adams and E. W. Coward (1971) *Small farmer Development strategies. A seminar report* (Columbus, Ohio State University), more broadly shared focus on farmers' initiatives are only relatively recent, for example W. M. Adams and D. M. Anderson (1988) "Irrigation before Development: Indigenous and Induced Change in Agricultural Water Management in East Africa," *African Affairs*, 87, 519–535; C. Reij and A. Waters-Bayer (2001) *Farmer innovation in Africa: A source of inspiration for agricultural development*

(London: Earthscan; M. Widgren and J. E. G. Sutton (eds) (2004) *Islands of intensive agriculture in Eastern Africa: Past & present* (London: British Institute in Eastern Africa; Athens: Ohio University Press).
6. For example, Barnett in the 1970s, who presents extremely useful field research on resource use and accessibility of tenants, but explains everything within the Neo-Marxist model of (under) development—see the Epilogue. T. Barnett (1977) *The Gezira Scheme: An illusion of development* (London: Frank Cass).
7. SA, 418/3/10: Letter from Bill to Manager SPS, dated 22/3/1928.
8. SA, 415/8/152: "Cotton from a Wilderness," The Daily Mail, dated 14/1/1925.
9. A. I. Clarkson (2005) *Courts, councils and citizenship: Political culture in the Gezira scheme in condominium Sudan* (PhD Thesis, Durham University), 13.
10. SA, 112/10/1: Minutes of Evidence Board of Trade Committee Growth of Cotton in the British Empire, dated 1/8/1917.
11. Clarkson (2005; 13).
12. SA, 112/3/1: Memorandum. The text includes the lovely quote that "the statistics are only approximate and are not the result of a detailed census," which does not entirely work with the numbers themselves.
13. SA, 112/10/1: Minutes of Evidence.
14. SA, 112/3/12: Native ownership in the 500,000 feddan area, Gezira Canal Scheme, dated 8/2/1913.
15. National Archives, FO141/578/1: Draft report to the Council of the British Cotton Growing Association on the possibilities of Cotton growing in the Anglo-Egyptian Soudan (undated, probably early 1912).
16. SA, 415/8/152: "Cotton from a Wilderness".
17. SA, 408/1/5: The Gezira Scheme, Reference Division Central Office of Information, dated 21/6/1950.
18. SA, 408/1/5: The Gezira Scheme.
19. See C. Lund (2013) "The Past and Space. On Arguments in African Land Control," *Africa*, 83, 14–35.
20. NA, FO14/579: Note of Financial Secretary on Gezira Policy 15/11/1929.
21. SA, 415/6/20: Letter from Sudan Government London Office to Eckstein, dated 26/11/1926.
22. NA, FO141/579: Osborne to Scott, dated 9/10/1923.
23. NA, FO14/579: Note on Gezira Policy, dated 15/11/1929.
24. The total number of Syndicate employees in Gezira (including support staff) came close to the total number of government officials in Sudan as a whole—about 300. This was recognized as a problem when the agreement between the Sudan Government and SPS was signed in 1919. "It is understood that in addition to the higher officials of the Syndicate who will probably be British for the most part the Syndicate will employ a considerable number of officials in direct contact with the natives who would number among them British and others of a lower grade of education. In the case of disturbance or serious difficulties arising between such an official and the tenants or natives the Government will have the right to make representations to the Syndicate and ultimately to insist on the removal of the official at least to a post where he would not be in contact with the natives."

SA, 415/3/18–20: Letter accompanying the Agreement between SPS and Sudan Government, dated 17/10/1919. See also SA: Notes on the Gezira Irrigation Project, January 1926.
25. NA, FO141/579: Governor Blue Nile to SPS, dated 20/1/1923.
26. NA, FO141/579: Governor Blue Nile to Director of Intelligence, dated 6/5/1923.
27. Clarkson (2005).
28. Clarkson (2005).
29. NA, FO141/579: Governor Blue Nile to Director of Intelligence, dated 6/5/1923.
30. NA, FO141/579: Osborne to Scott.
31. NA, FO141/579: Howell to First Secretary Cairo 8/11/1923. Perhaps Irrigation Adviser MacGregor was asked about his experience, as he had come from the Punjab Irrigation Service to Sudan in 1922. See Ali I. (1988) *The Punjab under imperialism 1885–1947* (Princeton, NJ: Princeton University Press), for the Punjab Canal Colonies.
32. NA, FO141/579: Stack to Allenby, dated 18/3/1924.
33. NA, FO141/579: Stack to Allenby.
34. SA, 415/10/59–62: Ordinary General Meeting SPS, dated 23/11/1921.
35. SA, 112/10/1: Minutes of Evidence.
36. SA, 417/1/56: 21st Ordinary General Meeting SPS, dated 14/11/1928.
37. SA, 418/3/3: Letter from inspector to Manager SPS, dated 14/3/1928.
38. SA, 418/3/3: Letter from inspector. "Even if not working personally the tenants would be a valuable type if he could be running his land as an interested & efficient master of labour. But with rare exceptions this is not the case. Absenteeism is far too general. [...] In many tenancies some wild West African is left in charge, incapable of understanding Inspectors advice or orders."
39. SA, 418/3/3: Letter from Inspector.
40. SA, 418/3/3, Letter from Inspector.
41. SA, 418/3/10: Bill to Manager.
42. SA, 418/3/10: Letter from Bill.
43. SA, 418/3/17. Letter from Poynz-Wright to Gaitskell, dated 25/3/1928. Emphasis in original. Is the remark "to say nothing of the questions which were asked in the House by the Labour members!" a first hint to Gaitskell's sympathy for Labor (or the position of his brother in the Labor Party)?
44. Clarkson (2005).
45. Clarkson (2005).
46. SA, 499/20/18: Bredin, The White Nile Alternative Livelihood Schemes, dated 15/3/1943.
47. SA, 499/20/14: Robertson, Definition of the duties of management in the Adbel Magid Area.
48. SA, 499/20/14: Robertson.
49. SA, 497/12/35: Dunn, Experimental Compact Holdings, dated 7/2/1952.
50. SA, 499/20/16: White Nile Alternative Livelihood Schemes, Conditions of Tenancy.
51. SA, 499/20/15: Robertson.
52. SA, 418/3/54–62: Handwritten Memorandum on the use of Native agents to assist in the administration of the Gezira Scheme, undated.

53. SA, 418/5/109–126: The future administration of the Gezira Scheme, Governor's Office, dated 1/2/1948; SA, 498/10/24–27: Note on Village Councils, undated, probably post World War II, as it mentions the "Wartime Samads Scheme" (498/10/26); nevertheless, the document will be representative of the instructions based on Schedule X. See Gaitskell A. (1959) *Gezira: A story of development in the Sudan* (London: Faber and Faber).
54. SA, 498/10/29: Governor Gezira Province, Notes on the Hosh experiment, dated 8/7/1941; SA, 418/5/109-126: The Future Administration. A quite similar note is dated 10/1/1948 entitled Gezira Scheme: Memorandum on some limitations at present affecting development (SA 418/7/1–21). The Governor basically confessed he copied much from Gaitskell: "You will see that I have quoted yours in extension to which Gaitskell answers "I am particularly delighted that you should have found so much common ground with me" (SA, respectively 418/5/108: Letter from Bredin (Governor Blue Nile Province) to Gaitskell, dated 28/1/1948; and 418/5/107 Letter from Gaitskell to Bredin; dated 3/2/1948). See also Clarkson (2005) and Willis J. (2011) "Tribal Gatherings: Colonial Spectacle, Native Administration and Local Government in Condominium Sudan," *Past & Present*, 211, 243–268.
55. Clarkson (2005).
56. The discussions do not always explain the difference...
57. SA, 498/10/28: Governor Gezira Province, Note on devolution in the Gezira Scheme. For information of tenants and local native authorities, dated 8/7/1941.
58. SA, 498/10/28: Note on devolution.
59. SA, 408/2/66: Morgan.
60. Clarkson (2005).
61. SA, 498/10/28: Note on Devolution.
62. SA, 498/10/29: Notes on the Hosh Experiment.
63. A misconception known from development corporation as well.
64. SA, 408/1/40: Letter from Robertson, Civil Secretary to all Governors and Head of Departments, Strictly Confidential, dated 1/1/1946.
65. SA, 408/1/43: Letter from Civil Secretary Robertson, dated 21/11/1946.
66. SA, 408/2/12: A broadcast on the Gezira Scheme from Omdurman Station, dated 18/1/1943.
67. SA, 498/10/2: SPS—Gezira Irrigation Scheme, dated 31/12/1938; there is one for the KCC as well.
68. SA, 418/3/94: Letter from John to Billy with Response of MacIntyre to Radio Message of 18/1/1943.
69. Clarkson (2005).
70. SA, 408/2/12: A broadcast on the Gezira Scheme from Omdurman Station, dated 18/1/1943; Gaitskell is not mentioned as the speaker, but the use of the term "corporate socialism" and SA, 418/3/94: Letter from John [Gaitskell] to Billy with response of MacIntyre to radio message of 18/1/1943 of Gaitskell, suggest he is.
71. SA, 408/2/11: A broadcast on the Gezira Scheme.
72. SA, 498/10/24: Note on Village Councils, not dated. As if nomadism is something easy or less valid. However, civilizations (or those that claim

they are) have often regarded nomadism as less worthy. See P. Heather (2010) *Empires and barbarians: The fall of Rome and the birth of Europe* (Oxford: Oxford University Press); J. M. Scott (2009) *The art of not being governed: An anarchist history of upland Southeast Asia* (New Haven, CT: Yale University Press); S. Stuurman (2009) *De uitvinding van de mensheid: Korte wereldgeschiedenis van het denken over gelijkheid en cultuurverschil* (Amsterdam: Bert Bakker).

73. SA, 498/10/24: Note on Village Councils, not dated.
74. Clarkson (2005).
75. SA, 498/8/10: Chinn—Gezira Irrigation, dated 15/2/1939. Irrigation design was already adapted: "In the old area large blocks of land served by several canals were placed in the Syndicates Inspector's direct control. In the new area this is never the case." Village based irrigation was also a good reason to ensure SID involvement, as the village model "shows the impossibility of a series of native run concessions, one striving against the other, unless the Irrigation department assumes the directive authority when such was required."
76. SA, 498/8/10: Gezira Irrigation.
77. SA, 498/3/10: Griffin—Water consumption and peak demand in the Gezira related to the crop area, dated 19/12/1936.
78. SA, 498/15/36: Letter from Griffin to Manager SPS, dated 3/3/1944. Interestingly enough, the SPS was represented by Gaitskell (!) who did not accept the SID position based on the Abdel Hakim experiment. It could "not be quoted as a precedent as no financial questions had been involved in practice" and "the main idea of the experiment was to educate." Mr. Gezira continued to combine training with business.
79. SA, 498/8/10: Gezira Irrigation.
80. Clarkson (2005).
81. Sudan Gezira Board (1967) *The Sudan Gezira Board: What it is and how it works*.
82. Clarkson (2005). The increased focus on social development in Gezira brought up questions of who should be the beneficiaries of these developments. Should all persons living in the Gezira profit or just the tenants—let alone those outside the scheme? I will return to this debate in chapter 6 and the Epilogue.
83. SA, 693/1/135: Hunt Diary, entry on 30/10/1945.
84. SA, 693/1/136: Hunt Diary, entry on 9/12/1945.
85. SA, 693/1/138: Hunt Diary, entry on 31/3/1946.
86. The original passage in SA, 418/5/109–126: The Future Administration of the Gezira Scheme, Governor's Office, dated 1/2/1948, reads "our staff;" the Note seems to be the same as in the note by Gaitskell dated 10/1/1948 entitled Gezira Scheme: Memorandum on some limitations at present affecting development (SA 418/7/1–21). One would expect, however, the "our staff" would be different for a governor or for Gaitskell...
87. The full list reads: Agriculture: Approach from the tenants viewpoint, as a full farmer; Experiments in mechanization; Experiments in planned labour supply; Experiments in animal husbandry, especially in fodder production, local concentrates; Experiments in stock breeding, folding, and protection; Experiments in poultry breeding; Experiments in fruit and vegetable growing; Experiments in fully satisfying village farming, rotations. Local

industries: Survey of local needs and encouragement of local handicrafts to meet them. Cooperation: Cooperative loan societies; Cooperative marketing societies. Seasonal and permanent labour force: Needs in housing, recreation, education, and technical training. Social amenities in villages: Housing; [On this] Research in cooperation with P.W.D. and perhaps with the Design Section of the Gordon College, & with Omdurman Technical School; Water supplies; Firewood ; [On this] Experiment in cooperation with Forestry Service. Social activities in villages: Educational tours & talks & Tours by research & other staff; Adult education; Cinemas; Travelling Book Clubs; Games; Youth Clubs; Farm Schools; Education in general; [On this] in cooperation with the Education Department, Research Institute, Gordon Memorial College, & other departments as required. Health: Review of health and; Application of remedial measures; [on this] in cooperation with the Medical Department. Publicity: Adequate presentation of all aspects of the needs and progress of the Gezira Scheme. SA, 418/5/109–126; Future Administration of the Gezira Scheme.
88. SA, 418/5/109–126: Future Administration of the Gezira Scheme.
89. SA, 498/10/33, A Directive on the Policy of Devolution in the Gezira, issued jointly by Governor Blue Nile Province and Manager Sudan Plantations Syndicate, dated 3/5/1947.
90. SA, 418/5/109–126: Future administration of the Gezira Scheme.
91. SA, 498/10/25, Note on Village Councils.
92. SA, 498/10/27, Note on Village Councils.
93. J. Robertson (1974) *Transition in Africa: From direct rule to independence*, London, Hurst & Company.
94. SA, 418/3/138: Draft Letter to the Editor Sudan Star, dated 1/3/1943.
95. SA, 418/3/138: Draft Letter to the Editor.
96. SA, 418/5/13: A consideration of the general political trend of the Gezira Scheme and its effect on our future interest here.
97. SA, 418/3/146: Letter from Griffiths to Gaitskell, dated 12/4/1943.
98. SA, 418/3/145: Letter from Gaitskell to Griffiths, dated 11/4/1943.
99. SA, 418/3/148: Letter from Billy to Gaitskell.
100. SA, 527/15/120–124: Letter from Bacon to Robertson, dated 8/3/1949.
101. SA, 408/2/66: Morgan.
102. SA, 693/1/127: Hunt Diary, entry on 8/4/1945. Clarkson (2005) allows the Syndicate a far too supportive role in developing Governmental policies. True, some inspectors were more helpful than others, but I would suggest that the Board of Directors—despite Gaitskell's influence—fully embraced the business model as most strongly defended by MacIntyre and Archdale. The Syndicate voluntary involvement in Gezira's social development was minimal. As much as the tenants cannot be described as a unity, I would suggest that Syndicate (and government for that matter) should not be treated as unity either, an idea I will elaborate upon in chapter 5 for the inspectors.
103. SA, 418/6/37–39: Letter from MacIntyre to Asquith, dated 15/08/1938.
104. SA, Gezira Scheme. A hand Book for new Personnel, undated, but from Sudan Gezira Board.
105. SA, 428/3/109: Culwick Diaries, entry on 14/12/1951.
106. SA, 428/3/130: Culwick Diaries, entry on 16/11/1952.
107. SA, 593/1/21: SID Annual Report 1952–1953.

108. SA, 498/2/27: R. J. Smith, The provision of canal foot-bridges for the Gezira Scheme, dated 17/12/1951.
109. SA, 408/1/33: Gezira Information Service, Report of a Press Conference, dated 7/2/1951. Although not for social development, the spoon-feeding theme had been used in the 1920s by Financial Secretary Schuster to characterize the SPS approach to train the tenants. How people initiative and pre-selected policies by the Board were related to each other was a debate ignored, but then, the Contested Development book had not been written yet (P. Barron, R. Diprose and M. Woolcock (2011) *Contesting development: Participatory projects and local conflict dynamics in Indonesia* [New Haven, CT: Yale University Press]).
110. SA, 415/8/19–21: Ordinary General Meeting SPS, dated 7/10/1910.
111. SA, 109/5/1: MacDonald A note on the Blue Nile Weir & the Gezira Scheme, dated 12/2/1915.
112. SA, 112/5/2: Wingate Note as to the effect of the war on raising of the Capital required for the Gezira Irrigation Scheme, dated 14/8/1916.
113. SA, 416/2/61–62: Ordinary General Meeting SPS, dated 28/10/1925. This comparison between different social groups is a continuous thread in the Sudanese colonial debate—and the African colonial debate in general, as I will argue in the Epilogue.
114. SA, 593/12/1: I. G. Simpson—Economic aspects of diversification in the Gezira, in: Agricultural development in the Sudan, dated 1965. For Simpson, this was good reason to start planning it: "There seems a case for concentrating additional vegetable production in those areas which are well placed to exploit the urban markets."
115. SA, 428/3/96: Diary Culwick, dated 12/3/1950. Some British staff had private boats, which were sometimes used for a larger picnic trip. British people met in the Clubs—discussed in the next chapter.
116. SA, 408/2/16: Morgan.
117. SA, 408/2/16: Morgan.
118. SA, 693/1/1: Hunt Diary, entry on 12/10/1934.
119. SA, 693/1/17: Hunt Diary, entry on 3/11/1938.
120. SA, 693/1/2: Hunt Diary, entry July 1936.
121. SA, 408/2/16: Morgan.
122. "A late lunch at the Grand, where we were joined by R McI (an officer of a Highland Regt—now with the Western Arab Corps of the S.D.F.), who talked about his job. He was astonished that we did not hit the nas. The troops enforce discipline often with corporal punishment at the hands of the Shawish (Sergeant). And very satisfactory it is too, if it is not abused, he said." (SA, 693/1/13: Hunt Diary, entry on 10/7/1938).
123. SA, 428/3/31: Diary Culwick, dated 15/5/1949. SA, 428/5/1; G. M. Culwick—Diet in the Gezira Irrigated Area, Sudan.
124. SA, 428/3/22: Diary Culwick, dated 26/4/1949.
125. SA, 428/3/74: Diary Culwick, dated 28/11/1949.
126. SA, 428/3/44: Diary Culwick, dated 30/6/1949.
127. SA, 428/3/22: Diary Culwick, dated 26/4/1949.
128. SA, 428/3/6: Diary Culwick, dated 4/7/1949.
129. SA, 428/3/33: Diary Culwick, dated 21/5/1949. Some of the "older S.P.S. men admitted that they had predicted complete failure for me, and had sat

back to watch me defeated by the non-cooperation and obstruction they expected."
130. SA, 428/3/37: Diary Culwick, dated 5/6/1949; emphasis in original. The women decided that Culwick was "going to marry an Arab husband."
131. SA, 428/3/22: Diary Culwick, dated 26/4/1949.
132. SA, 428/1/1: G. M. Culwick—A study of the human factor in the Gezira Scheme.
133. SA, 428/3/22: Diary Culwick, dated 26/4/1949. Although at one social event Culwick "saw dark shadows lurking on the outskirts." Weapons of the weak...See also SA, 428/1/1; The human factor.
134. SA, 428/1/1: The human factor.
135. SA, 418/3/18: Comments on the Gezira Scheme written from Abdel Hakim. This observer emphasized that it was more a case of "everyone for himself & the sons on their own."
136. SA, 428/1/1: The human factor. See P. F. M. McLoughlin (1962) "Economic Development and the Heritage of Slavery in the Sudan Republic," *Africa: Journal of the International African Institute*, 32, 355–391.
137. Clarkson (2005). See also P. Cross (1997) "British Attitudes to Sudanese labour. The Foreign Office Records as Sources for Social History," *British Journal of Middle Eastern Studies*, 24, 217–260; J. Spaulding (1988) "The Business of Slavery in the Central Anglo-Egyptian Sudan, 1910–1930," *African Economic History*, No. 17, 23–44.
138. SA, 428/3/93: Diary Culwick, dated 21/2/1950.
139. SA, 428/3/93: Diary Culwick, dated 21/2/1950.
140. Clarkson (2005).
141. SA, 408/2/16: Morgan. See J. Tait (1979) "Interner Kolonialismus und ethnisch-soziale Segregation im Sudan. Nigerianisch-Westafrikanische Arbeitsmigranten und das Arbeitsmarktsystem in der Gezira," *Africa Spectrum*, 14, 361–382.
142. Clarkson (2005). Sounds somehow quite a modern idea...
143. Much of the evidence for this is from the 1970s, especially from Barnett (1977), but there is no reason to assume similar differences were not present earlier. Culwick and the many discussions among the British suggest this.
144. Clarkson (2005).
145. SA, 428/1/1: The Human Factor.
146. "The Arabic word translated as 'slow, measured rhythmic tones' is *tarteel*." (Wikipedia).
147. SA, 418/6/37–39: MacIntyre to Asquith.

5 Another's Week's Toil: British SPS Inspectors and Their Idea(l)s

1. Sudan Archive, 693/1/84: Diary F. B. Hunt, entry on 9/3/1942.
2. Hunt's diary suggests that his main business was traveling around; he seems to have fired a gun only twice.
3. SA, 693/1/127: Hunt Diary, entry on 8/4/1945.
4. SA, 693/1/137: Hunt Diary, entry on 15/3/1946.
5. SA, 693/1/6: Hunt Diary; Hunt suggests that over a third of SPS staff came from Scotland. Many British in the Indian Colonial Service were Scottish too. In her diaries of the late 1940s, Ms. Johnson (wife of SPS inspector

Johnson) describes how a new staff member was recruited in the 1940s: "With my usual curiosity I asked him how he'd got the job & he described his interview with Wooding. Wooding 'Can you ride a horse?' Barrett 'Yes, Sir.' Wooding 'Can you drive a car?' Barrett 'Yes, Sir.' Wooding 'Have you got a dinner jacket?' B. 'Yes, Sir.' Wooding 'All right you're hired, good day'! & that was all!". SA, 751/10/13: Diary Ms. Johnson, 1946 August 12–September 22.

6. SA, 415/10/28: Twelfth Ordinary General Meeting SPS on 17/12/1919; see also SA, 416/1/58: Eighth Ordinary General Meeting SPS on 21/12/1915.
7. SA, 408/1/24: Gezira Information Service, Report of a Press Conference, dated 7/2/1951.
8. SA, 693/1/13: Diary Hunt, added comments.
9. The Sudan Gezira Board British staff register: incorporating the defunct Sudan Plantations Syndicate & Kassala Cotton Company. Sudan Gezira Board. Sudan Plantations Syndicate; Kassala Cotton Company. 1953.
10. SGB British staff register.
11. SA, 889/5/5: Letter from H. Huntington-Whitley, 26/5/1922; see also SA, 889/5/4: Letter from A. S. Ramsay, 20/5/1922.
12. SA, 889/10/1–7: Draft Manuscript on Sudan and SPS by Gibbons.
13. SA, 889/10/1–7: Draft Manuscript Gibbons.
14. SA, 403/12/2: Letter of employment E. A. P. Taylor, dated 17/11/1926. The diary of F. B. Hunt confirms that 400 EP was the starting salary, although by the time Hunt joined the SPS in 1934, his salary was reduced to 350 EP because of a salary reduction in the crisis of the 1930s (SA 693/1/7: Hunt Diary).
15. SA, 403/12/2: Letter of Employment.
16. SA, 849/12/13: Health Notes for Syndicate Inspectors, dated 6/1/1932.
17. SA, 403/12/2: Letter of Employment.
18. SA, 531/2/100: Letter Robertson 6/8/1923. Robertson became Civil Secretary in Sudan (1945–1953).
19. SA, 403/12/2: Letter of Employment.
20. SA, 880/12/47. Letter from Crompton to Edie dated 17/7/1905.
21. SA, 693/1/5: Hunt Diary, added comments.
22. SA, 417/1/59: 21st Ordinary General Meeting SPS, dated 14/11/1928.
23. SA, 889/10/1–7: Draft Manuscript Gibbons.
24. Culwick informs us that nothing would grow, but images of Gibbons' garden show a pretty sumptuous garden.
25. SA, 693/1/5: Hunt Diary, added comments. A suffrage would typically earn 2.50 EP per month, a cook 3 EP, a car boy and gardener about 2 EP, the bucket man 1.50 EP per month. The syce would earn 2 to 3 EP.
26. SA, 428/3/110: Culwick Diary, entry on 14/12/1951. The SGB did not provide those either. Culwick had five servants at the time. "To my mind, an extraordinary arrangement at the best of times—house the white men and the horses, but the black man can wait. (We used to think ourselves poorly housed often enough in Govt. Service in E. Africa, but at least our servants were included in the party when Govt. built the quarters)."
27. SA, 428/3/22: Culwick Diary, entry on 26/4/1949.
28. SA, 720/4: S. R. Simpson (1996) *Sudan Service*.
29. SA, 428/3/22: Culwick Diary, entry on 26/4/1949.

30. SA, 428/3/22: Culwick Diary, entry on 2/5/1949. "[Elise] got us well and truly lost in the chequerboard of the Gezira canals, until [someone] pointed out that the Southern Cross was now behind us whereas it ought to have been in front!"
31. SA, 693/1/13: Hunt Diary, entry on 11/7/1938.
32. SA, 693/1/8: Hunt Diary, entry on 17/3/1938.
33. "In this country no parcels are delivered. When one arrives at the local post office a chit is sent to the recipient. This chit has to be signed by the recipient who then has to fetch the parcel or have it fetched. [...] when the chit remained unanswered the Postmaster should have taken steps to find out why as one has to pay on a parcel left unclaimed for a number of weeks unless for some special reason." (SA, 895/7/30: Persis Aglen (wife of civil servant Edward Francis Aglen), dated 3/3/1943).
34. "It had been a Syndicate tradition right up to the second world war, that all new inspectors carried out their marour on horseback." (SA, 693/1/148: Hunt Diary, added comments).
35. SA, 693/1/8: Hunt Diary, entry on 12/3/1938.
36. SA, 889/10/1–7: Draft Manuscript Gibbons.
37. SA, 693/1/5: Hunt Diary, added comments. The SPS was alone in being hierarchical; the colonial service was pretty top-down too. What matters is how hierarchy was shaped and how inspectors responded.
38. SA, 693/1/7: Hunt Diary, added comments.
39. SA, 418/5/67: A Report on Zeidab May–June 1943, dated 19/7/1943, by Gaitskell.
40. SA, 693/1/7: Hunt Diary, entry on 2/3/1938. "though at the end of the year his general impression may knock an odd 5 or 10 EP of your rise."
41. SA, 693/1/7: Hunt Diary, entry on 9/3/1938.
42. SA, 693/1/7: Hunt Diary, entry on 2/3/1938.
43. SA, 693/1/1: Hunt Diary, entry in April 1940.
44. SA, 693/1/7: Hunt Diary, added comments.
45. SA, 693/1/7: Hunt Diary, added comments.
46. SA, 693/1/1: Hunt Diary, entry on 19/9/1945.
47. SA, 693/1/7: Hunt Diary, added comments.
48. SA, 418/3/7: Letter of inspector Gaitskell to Manager SPS, dated 14/3/1928.
49. SA, 418/3/14: Letter of senior inspector (Bill) to Manager SPS, dated 22/3/1928. Furthermore, the Senior Inspector considered it necessary to have a clear internal structure distinguishing between experienced and new inspectors. Gezira expansion had necessitated employing "a large number of young men" in need of supervision.
50. SA, 418/5/9: Letter from Manager SPS to Chairman and Managing Director SPS, dated 3/9/1932.
51. SA, 418/5/9: Letter Manager SPS.
52. SA, 418/5/10, Letter Manager SPS.
53. The note discussed leave arrangements as well. Although the author was "not a believer in the theory that a man cannot stay twenty months in the country without hurting his health," shorter periods between leaves would have the benefit of keeping everyone "fresher in their work" and giving junior inspectors "more opportunity of management." Needless to say that did not happen either (SA, 418/5/10: Letter Manager SPS, 3/9/1932).

54. SA, 751/9/22–23: Ms. Johnson Diary, August 1946.
55. SA, 752/2/19: Ms. Johnson Diary, August 1946; emphasis in original.
56. SA, 751/9/22–23: Ms. Johnson Diary, August 1946.
57. "This however did not apply to the so-called junior class of British employee, mostly artisans or working as hygiene officers. These never came to the clubs, or mixed socially. Individually they might be known to some, but on the whole they lived separate lives. In Khartoum they were members, not of the Sudan Club, but of the Khartoum Club." (SA, 693/1/9: Hunt Diary, added comments); "The distinction between the Sudan Club for senior officials and the Khartoum Club for junior officials persisted almost up to 1956. Membership of either club was determined by salary except in the case of the Political Service who, of course, and however poor, always belonged to the top club without question." (Kenrick R. [ed] 1987 *Sudan Tales: Recollections of some Sudan Political Service wives 1926–1956*, Cambridge, The Oleander Press, 28.)
58. SA, 720/4: Simpson (1996).
59. SA, 851/6/24: Newspaper Clipping.
60. SA 889/5/8–10: Letter from Gaitskell, 15/7/1950.
61. SA, 720/4: Simpson (1996).
62. SA, 889/10/1–7: Draft Manuscript Gibbons
63. SA, 531/2/100: Letter Robertson 6/8/1923. Even though they would be "better off not to be married" given the need "to rough it a good deal—in the cotton field!."
64. SA, 403/12/2: Letter of Employment.
65. SA, 693/1/131, Diary of Hunt. According to Ms. Johnson, Gaitskell may have spoken out of personal experience: "Stephanie G. [Gaitskell] hates every moment of life here" (SA 751/14/5: Ms. Johnson diary, December 1946). One of the issues may have been that Ms Gaitskell had "[...] an intense dislike of touching any hand that isn't her own colour" (SA 751/8/5: Ms. Johnson diary, December 1946), as she made '[...] no secret of how she dislikes natives as she calls them." (SA 751/16/25: Ms. Johnson diary, March 1947). Culwick confirms that "his [Gaitskell's] wife hates the Sudan and won't come out any more" (SA, 428/3/74: Culwick Diary, entry on 28/11/1949). We will return to British-Sudanese relations in Chapter 6 and in the Epilogue, as Ms Gaitskell was certainly not the only one with apparent issues on such relations.
66. SA 889/5/8–10: Letter from Gaitskell.
67. SA, 889/10/1–7: Draft Manuscript Gibbons.
68. SA, 428/3/25: Culwick Diary, entry on 23/4/1949.
69. SA, 693/1/1: Hunt Diary, entry on 23/2/1936.
70. SA, 889/10/1–7: Draft Manuscript Gibbons.
71. SA, 428/1/40: Culwick Letter to "Folks," entry on 20/5/1947.
72. SA, 428/3/32: Culwick Diary, entry on 13/5/1949.
73. "The mess in the house after such a storm has to be seen to be believed. Even with the windows shut, it gets in at doors and windows." SA, 428/3/32: Culwick Diary, entry on 13/5/1949.
74. Although it depended on the car one drove. Bedfords were "bitches" even in little mud and would not "steer in it at all." Land Rovers "guaranteed to go through anything." SA, 428/3/40: Culwick Diary, entry on 21/6/1949.

75. SA, 428/3/32: Culwick Diary, entry on 13/5/1949; "Hear that a lot of people got caught at the 88 Club last night." (SA, 428/3/10: Culwick Diary, entry on 23/9/1949).
76. SA, 693/1/6: Hunt Diary, added comments
77. SA, 693/1/13: Hunt Diary, entry on 10/7/1938. SPS employees were not that different from other colonial British. On a trip to Egypt on the Mediterranean, civil servant Simpson met British people travelling further east with which "[...] 'social drinking' took on altogether a new dimension." (SA, 720/4: Simpson (1996).
78. SA, 693/1/8: Hunt Diary, entry on 12/3/1938.
79. Someone, probably Mellor himself, made the poster from the text of this song with additional drawings.
80. SA, 693/1/6: Hunt Diary, entry on 27/2/1938.
81. SA, 751/8/3: Ms. Johnson diary, May 1946.
82. SA, 693/1/6: Hunt Diary, entry on 27/2/1938.
83. The position of Sudan in war affairs was a little complex, as Egypt—part of the Condominium—was neutral; whether SPS inspectors should join the forces or stay on the job was unclear. This confusion was solved once Sudan was at war with Italy in June 1940. Just for World War II, a majority of inspectors had ten years or more of SPS service, as there were only a very few post-Depression recruits. (SA, 889/10/1–7: Draft Manuscript Gibbons).
84. SA, 895/6/38: Diary of Aglen. Those who stayed in Gezira had to have trenches in the gardens. "Some people have very grand trenches with seats therein and other amenities—ours is just a trench and no more but so long as it will fulfil its purpose I shall be quite satisfied." Nor did class differences or hierarchy necessarily disappear during the war. "During the War, one Governor of Darfur wired the US Air Force that they could not land at the particular time they requested as he would be playing polo that afternoon on the landing strip." (Kenrick 1987).
85. SA, 501/2/5–14: SID Annual Report for 1944, dated 6/4/1945.
86. SA, 889/9/2: Letter of Gibbons to Civil Secretary, via Manager SPS, Barakat, dated 16/8/1943.
87. SA, 428/3/13: Culwick Diary, entry on 26/4/1949.
88. Some improved after resigning, as apparently the Davidsons in Nairobi: "What a difference in them since they escaped from the S.P.S! They are once again normal people." (SA, 751/17/8: Ms. Johnson diary, March 1947).
89. SA, 751/20/6: Diary Ms. Johnson, October 1947.
90. SA, 895/6/27: Diary of Aglen, July 1940.
91. SA, 751/11/11: Ms. Johnson Diary, October 1946.
92. SA, 428/3/13: Culwick Diary, entry on 2/12/1949; SA, 428/3/77: Culwick Diary, entry on 18/12/1949.
93. SA, 428/3/47: Culwick Diary, entry on 16/7/1949.
94. SA, 428/3/67: Culwick Diary, entry on 1/11/1949.
95. SA, 428/3/13: Culwick Diary, entry on 26/4/1949; "and they are amazed but delighted that I will live in the village, because they think it will mean a lot to the village to have a European really among them like that."
96. SA, 751/19/39, Ms. Johnson Diary, September 1947.
97. SA, 751/11/18, Ms. Johnson Diary, October 1946.
98. SA, 751/8/9, Ms. Johnson Diary, May 1946.
99. SA, 752/8/15, Ms. Johnson Diary, June 1946.

100. SA, 752/8/1, Ms. Johnson Diary, May 1946.
101. SA, 752/8/5, Ms. Johnson Diary, May 1946.
102. SA, 693/1/7: Hunt Diary, entry on February 1942.
103. SA, 889/4/15: Letter from Gaitskell to Gibbons, dated 1/7/1950.
104. SA, 889/4/19: Letter to Gibbons, dated 28/5/1953.
105. SA, 889/4/20: Letter to Gibbons.
106. SA, 889/4/20: Handwritten on Back of Letter to Gibbons, dated 3/6/1953.
107. SA, 889/4/21–24: Letter from Watt to "Dear Roddy" Dew, dated 5/8/1953.
108. SA, 889/4/28: Letter from Roddy to Gibbo, dated 12/8/1953; SA, 889/4/29: Letter from Roddy to Gibbo, dated 26/8/1953.
109. SA, 889/4/30: Letter from General Manager SGB to Gibbons, dated 3/9/1953.
110. SA, handwritten on back of 889/4/24.
111. SA, 889/4/25–27: Terms of Service for British Staff Employed on Fixed Long Contract, dated 8/8/1953.
112. SA, 889/4/50: Letter from Gibbons to Waite, dated 13/4/1955.
113. SA, 889/4/52: Letter from Waite to Gibbons, dated 2/5/1955.
114. SA, 889/4/54: Letter from Gibbons to General Manager, dated 9/5/1955, with reference to 889/4/52 (letter of 2/5/1955); "I am now doubling the jobs of field and Development Advisor."
115. SA, 889/4/59: Letter from Financial Controller to Gibbons, dated 31/5/1955.
116. SA, 889/4/69: Letter from Waite to Gibbons, dated 31/3/1956.
117. SA, 889/4/75: Letter from Abdulla to Gibbons, dated 20/9/1958.
118. SA, 889/9/22: Letter from Gibbons to Governor Blue Nile Province, undated.
119. SA, 889/9/22: Letter from Gibbons to Manager Barclays Bank, Wad Medani, undated.
120. SA, 889/7/30: Letter from Gaitskell to Gibbons, dated 27/4/1959.
121. SA, 889/8/13: Letter to Alison Gibbons dated 1/7/1974.
122. SA, 889/8/14–15. Letter from Leigh to Alison Gibbons, dated 1/7/1974.
123. SA, 889/4/43: Letter from Hood, dated 28/6/1978.
124. SA, 889/8/45: Small Note.
125. There is some confusion because of two different years (1981 and 1982) being mentioned in the archival material. I tend to think that 1981 is the correct date, as it appears more often in the material.
126. SA, 889/8/51: Letter from Lesley, dated 1/7/1980.
127. SA, 889/8/79: Letter from Sara Taffinder, dated 8/7/1981.
128. *The Managil Extension: An achievement of the Republic of Sudan after independence*, Information Centre of the Ministry of Social Affairs, undated.
129. SA, 752/10/5: Diary Ms. Johnson, dated 5/12/1950.
130. National Archives, LAB 8/1890: Minutes meeting 25/3/1955; see also NA, LAB 8/1890: Letter from Waite to Luce, dated 15/2/1955. "When Sir Donald Perrott, K.B.E. (now one of the Atomic Energy big shots) and I were running down the Overseas Food Corporation three years ago, we enlisted the support of your Ministry to help us find alternative employment for redundant personnel in Tanganyika. [...] I have something of a similar situation to handle now in the Sudan but on a very much smaller scale,

and I am writing to ask if you think there would be any chance of finding employment for about 100 British Agricultural Inspectors and Supervisors whose posts will be Sudanised in about six months time. Some would be of use in this Country [UK, letter is sent from London] but in the main the whole of their experience has been in the Sudan engaged on growing cotton and fodder crops in the largest irrigated scheme in the world, and in this sort of work they are acknowledged experts. There is, of course, no responsibility whatever attaching to the Sudan Gezira Board to find jobs for the British staff whose services are being dispensed with on account of Sudanisation, for each member is being generously compensated according to his length of Service." (NA, LAB 8/1890: Letter from Raby to James, dated 21/10/1954).

131. NA, LAB 8/1890: Letter from Raby to James, dated 21/10/1954.
132. NA, LAB 8/1890: Letter from James to Wyer, dated 25/3/1955.
133. NA, LAB 8/1890: Letter from Wyer to James, dated 3/11/1954.

6 Move from the Old Grooves: Gezira Continuity and Change after World War II

1. Sudan Archive, 498/6/49: Irrigating the Sudan. Benefits of a nationalized Gezira Scheme, The Times 3/7/1950.
2. National Archives, FO371/41423: Letter from Huddleston to MacIntyre, dated 25/3/1944.
3. SA, 416/5/72: Meeting SPS 1945.
4. See R. O. Collins (2008) *A history of modern Sudan* (Cambridge: Cambridge University Press); M. W. Daly (1986) *Empire on the Nile: The Anglo-Egyptian Sudan, 1898–1934* (Cambridge: Cambridge University Press); J. Ryle, J. Willis, S. Baldo and J. M. Jok (2011) *The Sudan Handbook* (Woodbridge: James Curry). There was a colonial power at the time criticizing the new policy of the British, Belgium, or rather the Governor-General in the Belgian Congo. The Sudan Civil Secretary visited the Congo in 1947 and reported to his own Governor-General that "[t]heir new palace is to have forty rooms for the Governor-General and another forty to accommodate the Belgian royal family when it visits the Congo. He was rather scornful of British ideas on education filling les negres with dreams of independence etc. long before they were ready for it." (SA, 527/7/10–11: Letter from Robertson to Governor-General, dated 19/8/1947). As we all know, the Belgian Congo does not provide the most enlightened example of colonial government and decolonization: see D. G. Van Reybrouck (2010) *Congo: Een geschiedenis* (Amsterdam: De Bezige Bij) (or (2012) *Congo: Une histoire* (Arles: Editions Actes Sud) or (2014) *Congo: The epic history of a people* (New York: HarperCollins).
5. SA, 418/3/206–208: The Future of the Gezira Scheme. A talk at the Official's Club, Medani, dated 8/12/1944.
6. SA, 416/5/26: SPS Board Minutes meeting 261, dated 28/4/1938.
7. SA, 416/5/27–29: SPS Board Minutes meeting 262, dated 28/7/1938.
8. SA, 416/5/41: SPS Board Minutes meeting 266, dated 10/10/1939.
9. SA, 416/5/47: SPS Board Minutes meeting 268, dated 26/3/1940.

10. SA, 501/6/10–12: Broadcast from Omdurman on Local Government and the Individual by the Assistant Civil Secretary, dated 26/8/1943. Perhaps the ACS was Philip Brown Broadbent.
11. SA, 416/5/66: SPS Ordinary General Meeting 35, dated 5/5/1943.
12. NA, FO371/41423: Letter from MacIntyre to Governor-General Sudan, dated 21/4/1944.
13. NA, FO371/41423: Huddleston to MacIntyre: see also NA, FO371/41423: Telegram from Malisudan aan Rugman, dated 14/5/1944.
14. NA, FO371/41423: Letter from Governor-General Sudan to Lord Killearn, dated 10/6/1944: in the words of Sir Humphrey Appleby (Yes Minister) it would have been "a courageous decision" to continue with the SPS.
15. SA, 416/5/69: SPS Board Minutes meeting 277, dated 13/8/1944.
16. SA, 408/1/1: Letter from SPS to shareholders, dated 24/8/1944.
17. Original quote in SA, SUD.A.PK1569.2SUDAN: Reprint BCGA April 1950.
18. SA, 416/5/72: Report in The Financial Times on SPS Ordinary General Meeting 37 (held at 19/4/1945), 20/4/1945.
19. Which was not exactly what the government intended, as we will see below. Gezira had to stay the money maker…
20. SA, 416/5/72: *The Financial Times.*
21. SA, 408/2/75: Gezira News in London, dated 4/9/1944.
22. SA, 408/2/74: Item from Sudan Times, dated 31/8/1944.
23. SUD.A.PK15691KAS: SUD.A.PK1569.2SUD: Minutes 279, dated 21/9/1945: Minutes 280, dated 13/2/1946: Minutes 288, dated 4/5/1949.
24. SA, 416/5/120: SPS Ordinary General Meeting 43, dated 4/4/1951.
25. See Mollan (2008) for the financial affairs of the SPS. Gaitskell apparently once said that the SPS had offered to run "a scheme in West Africa, and so far it is now known if this will be accepted or not," but I have not found more on this. (SA, 418/3/210–214: Translation of an article published in "El Rai El Aam," Issue 975, dated 2/7/1948.)
26. SA, 418/3/210–214: Translation of an article published in "El Rai El Aam," Issue 976, dated 2/7/1948.
27. SA, 752/18/15, Sudan Star article "Fears for future of SPS," dated 10/1/1949. I would agree with Rai El Amm that not knowing how to run a scheme like Gezira considerable time in advance before the turnover from SPS to government would not win the beauty contest anywhere, but I have not found any reference to Sudanese business men or handing scheme responsibility back to the SPS as options being discussed.
28. Robertson J. (1974) *Transition in Africa* (London: Hurst & Company).
29. Robertson (1974).
30. SA, 693/1/133: Hunt Diary, entry on 17/9/1945.
31. SA, 517/10/18–48: Report on the future of the Gezira Scheme, dated 27/7/1946: see also SA: Report of the Gezira Special Committee and the record of the debate thereon, undated.
32. SA, 517/10/18–48: The Future of the Gezira Scheme.
33. SA, 517/10/18–48: The Future of the Gezira Scheme.
34. SA, 517/10/1–17: Financial Secretary's Statement on Report Gezira Special Committee 1944, undated. Although undated, the response by the Financial Secretary to the report of the Special Committee would have to

be dated between July 27 and August 18, 1947, as on this last date another document—partially in response to the Financial Secretary—in the discussion was prepared.
35. Remember the less than positive evaluation of the Financial Secretary Schuster in chapter 3?
36. SA, 517/10/1–17: Financial Secretary on Report Gezira.
37. SA, 527/7/22: Letter from Robertson to Gaitskell, dated 28/8/1947. In his memoirs, Robertson remembers these events like this: "After some hesitation I produced a memorandum advocating that part of the money, which had previously gone to Syndicate shareholders, should be used to improve social services in the Gezira area: local authorities in the Gezira itself went further and advocated a sort of Tenessee Valley Authority, whereby profits arising from the Gezira Scheme should be used in the Gezira. The Advisory Council recommended that while the Scheme should benefit the country as a whole, it should more particularly cater for the Gezira area. The Financial Secretary, however, pointed out that the Scheme had been supported by the general taxpayer in bad years when it has made no profits, and in the payment of the interest on the loans which financed the dam and the canalisation, and these would not be finally paid off until the 1970s" Robertson (1974).
38. The booklet was D. E. Lilienthal (1944) *Tennessee Valley Authority (TVA): Democracy on the March* (New York: Penguin).
39. SA, 527/7/44: Letter from Robertson to Hawkesworth, dated 14/9/1947.
40. SA, 527/7/51–21: Letter from Gaitskell to Robertson, dated 12/9/1947.
41. SA, 527/7/23–25: Draft Note on Gezira from Robertson, dated 28/8/1947.
42. SA, 527/7/23–25: Draft Note on Gezira.
43. SA, 418/5/108: Letter from Bredin (Governor Blue Nile Province) to Gaitskell, dated 28/1/1948: and SA, 418/5/107: Letter from Gaitskell to Bredin, dated 3/2/1948.
44. SA, 418/5/109–126: The future administration of the Gezira Scheme, Governor's Office, dated 1/2/1948. A quite similar note dated 10/1/1948 is entitled Gezira Scheme: Memorandum on some limitations at present affecting development (SA 418/7/1–21).
45. SA, 418/5/109–126: Future Administration of Gezira.
46. SA, 418/7/22–24: Letter from Gezira Research Farm, dated 22/3/1948 on the Future administration of the Gezira Scheme.
47. One could actually question the relevance of TVA for Gezira. For US standards, the TVA may have been an example of direct and integrated project management, but it was nothing compared to Gezira in detailed control of daily life.
48. SA, 418/7/26–28: Draft Minutes of the 32d Meeting of the Gezira Advisory Board, dated 8/4/1948.
49. SA, 418/7/26–28: Draft Minutes Gezira Advisory Board.
50. SA, 418/7/26–28: Draft Minutes Gezira Advisory Board.
51. SA, 418/7/26–28: Draft Minutes Gezira Advisory Board. In his own notes, Gaitskell summarized the differences between the two positions. "Difference between us and DofA. Separation v Coordination […] Our reasons for preferring coordination (1) past experience (2) British staff vital to preach new aspects (3) Mangt 2 councils too cumbersome (4) [not clear for a tenant]

anyway (5) B.I. does not form part of local Govt. He merely general feudal father of all interests." SA, 418/7/47–48: Notes from Gaitskell at the 32d Meeting of the Gezira Advisory Board, dated 8/4/1948.

52. SA, 418/7/26–28: Draft Minutes Gezira Advisory Board, dated 8/4/1948.
53. "Under the existing arrangements, the principal functions of the Companies are: (a) The general agricultural management of the Scheme and the provision of the necessary staff for that purpose: (b) The collection and storage of the cotton crop, involving the provision and maintenance of the Gezira Light Railway: (c) The ginning of the crop, involving the provision and operation of ginning factories: (d) The provision and maintenance of sufficient machinery to carry out the heavy agricultural processes: (e) The marketing of the cotton crop including the seeds: (f) The provision of such loans to tenants as they may need to carry out their agricultural operations: (g) The provision and maintenance of such houses, stores, offices, workshops, and materials as are necessary for the above purposes." SA, 408/1/2: Letter of A. L. Chick, Financial Secretary, accompanying "The future administration of the Gezira Scheme," Note from Financial Department, Khartoum, dated 16/7/1949.
54. SA, 408/1/2: Letter of Chick: see also SA, 408/1/5–14: The Gezira Scheme, Reference Division (Colonial Section), Central Office of Information, London, dated 21/6/1950.
55. SA, 408/1/2: Letter of Chick.
56. SA, 408/1/5–14: The Gezira Scheme.
57. SA, 498/10/34: Gezira Tenants Representative Body, dated 7/1947.The 1949 note discussed the option to appoint one last body, a Gezira Advisory Council which would review "all matters concerning the operation and welfare of the Scheme" as well. It would be a large body, basically including everyone with an interest in the scheme: the note clearly doubted the need for this council. SA, 408/1/2: Letter of Chick.
58. SA, 408/1/3–4: The Future Administration of the Gezira Scheme, Note from Financial Department, Khartoum, dated 16/7/1949. To ensure that the correct message on Gezira would reach the broader public, especially "the part which has been played by the Sudanese people themselves in formulating the new proposals," an official history of the scheme became important. (SA, 408/1/5–14: The Gezira Scheme). See chapter 7 for more on this aspect.
59. SA: Select Committee of the Legislative Assembly on the Future Administration of the Gezira Scheme, dated 12/2/1950, Committee appointed at 16/7/1949: SA, 408/1/5–14: The Gezira Scheme.
60. SA, 408/1/5–14: The Gezira Scheme.
61. SA, 498/10/37: The Gezira Scheme Bill, dated 19/4/1950.
62. "The only differences are in respect of the costs of social development and, so far as concerns extensions, 'abu Ishreens,'" which originally had been paid by the government. (SA, 498/10/37: The Gezira Scheme Bill.)
63. SA, 498/10/37: The Gezira Scheme Bill: Explanatory Memorandum, dated 19/4/1950: "Although after the 1949/50 crop the bales of cotton from the Gezira will not bear the marks well known for many years—"S.P.S. or K.C.C."—there is every hope that the high quality of the cotton

which Lancashire has known and come to expect, will continue." (SA, SUD.A.PK1569.2SUDAN: Reprint BCGA April 1950).
64. SA, 408/1/5-14: The Gezira Scheme. See H. J. Sharkey (2003) *Living with colonialism: Nationalism and culture in the Anglo-Egyptian Sudan* (Berkeley: University of California Press).
65. Robertson (1974). It was clear from an early date the Sudan Government did much to ensure Gaitskell continuing service for Gezira. "I saw quite a bit of Arthur Gaitskell, and I was disappointed that he did not express himself more definite about staying on. You seem to have gone all out to draw him in by making him a member of the Cabinet and Chairman of the Gordon College Council." Some texts almost read like a job description specifically for Gaitskell, for example when a need was mentioned for a "man with ability and personality" to "build up an organisation in the Gezira which would fire the tenants, villagers, schoolmasters, petits-functionaires and other Gezira inhabitants with a joint enthusiasm to create better social conditions and greater economic prosperity than they have ever known." SA, 527/7/23-25: Draft Note on Gezira.
66. SA, 993/7/99-100: Letter to J. Carmichael, dated 30/11/1952.
67. SA, 529/8/36-37: Letter from Higgin to Robertson, dated 30/11/1951.
68. SA, 408/1/22: Gezira Information Service, Report of a Press Conference, dated 7/2/1951. The Sudan Government did agree with the SGB that "one partner should be definitely responsible for marketing whose decision on the subject should be final, rather than leave some confusion of functions" (SA, 408/1/22: Gezira Information Service, Report of a Press Conference, dated 7/2/1951: see also SA, 991/1/12-15: Carmichael, Note on Cotton marketing, dated 23/11/1954). "The marketing of cotton is a complicated task and should be left in the hands of experts subject only to the general supervision of the Council," but it simply wanted to have more influence on that partner's behavior and decisions. For example, we learn that in 1954, the (Sudanese) Council of Ministers "directed the Board" to sell cotton by auctions only, because the Council "had to reckon with public opinion and the tenants." (SA, 991/1/12-15: Note on Cotton marketing).
69. SA, 993/7/99-100: Letter to Carmichael.
70. In Ms. Johnson's diary, the succession of Gaitskell reads like a good gossip story—or as the return trip to Paris from Napoleon once escaped from Elba. At the end of 1950, "Bill [Beer] very perturbed at Inspectors & tenants going to him & asking if it's true John Gaitskell is leaving next april. [...] Bill said that if the tenants started getting worried about the future régime it was going to lead to trouble. [...] Bill had warned Central Government who poo pooed the idea & said 'What do these people mean by all this, they've no right to spread rumours.' (That's the British staff.)" (SA, 752/9/3: Diary Ms. Johnson 1950). On September 1, 1951: "[Someone] thinks J.G. might go next year. Feels he J.G. would like to go into politics, Labour of course." (SA, 752/11/19: Diary Ms. Johnson, dated 1/9/1951). On October 12 1951: "J.G. definitely going in April [...]. Carmichael, Bacon, Hillard all having been mentioned as successors to J.G." (SA, 752/11/25: Diary Ms. Johnson, dated 12/10/1951). On October 19 1951: "Heard J. G. leaving April next week. Selling house. Advertisement in English & Scottish paper for new Managing Director for here. Three to four thousand a year." (SA, 752/12/3: Diary Ms. Johnson, dated 19/10/1951). On November 3

1951: "New Managing Director has been chosen." (SA, 752/12/6: Diary Ms. Johnson , dated 3/11/1951).
71. SA, 752/12/10: Diary Ms. Johnson, dated 14/11/1951.
72. SA, 752/12/12: Diary Ms. Johnson, dated 29/11/1951.
73. SA, 752/12/13: Diary Ms. Johnson, dated 2/12/1951.
74. Ms. Johnson remarks about that decision: "[Raby] is not interested in politics. Knew Smuts who asked him to reorganise the works. Left because after death politics entered into business too much in South Africa. Feels they are too many conflicting racial troubles. One couldn't start a business at Springs because one was expected to start it as so & so to please so & so" (SA, 752/13/23-24: Diary Ms. Johnson, dated 20/4/1952).
75. Which would not necessarily be a recommendation, see chapter 7.
76. SA, 752/18/2-3: Sudan Star, article "Possible successor to Mr. Gaitskell," dated 3/12/1951. From the same article we learn that "Mr. Raby is the man who discovered how the Germans made the V.2. bomb. He has an outstanding war record and extensive knowledge of works."
77. SA, 752/12/16: Diary Ms. Johnson, dated 14/12/1951.
78. SA, 752/12/19: Diary Ms. Johnson, dated 4/1/1952.
79. SA, 752/12/21: Diary Ms. Johnson, dated 7/1/1952.
80. SA, 752/12/21: Diary Ms. Johnson, dated 7/1/1952.
81. SA, 418/9/1-2: Letter from Chick, Financial Secretary, to Gaitskell, dated 19/3/1952. The final book became A. Gaitskell (1959) *Gezira. A story of development in the Sudan* (London: Faber and Faber). Apparently Gaitskell also had other options for work: "Ministry of Materials are contemplating offering A. Gaitskell post as independent member of Raw Cotton Commission at £3,000 per annum. Before doing so they would like a personal estimate of his character and abilities and some account of the nature of his functions as General Manager of the Gezira Board." (NA, FO371/9700: Foreign Office to Khartoum, dated 13/5/1952). In the many positions that Gaitskell has fulfilled after his Gezira work, however, I have not found this position. According to Ms. Johnson not everyone in the government was that positive about Gaitskell staying on: "J.G. is supposed to be rather toying with the idea of continuing to come out for 3 months of the year as the locals direct him so but as far as we hear the Government don't!" (SA, 752/12/16: Diary Ms. Johnson, dated 14/12/1951).
82. SA, 529/8/41-42: Letter from Luce to Chick, dated 8/12/1951.
83. "Had very trying time with J. G. who would moan & ask whether he ought to go & leave the Sudanese & whether he ought not to put off his plane date! Was told by Raby that he Raby was now in the Chair & that he J.G. had had it!" (SA, 752/13/23-24: Diary Ms. Johnson, dated 20/4/1952).
84. SA, 752/12/16: Diary Ms. Johnson, dated 14/12/1951.
85. SA, 752/13/16: Diary Ms. Johnson, dated 20/3/1952.
86. SA, 752/16/8: Diary Ms Johnson, dated 26/3/1953.
87. SA, 752/16/9: Diary Ms Johnson, dated 27/3/1953.
88. SA, 993/7/101: Letter from Raby to Carmichael, dated 15/1/1953: SA, 993/7/102: Letter from Carmichael to Raby, dated 19/1/1953.
89. SA, 993/7/157: Letter from Carmichael to Raby, dated 17/12/1953.
90. SA, 991/1/16-26: A verbatim report of Mr. G. W. Raby's recorded speech, including handwritten comment. On the SGB mission to France, which was to study the options of fiber from cotton, Raby said that he "particularly

asked Mr. Mekki Abbas to head this mission because it is his country and he had got to see to the scheme being successful over the years." Indeed, Mekki Abbas would succeed Raby.
91. SA, 991/1/16–26: A Verbatim Report.
92. SA, 993/7/98: Letter from C. G. Davies to J. Carmichael, dated 26/11/1952.
93. SA, 993/7/99–100: Letter to J. Carmichael, dated 30/11/1952.
94. SA, 991/1/16–26: A Verbatim Report.
95. SA, 991/1/16–26: A Verbatim Report.
96. Apparently, Raby showed Gaitskell-type behavior as he tried to return to Gezira after deciding (forced to decide?) to leave. Carmichael writes about this with some malicious delight: "I enclose copies of correspondence which have recently been passed between Raby and the Sudanese Government, from which you will see he is trying hard to get back to the Sudan. [...] his position in the United Kingdom Atomic Energy Authority was going to be at a much lower level than Raby has anticipated when he accepted the appointment." (SA, 994/1/82: Letter from Carmichael to Luce, dated 3/11/1955).
97. SA, 994/1/15: Letter from Hammad Tewfik Hammad, Minister of Finance & Economics to Raby, dated 19/3/1955.
98. Gaitskell (1959).
99. SA, 498/10/54: Letter from Financial Secretary on Sudan Gezira Board Publicity, dated 25/3/1953.
100. SA: Report of the Gezira Special Committee and the record of the debate thereon, undated.
101. SA: Select Committee of the Legislative Assembly on the Future Administration of the Gezira Scheme, dated 12/2/1950, Committee appointed at 16/7/1949.
102. SA, 408/1/27: Report of a Press Conference. The 140 British earned £147,254, the 1,800 non-British £272,547. "In case someone may deduce from this that the average Britisher's pay was over £1000 while the average Sudanese was £150 and that there was some Nationality bias in this, it must be pointed out that the figures are not comparable because on the one side you have all the chief executive posts including the Manager, while on the other side you have everyone else down to the Office Murasla. It is not possible to give a comparison, such as is given in Government statistics, between the numbers and salaries of British and Sudanese in the same Scale, because in the Gezira Scheme until 1st July 1950 no Sudanese were employed in the same jobs as Britishers, who, apart from a few Headquarters and Engineering posts, comprised the field staff."
103. SA, 408/1/24: Report of a Press Conference.
104. SA, 752/8/9: Diary Ms Johnson: "Found the Sudanese who came for Barakat interview very full of themselves as if they owned the earth & all felt they had only come to sign on dotted line!"
105. SA, 752/10/13: Diary Ms Johnson, dated 7/1/1951.
106. SA, 752/10/13: Diary Ms Johnson, dated 7/1/1951: "Asked by Husui why he'd applied he said well he didn't see why he couldn't in the advertisement it said 'No qualifications needed'! (which is quite true thus it left out that one had to be honest & trustworthy & all the other things needed to be an Inspector)."

107. SA, 752/10/15: Diary Ms Johnson, dated 14/1/1951.
108. SA, 428/3/40: Diary Culwick, dated 21/6/1949: underlining in original.
109. SA, 428/3/102: Diary Culwick, dated Easter Day 1950.
110. SA, 752/13/5: Diary Ms Johnson, dated 9/2/1952.
111. SA, 752/9/3: Diary Ms Johnson, July 1950.
112. SA, 752/10/13: Diary Ms Johnson, dated 7/1/1951.
113. SA, 752/15/8: Diary Ms Johnson, dated 8/11/1952.
114. NA, FO371/102759: Letter from Riches to Allen, dated 26/8/1953.
115. SA, 992/1/121: Carmichael, Sudanisation of the Ministry of Finance, dated 17/2/1955.
116. The full quote is "In principle, I believe that it's good for INGOs to be staffed by nationals from the countries they're functioning in. There are many reasons why this is a good thing which I won't list here. Our current national staff are brilliant, they've been educated in other neighbouring countries or refugee camps, and I believe that given time they'll have built the capacity to move up and thrive in the Manager roles. As an international NGO, this is what we want: to improve the capacity of the work force so that the internationals can step back and take their resources elsewhere. However, South Sudan has been engaged in 2 civil wars since 1955 [...] and I feel like it's a bit too early to be kicking out the expat human resources. Many NGOs wouldn't want to risk their reputations by forcibly taking on staff who are not able to do the roles to international standards, or who need a lot of capacity building to meet these standards, and many donors will be reluctant to donate to organisations where International HQ's are made to manage from afar. Also, that the expat staff are asked to leave with little or no notice period means that any sort of meaningful hand-over of roles, or capacity building of the new staff members within the organisation will be impossible." (http://gemmakatehay.blogspot.de/2011/11/sudanisation.html: accessed on June 23, 2012).

7 The Everlasting Rectangles: Gezira and International Development

1. See M. Al-Rahim (1969) *Imperialism and nationalism in the Sudan: A study in constitutional and political development, 1899–1956* (Oxford: Oxford University Press).
2. Sudan Archive, 993/7/111–113: Letter plus note from Luce to Carmichael, dated 7/8/1953.
3. SA, 993/7/139–140: Letter from Howe to Secretary of State, dated 17/9/1953. See K. M. Barbour (1959) "Irrigation in the Sudan. Its Growth, Distribution and Potential Expansion," *Transactions and Papers (Institute of British Geographers)*, No. 26, 243–263.
4. R. Chambers (1969) *Settlement schemes in tropical Africa: A study of organisations and development* (London: Routledge); see J. M. Hodge, G. Hödl, and M. Kopf (eds) (2014) *Developing Africa: Concepts and practices in twentieth century colonialism* (Manchester: Manchester University Press).
5. See E. Akyeampong, R. H. Bates, N. Nunn, and J. A. Robinson (2014) *Africa's development in historical perspective* (New York: Cambridge University Press); M. Black (2007) *The No-Nonsense guide to international*

development (Oxford: New Internationalist Publications); S. Chari and S. Corbridge (eds) (2008) *The development reader* (London: Routledge); N. Cullather (2000) "Development? It's history," Diplomatic History, 24, 641–653; P. Hopper (2012) *Understanding development* (Cambridge: Polity Press); U. Kothari (ed.) (2005) *A radical history of development studies: Individuals, institutions and ideologies* (London: Zed Books); M. F. S. Kunkel (2011) "Writing the History of Development. A Review of the Recent Literature," *Contemporary European History*, 20, 215–232.

6. The SGB complained that some British firms were not "as forthcoming in assisting the furtherance of development schemes as firms in other countries;" the Sudan Government agreed to "impress on such firms the political desirability of close cooperation and assistance." SA, 993/7/114: Letter from Carmichael to Davies, dated 8/8/1953.
7. SA, 993/7/115: Letter from Carmichael to Luce, dated 20/8/1953. Apparently Luce insisted in his own views, as we learn from a second letter from Carmichael that Luce had sent his ideas along to others—something that Carmichael and others in Khartoum found "hard to understand," as they had not been consulted in this matter that is "of such vital importance and which is never far away from our thoughts." (SA, 993/7/17: Letter from Carmichael to Luce, dated 25/8/1953).
8. SA, 993/7/122: Letter and Note from Luce, dated 18/9/1953.
9. SA, 993/7/122: Letter and Note from Luce.
10. SA, 993/7/144–147: Note from Carmichael, dated 6/10/1953.
11. SA, 993/7/163–164: Letter from Allen, dated 26/11/1953.
12. SA, 993/7/151–153: Note from Carmichael.
13. SA, 993/6/8–15: The Interest of Great Britain in theSudan, dated 11/12/1956.
14. SA, 593/1/1–28: Annual Report of the Ministry of Irrigation & H.E.P 1957–1958, undated.
15. SA, 993/6/36–49: Sudan. Items of topical interest, Carmichael, dated 8/4/1959.
16. Despite the Russians being "of course, very clever in timing their visit over Ramadan when Sudanese intellects are not at their sharpest" (SA, 993/6/36–49: Sudan. Items of topical interest).
17. SA, 992/2/62–65: Development in the Sudan, Carmichael, dated 14/4/1956.
18. SA, 992/2/68–69: Development Circular Letter No. 5–1956, dated 20/6/1956.
19. SA, 992/2/100–102: Development Circular Letter No. 9–1956, dated 16/10/1956.
20. SA, 991/1/34–40: Speech for Acting Minister, undated.
21. SA, 992/2/12–15: Main Features of Development Possibilities and Their Relation to Budget Surpluses, dated 30/10/1955.
22. SA, 992/2/20–23: Planning future development in the Sudan, dated 26/9/1955 (likely). The lack of staff and resources was also an international issue: "The Colombo Plan is very much in arrears, not for lack of money but for lack of men and machines and designs to spend it." A multilateral fund to assist states in southeast Asia, the Colombo Plan was initiated on a Commonwealth Conference in Colombo, Sri Lanka, January 1950.

23. Draft minutes of a meeting to consider certain aspects of the Gezira Southwest (Managil) Extension, dated 15/9/1955.
24. SA, 990/2/9–10: Note for Consideration by Ad hoc Committee on Managil Management of South West Gezira (Managil) Extension, Development Branch, Ministry of Finance & Economics, dated 10/9/1955.
25. SA, 990/2/38–46: Minutes of the Second Meeting of the Ad Hoc Committee on Managil, undated.
26. SA, 990/2/50–57: Approval by Development Committee of the Recommendations of the Ad Hoc Committee on Managil, dated 29/3/1956.
27. An IBRD mission on Managil even stated that "the Gezira Scheme can be regarded as a highly successful pilot scheme for the Managil Extension Scheme" (SA, 594/5/1–32: The Managil Extension Irrigation Project, International Bank Report).
28. SA, 990/2/38–46: Second Meeting Ad Hoc Committee.
29. SA, 990/2/50–57: Approval by Development Committee.
30. Gezira counted at the time 14,000 5-feddan tenancies and 16,000 10-feddan tenancies, in a 1:4 rotation (SA, 990/2/38–46: Minutes of the second Meeting).
31. SA, 990/2/22: SGB Report on Managil Extension, dated 16/11/1955.
32. SA, 990/2/38–46: Second Meeting Ad Hoc Committee.
33. SA, 593/1/1–28: SID Annual Report 1953–1954.
34. SA, 501/5/1–34: Annual Report of the Ministry of Irrigation & H.E.P 1954–1955, undated.
35. SA, 593/1/1–28: Annual Report of the Ministry of Irrigation & H.E.P 1955–1956, undated.
36. SA, 593/1/1–28: Annual Report of the Ministry of Irrigation & H.E.P 1957–1958, undated.
37. SA, 593/1/1–28: Annual Report of the Ministry of Irrigation & H.E.P 1959–1960, undated. See Shaw D. J. (1965b) "The Managil South-Western Extension: An Extension to the Gezira Scheme. An Example of an Irrigation Development in the Republic of the Sudan," *Koninklijk Nederlands Aardrijkskundig Genootschap*, 82.
38. The International Bank for Reconstruction and Development (IBRD) is one of five institutions of the World Bank Group. It was established in 1944 to finance reconstructing Europe after World War II. The World Bank was formal established on December 27, 1945, within the context of the Bretton Woods agreements, with a similar mandate.
39. SA, 593/8/1–61: International Bank for Reconstruction and Development, The Economy of the Sudan, Main Report, dated 25/2/1958.
40. SA, 593/8/1–61: IBRD, The Economy of the Sudan.
41. SA, 990/2/162–164: Meeting on Managil Extension, dated 6/2/1958.
42. SA, 594/5/1–32: The Managil Extension Irrigation Project, International Bank Report.
43. SA, 424/6/1–20: The Economics of Major Irrigation Schemes in the Sudan, Ministry of Finance & Economics, J, Carmichael, Financial Adviser, dated 13/2/1958.
44. SA, 593/8/1–61: IBRD, The Economy of the Sudan.
45. SA, 424/6/21–37: Uses for Water Made Available by the Roseires Dam, Carmichael, dated 23/2/1958.

46. SA, 424/6/38–43: Productive Development in the Period between the Completion of the Managil Extension and the Completion of the Roseires Dam, Carmichael, dated 26/2/1958; SA, 424/6/21–37: Uses for Water by Roseires.
47. SA, 424/6/38–43: Productive Development between Managil and Roseires.
48. SA, 424/6/44–48: A Possible Programme of Irrigation Development in the Sudan Relying on Internal Resources, Carmichael, dated 6/3/1958.
49. SA, 593/10/1–123: International Bank for Reconstruction and Development, Report on the Technical Mission on Sudan Irrigation, dated 17/4/1959.
50. SA, 593/10/1–123: IBRD, Technical Mission Sudan Irrigation.
51. SA, 593/10/1–123: IBRD, Technical Mission Sudan Irrigation; Handwritten in the margin to this "consideration" is: "S.I.D. has tried this for 20 years, only Abdel Magid & 2 blocks in N.W."
52. SA, 593/10/1–123: IBRD, Technical Mission Sudan Irrigation.
53. SA, 594/5/33–37: Preliminary Comments on the Report of the Technical Mission of the International Bank on Sudan Irrigation, dated 5/5/1959.
54. SA, 593/12/1–410: Taha T. J. (1965) "Optimisation of production in the Gezira Scheme," in *Agricultural Development in the Sudan* (Khartoum: Philosophical Society of the Sudan, Sudan Agricultural Society); see T. Tvedt (2004) *The river Nile in the age of the British: Political ecology and the quest for economic power* (London: I. B. Taurus).
55. SA, 990/2/170–171: Note from Carmichael, dated 14/6/1959.
56. See also SA, 851/1/5–10: Note on the Activities of the Village Farming Exp, Mohd. Abdel Gadir El Mufti, The Sudan Gezira Board—Social Development Department Village Farming Experiments, dated 27/11/1960.
57. SA, 849/13/1–12: Review of Future Agricultural Policy, dated 30/3/1953.
58. SA, 849/13/1–12: Future Agricultural Policy.
59. SA, 991/1/51: Letter from Carmichael to Mekki Abbas, dated 28/11/1955.
60. SA, 593/12/1–410: I. G. Simpson (1965) "Economic aspects of diversification in the Gezira," in *Agricultural Development in the Sudan* (Khartoum: Philosophical Society of the Sudan, Sudan Agricultural Society).
61. SA, 991/1/81–84: Report on the Marketing of Sudan Cotton, A. N. Shimmin, dated 2/2/1956.
62. SA, 755/1/6–135: H. Ferguson, Crops other than Cotton in the Sudan Gezira, undated; "A few crops such as dura and lubia may grow as well in the Gezira as elsewhere, but the vast majority of 'alternative' crops grow better elsewhere in the Sudan, often without the necessity and expense of irrigation." Cotton was a Gezira crop; food crops should come from elsewhere, and mixed farming with livestock could be beneficial for social reasons.
63. SA, 593/10/1–123: IBRD, Technical Mission Sudan Irrigation.
64. SA, 593/12/1–410: Simpson (1965).
65. SA, 593/12/1–410: S. E. L. Noah (1965) "Agricultural extension—its role and importance in agricultural development in the Sudan" in *Agricultural development in the Sudan* (Khartoum: Philosophical Society of the Sudan, Sudan Agricultural Society).

66. SA, 593/12/1–410: Taha (1965). The areas were Hamad el Nil (South Group), Tebub (Centre Group), Abdil Gilal (Messellamia Group) and Feteis (Wad Shair Group).
67. SA, 756/7/7–109: The Development of Agriculture in the Main Gezira Area. Interim Report by the Working Party. Agricultural Research Division. Department of Agriculture. Ministry of Agriculture, Sudan, May 1965.
68. SA, 593/12/1–410: Taha (1965).
69. SA, 756/7/7–109: Interim Report Working Party.
70. SA, 756/7/7–109: Interim Report Working Party.
71. SA, 756/7/7–109: Interim Report Working Party. Around this time the issue of "proper" use of natural resources was high on the international agenda (M. Clawson (ed.) (1964) *Natural resources and international development* (Baltimore: Johns Hopkins Press); UNESCO (1964) "Social Aspects of African Resource Development," *International Social Science Journal*, 16). See C. S. Wright (ed.) (1962) *Africa and irrigation* (Salisbury: Proceedings of an International Symposium).
72. SA, 756/7/7–109: Interim Report Working Party.
73. SA, 756/7/5–6: Comments on the Working Party's Interim Report.
74. SA, 756/7/7–109: Interim Report Working Party.
75. SA, 756/7/7–109: Interim Report Working Party. See I. G. Simpson (1970) "New Approaches to Irrigated Farming in the Sudan: Organization and Management," *Land Economics*, 46, 287–296.
76. SA, 756/7/7–109: Interim Report Working Party.
77. SA, 756/7/5–6: Comments on the Working Party's Interim Report.
78. SA, 756/7/7–109: Interim Report Working Party.
79. SA, 593/12/1–410: Simpson (1965).
80. SA, 756/7/7–109: Interim Report Working Party; emphasis in original.
81. SA, 756/7/7–109: Interim Report Working Party.
82. Gezira Study Mission, Main Report, International Bank for Reconstruction and Development 1966.
83. Gezira Study Mission, Main Report, IBRD 1966.
84. SA, 593/8/1–61: IBRD, The Economy of the Sudan.
85. SA, 498/6/49: The Times, Irrigating the Sudan. Benefits of a nationalized Gezira Scheme, dated 3/7/1950.
86. SA, 408/1/5–14: The Gezira Scheme.
87. SA, 501/5/1–34: Annual Report of the Ministry of Irrigation & H.E.P 1954–55, undated. Compare with the claims by the Dutch irrigation engineers after World War II, who redefined their Javanese approach to an international smallholders approach (M. W. Ertsen (2010) *Locales of Happiness: Colonial irrigation in the Netherlands East Indies and its remains, 1830—1980* [Delft: VSSD Press]).
88. SA, 408/1/5–14: The Gezira Scheme, emphasis in original.
89. SA, 408/2/11–13: A Broadcast on the Gezira Scheme from Omdurman Station, dated 18/1/1943.
90. SA, 593/12/1–410: D. J. Shaw (1965a) "The development and contribution of irrigated agriculture in the Sudan," in *Agricultural Development in the Sudan* (Khartoum: Philosophical Society of the Sudan, Sudan Agricultural Society).
91. C. C. Reining (1966) *The Zande scheme: an anthropological case study of economic development in Africa* (Evanston: Northwestern University Press).

See also W. A. Hance (1955) "The Zande Scheme in the Anglo-Egyptian Sudan," *Economic Geography*, 31, 149–156.
92. Chambers (1969).
93. In 1955, Hance compared the Zande management model (the Equatoria Projects Board) with the Sudan Gezira Board, Gash was out of the picture. For Hance, Zande "leaned heavily upon the highly successful experience of the Gezira, particularly in the application of science and in the organizational arrangements." (Hance [1955]).
94. Reining (1966).
95. Chambers (1969.)
96. Reining (1966).
97. SA, 994/1/109–110: Extracts from letter to Carmichael from McCall, dated 4/12/1955; "You will remember John Smith's ideas on the subject, with large new areas being opened up and seven Canadian engineers being recruited to run the areas." See also S. Serels (2013) *Starvation and the State: Famine, slavery, and power in Sudan, 1883–1956* (New York: Palgrave Macmillan).
98. SA, 593/10/1–123: IBRD, Technical Mission Sudan Irrigation. See also J. B. Bascom (1990) "Food, Wages, and Profits. Mechanized Schemes and the Sudanese State," *Economic Geography*, 66, 140–155.
99. National Archives, CO111/784/10: Letter from Lethem to Beckett, dated 11/12/1944.
100. NA, CO111/784/10: Letter from Lethem to Beckett, dated 25/4/1945.
101. NA, CO111/784/10: Letter from Lethem to Beckett, dated 11/12/1944. See also J. D. N.Versluys (1953) "The Gezira Scheme in the Sudan and the Russian Kolkhoz: A Comparison of Two Experiments," *Economic Development and Cultural Change*, 2, 32–59; J. D. N. Versluys (1954a) "The Gezira Scheme in the Sudan and the Russian Kolkhoz: A Comparison of Two Experiments, Part II," *Economic Development and Cultural Change*, 2, 120–135; J. D. N. Versluys (1954b) "The Gezira Scheme in the Sudan and the Russian Kolkhoz: A Comparison of Two Experiments, Part III," *Economic Development and Cultural Change*, 2, 216–235.
102. NA, CO111/784/10: Letter from Tempany to Mayall, January 1945; A comment dated 22/1/1945 even suggested that the Kassala Board would show that a partnership with a private firm was no longer the model in Sudan and that "means have apparently been found of eliminating this aspect." (NA, CO111/784/10: Circulation Notes with file). A letter suggested something similar: "Yes, the point you make about Kassala is interesting. I was at Kassala in about May 1925 and at that time the whole show was run by a Company, though I cannot remember the actual details of the arrangement. Nor do I remember the date at which the whole thing was taken over by the Government directly. As you say this has no doubt been the precedent and experiment which has brought about the change in the Gezirah." (NA, CO111/784/10: Letter from Lethem to Beckett, dated 25/4/1945). I think both persons missed the particular political reasons why a private firm like the KCC moved out and the government moved in.
103. NA, CO111/784/10: Letter from Beckett to Lethem, dated 22/1/1945.
104. NA, CO111/784/10: Circulation Notes with File.
105. NA, CO111/784/10: Circulation Notes with File.
106. NA, CO111/784/10: Letter from Beckett to Lethem; dated 26/2/1945.

107. http://www.worldipcomgroup.com/theworlddiplomat/africaninstitutions.html; "The mission of the ECA was 'to initiate and participate in measures for facilitating concerted action for the economic development of Africa, including its social aspects, with a view to raising the level of economic activity and standards of living in Africa and for maintaining and strengthening the economic relations of countries and territories in Africa.'"
108. SA, 418/9/15: Letter from Roger Lubborn to Gaitskell, dated 28/1/1953. Gaitskell has also produced pages and pages on the Gezira and why it was a model to follow in international development, particularly in Africa—but I will discuss those in more detail in the Epilogue.
109. SA, 889/7/33–34: Letter from Junken (?) to Gibbons, dated 5/5/1959.
110. SA, 889/7/31: Letter from Dew from Nairobi to Gibbons.
111. SA, 851/6/22: Letter of Recommendation from Dew for Smith, dated 29/3/1955.
112. SA, 850/5/1–27: Cotton Production in Private Estates, Blue Nile Province, Sudan. Final Draft, issued by Barclays Bank, August 1958.
113. SA, 994/3/1–15: Impressions Formed on a Tour of Our Companies in Africa, Carmichael, dated March 1967.
114. SA, 759/3/1–6: CV of Hugh Ferguson, undated.
115. SA, 759/3/7–9: Letter from Ferguson to India Tea Association, dated 11/10/1954.
116. SA, 759/3/10: Letter from Ferguson to Hawkesforth, dated 14/10/1954; SA, 759/3/24: Letter from Hawkesforth to Ferguson, dated 24/12/1954.
117. SA, 759/3/19–23: Letter from Ferguson to Fisons Limited.
118. SA, 759/3/24: Letter from Hawkesforth to Ferguson, dated 24/12/1954.
119. SA, 755/5/20: Letter from Ferguson to Gaitskell, dated 5/5/1961.
120. SA, 755/5/23: Letter from Bedford to Ferguson, dated 10/5/1961.
121. Gaitskell promised he would mention Ferguson to his Ethiopian contacts, but in the same letter he expressed the wish that "someone could persuade [Ferguson] you to take over the Research station in the Gezira!" It would ensure the post would not go to "the Germans—so much more active than we are overseas now" and prevent "German sprays and fertilizers" in Gezira. (SA, 755/5/24: Letter from Gaitskell to Ferguson, dated 16/5/1961).
122. SA, 755/5/20: Letter from Ferguson to Sayed Wadie Habashi, dated 2/8/1961.
123. SA, 755/5/20: Letter from Gaitskell to Ferguson, dated 28/10/1961.
124. SA, 759/1/46: Letter from Ferguson to Eric (IBRD), dated 25/2/1967; SA, 759/1/47: Letter from E. M. (Eric) Sicely yo Ferguson, dated 9/11/1967.
125. SA, 759/1/52: Letter from L. C. Phillips to Ferguson, dated 23/12/1970.
126. SA, 593/1/1–28: SID Annual Report 1953–1954. In 1935/1936 engineer Williams from the SID had toured India, but his main conclusions at that time had been that there was not much to take from India. Perhaps the working procedures, a crop, and the need for hydraulic research (see later in this chapter). (W. M. Williams (1936) *A report on a study of India Irrigation Works*, Irrigation Department, Sudan Government). ICOLD was established in 1928; ICID was established in 1950. Both act as intergovernmental

forums on their respective topics, with a membership focus on public works departments.
127. SA, 593/1/1–28: SID Annual Report 1953–1954.
128. SA, 593/1/1–28: SID Annual Report 1952–1953.
129. SA, 593/1/1–28: SID Annual Report 1953–1954.
130. SA, 501/5/1–34: Annual Report of the Ministry of Irrigation & H.E.P 1954–1955, undated.
131. SA, 639/13/1–23: Report by Sir Claude Inglis on Irrigation problems in the Sudan and Recommendations as to How to Deal with Them, Wallingford, January 1954. India remained a thankful country for comparison, see, for example, D. S. Thornton (1966) "Contrasting Policies in Irrigation Development, Sudan and India," *Development Studies*, No 1, University of Reading.
132. SA, 593/1/1–28: SID Annual Report 1953–1954. Using expertise in the way that suits the client is not new either...
133. SA, 639/13/1–23: Report by Sir Claude Inglis. In 1951, a Hydraulic Research Station at Sennar had been proposed, with reference to India, where "no responsible engineer would dream of deciding upon a design for an irrigation structure of any importance without first having it carefully tested at a Research Station for possible hydraulic weaknesses." (SA, 499/13/34–40: A Note on the Urgent Need for a Hydraulic Research Station in the Sudan, H. A. Morrice, dated 16/4/1951).
134. SA, 501/5/1–34: Annual Report of the Ministry of Irrigation & H.E.P 1954–1955, undated.
135. Like the visit of Dr. Khoshla from India for Roseires (SA, 593/1/1–28: SID Annual Report 1956–1957).
136. As I explain in Ertsen (2010) and M. W. Ertsen (2007) "The Development of Irrigation Design Schools or How History Structures Human Action," *Irrigation and Drainage*, 56, 1–19.
137. K. M. Baker (1989) *Agricultural change in Nigeria* (London: John Murray), 41.
138. SA, 418/4/2–7: The Gezira Scheme, looking backwards and looking forwards. Broadcast by Mr. Arthur Gaitskell C. M. G., Managing Director of the Sudan Gezira Board, from Omdurman, dated 15/7/1950.

Epilogue

1. The ideas move from ecological disaster (N. Pollard [1984] "The Sudan's Gezira Scheme. A study in failure" in E. Goldsmith and N. Hildyard [eds] *The social and environmental effects of large dams. Volume 2: Case studies* [Camelford: Wadebridge Ecological Center]) to sustainable hopes (A. M. Eldaw [2004] *The Gezira Scheme: Perspectives for sustainable development* [Bonn: German Development Institute]).
2. Euroconsult, Sir Alexander Gibb and Partners and TCS Sudan (1982) *Sudan Gezira Rehabilitation Project* (6 volumes); M. R. Francis and O. Elawad (1986) *A study of the management of minor canals in the Gezira Irrigation Scheme, Sudan* (Wallingford: Wallingford Hydraulics Research); M. R. Francis and O. M. A. Elaweed (1989) *Diagnostic investigations and rehabilitation of canals in the Gezira Scheme, Sudan* (Manila: Asia Regional Symposium). A more general overview typical for the time can be found

in L. F. Kortenhorst, P. N. G. Van Steekelenburg and L. H. Sprey (1989) "Prospects and Problems of Irrigation Development in Sahelian and Sub-Saharan Africa," *Irrigation and Drainage Systems*, 3, 13–45. See for a brief Sudan's independence history K. M. Barbour (1980) "The Sudan since Independence," *The Journal of Modern African Studies*, 18, 73–97.
3. T. Barnett (1977) *The Gezira Scheme: An illusion of development* (London: Frank Cass). Reviews include M. E. Adams (1980) Review of T. Barnett (1977) "The Gezira Scheme: An Illusion of Development," *Economic Development and Cultural Change*, 28, 633–636; G. Ellis (1979a) Review of T. Barnett (1977) "The Gezira Scheme: An Illusion of Development," *ASA Reviews of Books*, 5, 34–36; G. Ellis (1979b) Review of Barnett T. (1977) "The Gezira Scheme. An Illusion of Development," *The Journal of Modern African Studies*, 17, 340–342; P. C. Salzman (1978) Review of T. Barnett (1977) "The Gezira Scheme: An Illusion of Development," *Annals of the American Academy of Political and Social Science*, 440, 174–175; J. Samoff (1978) T. Review of Barnett (1977) "The Gezira Scheme: An Illusion of Development," *The American Historical Review*, 83, 1068–1069; D. Seddon (1978) Review of T. Barnett (1977) "The Gezira Scheme: An Illusion of Development," *International Affairs*, 54, 161–163. See also A. Abdelkarim (1985) "Why Do Gezira Tenants Withhold Their Households' Labour?," *Journal of Asian and African Studies*, 20, 72–88; A. Abdelkarim (1986) "Wage Labourers in the Fragmented Labour Market of the Gezira, Sudan," *Africa: Journal of the International African Institute*, 56, 54–70. See also J. Tait (1978) "Diversifizierung, Mechanisierung und Kapitalisierung der Produktion im Sudan Gezira Scheme: Auf dem Weg zur Überwindung kolonial deformierter Agrarstrukturen?', *Africa Spectrum*, 13, 165–178.
4. M. R. Francis and R. Hinton (1987) *Minor Canal Management in the Gezira Irrigation Scheme, Sudan: An Interim Report on Diagnostic Investigations During 1986–1987* (Wallingford: Wallingford Hydraulics Research). My guess would be that the handwritten comments were by Farbrother, who published a detailed study on irrigation in Gezira in 1974. H. G. Farbrother (1974) *Irrigation practices in the Gezira* (Khartoum: Cotton Research Report). See also H. G Farbrother (1979) *Irrigation practices in the Gezira* (London: Cotton Research Corporation); H. G. Farbrother (1976) *Technical notes on water use* (Agricultural Research Corporation). See also A. Ahmed and M. Tiffen (1986) *Water management in the Gezira Scheme, Sudan: A survey of farmers attitudes on two minor canals* (Wallingford: Wallingford Hydraulic Research).
5. World Bank (1983) *Sudan Gezira Rehabilitation Project*, Staff Appraisal Report No. 4218.SU; World Bank (1983) *Sudan Gezira Rehabilitation Project: Implementation*, Volumes I and II. See also H. Pluesquellec (1990) *The Gezira Scheme in Sudan: Objectives, design and performance* (Washington: World Bank). See M. Goldman (2005) *Imperial nature: The World Bank and struggles for social justice in the age of globalization* (New Haven, CT: Yale University Press) for an analysis of the Bank.
6. Sudan Archive, 428/1/1: G. M. Culwick—A study of the human factor in the Gezira Scheme.
7. SA 501/2/34: Sudan Irrigation Department Annual Report 1946.

8. SA, 415/10/60: 415/10/73-75; Ordinary General Meeting SPS, dated 23/11/1921.
9. SA, 415/10/60: Ordinary General Meeting SPS.
10. SA, SUD.A.PK 1569-2SUDAN: "The Sudan Plantations Syndicate Ltd and the Kassala Cotton Co. Ltd. and their work in the Anglo-Egyptian Sudan," reprint from The forty-fifth annual report of the British Cotton Growing Association (1950). Perhaps this tone was to be expected in 1950, but as late as 1982 we encounter a "salute" to SPS inspectors and "their gallant wives" who had been a major part of the success of "this great enterprise." SA, 851/6/1-2: G. Bradin 1982 Foreword to Smith's Note on Gezira, dated 23/9/1982. The BCGA also praised the "British women, the wives of the European officials and staff," as they had offered "that little extra comfort for the tired, and at times cross, worker, to help him forget his worries." Indeed, the position of British women in the Gezira was that of a dependent—social researcher Culwick was the first women in the service of the SPS, as late as 1949 (!).
11. SA, 415/10/60: Ordinary General Meeting SPS.
12. SA, 418/3/94: Letter from John to Billy with Response of MacIntyre to Radio Message of 18/1/1943 of Gaitskell.
13. See for example the *Report of the East Africa Commission presented to the British Parliament by the Secretary of State for the Colonies* (1925) (London: His Majesty's Stationary Office).
14. SA, 499/14/2: Five Year Plan for Postwar Development.
15. SA, 495/2/26: R. J. Smith SID Memorandum No. 1, dated 23/12/1950.
16. M. Adas (1989) *Machines as the measure of men: Science, technology, and ideologies of Western dominance* (Ithaca, NY: Cornell University Press).
17. R. Chambers (1969) *Settlement schemes in tropical Africa. A study of organisations and development* (London: Routledge). Raby, the successor of Gaitskell in the SGB, briefly worked for the Groundnut Scheme. See for the mechanized durra schemes, S. Serels (2013) *Starvation and the State: Famine, slavery, and power in Sudan, 1883-1956* (New York: Palgrave Macmillan). See for general UK development policies J. M. Hodge (2007) *Triumph of the expert: Agrarian doctrines of development and the legacies of British colonialism* (Athens: Ohio University Press); J. M. Hodge, G. Hödl and M. Kopf (eds) (2014) *Developing Africa: Concepts and practices in twentieth-century colonialism* (Manchester: Manchester University Press). N. Farson (1949) *Last chance in Africa* (London: Victor Gollancz) describes the British feelings quite well.
18. See J. S Hogendorn. and K. M. Scott (1981) "The East African Groundnut Scheme. Lessons of a Large-scale Agricultural Failure," *African Economic History*, No. 10, 81-115, and A. Wood (1950) *The groundnut affair* (London: The Bodley Head).
19. K. D. S. Baldwin (1957) *The Niger Agricultural Project* (Cambridge, MA: Harvard University Press).
20. Baldwin (1957).
21. T. Forrest 1981 "Agricultural policies in Nigeria, 1900-78" in J. Heyer, P. Roberts and G. Williams (eds) *Rural development in tropical Africa* (New York: St. Martin's Press), 233.
22. National Archives, CO822/1106: Letter dated 6/7/1954. Somehow material from the archives offers room for so many names from movies, like

Mr. Smith (The Matrix) and Mr. Brown, who could have been a character in Reservoir Dogs. Then again, Smith and Brown are not really the most special names...

23. National Archives, CO822/1106: Letter 6/7/1954. See also Chambers (1969).
24. NA, CO822/1106: Letter from Secretary of State for the Colonies to the Officer Administering the Government of Kenya, dated 1/9/1954.
25. NA, CO822/1106: Letter from Brown, dated 21/8/1954.
26. NA, CO822/1106: Letter from Kenya, dated 16/9/1954; NA, EAF209/291/02: Proposed Colonial Development and Welfare rice investigation scheme in favour of the Government of Kenya involving a free grant of £1,000 to meet the cost of a visit by an expert to advise on rice irrigation, especially the land settlement aspect, in Kenya, dated 13/10/1954. "Mr. Brown's strong point is land settlement rather than rice and I am sure he would welcome any advice Mr. Rhind can give him on rice cultivation." NA, EAF209/291/02: Note, dated 8/10/1954.
27. G. Diemer (1990) *Irrigatie in Afrika: Boeren en ingenieurs, techniek en* kultuur (Amsterdam: Thesis Publishers).
28. J. Moris (1973) "The Mwea environment" in Chambers R. and Moris J. (eds) *Mwea, an irrigated rice settlement in Kenya* (München: Weltforum Verlag).
29. See T. Kanogo (1987) *Squatters and the roots of Mau Mau* (London: James Currey).
30. NA, CO822/952: Recent developments affecting the standard of living of the African and his sense of security, dated 16/12/1953. See also J. C. De Wilde (1967a) *Experiences with agricultural development in tropical Africa. Vol I: The synthesis* (Baltimore, MD: Johns Hopkins Press). J. C. De Wilde (1967b) *Experiences with agricultural development in tropical Africa. Vol II: The case studies* (Baltimore, MD: Johns Hopkins Press); Diemer (1990).
31. NA, CO822/1554: Accelerated development of African agriculture. Proposals of the Ministry of Agriculture, Kenya, dated 21/2/1957; NA, CO822/1554: Note of a meeting held on 28th February, 1957, in Mr. Nye's office, Colonial Office.
32. NA, CO822/213: Press Office Handout, No. 773, dated 1/12/1953. Some years later costs pops up: "In the Mwea I am anxious to see a large number of families settled, if only to avoid the crippling expense to the scheme of employing large numbers of landless whites and detainees." (NA, CO822/1730: Letter from Browris, dated 27/2/1957).
33. NA, CO822/1730: Kenya News, Press Office Handout No. 109, dated 10/2/1958.
34. NA, CO822/156: Despatch from the Governor of Kenya, No. 81, dated 10/5/1951. See for a more general overview of development policy in G. Kenya Holtham and A. Hazlewood (1976) *Aid and inequality in Kenya: British development assistance to Kenya* (London: Croom Helm).
35. NA, CO822/550: Colony and Protectorate of Kenya. Development and Reconstruction Authority. Quarterly report for the period 1st January to 31st March, 1952; Quarterly report for the period 1st April to 30th June, 1952 : Quarterly report for the period 1st July to 30th September, 1952
36. NA, CO822/193: Employment Schemes for Kikuyu Repratriates, Appendix A—A description of Kikuyu employment schemes Central Province,

dated 28/10/1953; NA, CO822/193: Intensification of Agricultural Development in the African areas of Kenya, dated 2/11/1953.
37. NA, CO822/1105: Telegram from the Officer administering the Government of Kenya to the Secretary of State for the Colonies, dated 9/7/1954.
38. Chambers (1969; 89).
39. Chambers (1969; 85). See also R. Chambers and J. Moris (eds) (1973) *Mwea, an irrigated rice settlement in Kenya* (München: Weltforum Verlag).
40. Chambers (1969; 72).
41. Chambers and Moris (1973; 70).
42. J. J. Veen (1973) "The production system" in Chambers R. and Moris J. (eds) *Mwea, an irrigated rice settlement in Kenya* (München: Weltforum Verlag), 127.
43. De Wilde (1967b; 236–237). John C. de Wilde was Acting Director of the Economic Staff of the International Bank for Reconstruction and Development at the time.
44. Chambers (1969; 94).
45. Chambers (1969; 112). See also Golkowsky R. (1969) *Bewässerungslandwirtschaft in Kenya: Darstellung grundsätzlicher Zusammenhänge am Beispiel des Mwea Irrigation Settlement* (München: Weltforum Verlag).
46. Veen (1973).
47. G. G. Manig (1973) "Engineering aspects and water control" in Chambers R. and Moris J. (eds) *Mwea: An irrigated rice settlement in Kenya* (München: Weltforum Verlag), 91. After 35 years of experience in Mozambique and Tanzania, George G. Manig joined the Public Works Department in Nairobi in 1961. He started working in Mwea in 1963 and became the Senior Irrigation Engineer of the National Irrigation Board in Kenya in 1966.
48. Forrest (1981; 241). These years increasing food production was high on the international agenda (R. Dumont and B. Rosier [1969] *The hungry future* [London: Andre Deutsch]), but on the Sudanese agenda as well (P. Oesterdiekhoff and K. Wohlmuth (eds) (1983) *The development perspectives of the Democratic Republic of Sudan: The limits of the Breadbasket Strategy* [München: Weltforum Verlag]). See as well J. O'Brien (1981) 'Sudan. An Arab breadbasket?', *MERIP Reports*, No. 99, 20–26 for a slightly later contribution to the food debate.
49. K. M. Baker (1989) *Agricultural change in Nigeria* (London: John Murray).
50. T. Wallace (1979a) "The impact of a large-scale irrigation scheme on two wards in a small town in Kano state, Nigeria. Its implications for rural development," in Southall A. (ed.) *Small urban centers in rural development in Africa* (Madison: University of Wisconsin).
51. T. H. Strong and P. Paton (1968) *Economic appraisal and analysis of requirements for development of the Talata Mafara area, Sokoto Rima Drainage Basin* (Rome: United Nations Development Program and Food and Agriculture Organization), 63. See also G. Andrae and B. Beckman (1985) *The wheat trap: Bread and underdevelopment in Nigeria* (London: Zed Books).
52. NEDECO (1970) *Kano River Project feasibility study* (Amsterdam: NEDECO), 123.

53. Andrae and Beckman (1985; 114).
54. NEDECO (1974) *Some considerations and recommendations on the management of the Kano River Project* (Amsterdam: NEDECO), 1–2.
55. Andrae and Beckman (1985; 113).
56. A. C. Bird (1983) "The land issue in large scale irrigation projects, some problems from Northern Nigeria" in W. M. Adams and A. T. Grove (eds) *Irrigation in tropical Africa: Problems and problem solving* (Cambridge: Cambridge University Press), 79. See also W. M. Adams (1991) "Large Scale Irrigation in Northern Nigeria: Performance and Ideology," *Transactions of the Institute of British Geographers, New Series,* 16, 287–300; R. W. Palmer-Jones (1981) "How Not to Learn from Pilot Irrigation Projects: The Nigerian Experience," *Water Supply and Management,* 5, 81–105; S. Pierce (2005) *Farmers and the state in colonial Kano: Land Tenure and the Legal Imagination* (Bloomington: Indiana University Press).
57. J. M. Siann (1983) "Labour constraints in the implementation of irrigation projects" in W. M. Adams and A. t. Grove (eds) *Irrigation in tropical Africa: Problems and problem solving* (Cambridge: African Studies Center), 91.
58. De Wilde (1967a; 63–64); Chambers and Moris (1973).
59. J. Heyer, P. Roberts, and G. Williams (1981) "Rural development" in J. Heyer, P. Roberts, and G. Williams (eds) *Rural development in tropical Africa* (New York: St. Martin's Press), 8.
60. Wallace (1979a; 243).
61. Many settlement schemes in tropical Africa did not meet expectations of their planners because economic benefits were not enough to attract settlers—as for example in the Office du Niger in Mali for a long time. See E. Schreyger (1984) *L'Office du Niger au Mali 1931 à 1982: La problématique d'une grande enterprise agricole dans la zone du Sahel* (Stuttgart: Beitrage zur Kolonial- und Überseegeschichte 27); G. Spitz (1949) *Sansanding: Les irrigations du Niger* (Paris : Société d'Etitions Géographiques, Maritimes & Coloniales) ; M. M. Van Beusekom (2002) *Negotiating development: African farmers and colonial experts at the Office du Niger, 1920–1960* (Portsmouth: Heinemann). M. M. Van Beusekom (1989) *Colonial rural development. French policy and African response at the Office du Niger, Soudan Français (Mali), 1920–1960* (PhD Thesis, Johns Hopkins University).
62. Veen (1973; 124).
63. Baker (1989; 38; emphasis added).
64. Andrae and Beckman (1985; 114).
65. Quoted in T. Wallace (1981) "The Kano River Project, Nigeria: The impact of an irrigation scheme on productivity and welfare" in J. Heyer, P. Roberts, and G.Williams (eds) *Rural development in tropical Africa* (New York: St. Martin's Press), 287; emphasis in original.
66. T. Wallace (1979b) *Rural development through irrigation: Studies in a town on the Kano River Project*, Research Report 3, Ahmadu Bello University, 5. Original quote J. C. Wells (1974) *Agriculture policy and economic growth in Nigeria, 1962–8* (Ibadan: NISER).
67. J. A. Sagardoy, A. Bottrall and G. O. Uittenbogaard (1986) *Organization, operation and maintenance of irrigation schemes* (Rome: Food and Agriculture Organization). Recently Mwea farmers and human rights organizations in Kenya have shown that the FAO was correct in doubting

the validity of the Mwea model. Its original colonial principles were not untouchable anymore. See for example Kenya Human Rights Commission (2000) *Dying to be free: The struggle for rights in Mwea* (Nairobi).
68. BBC's "Yes Minister," Series 2, Episode 2 (1981). Bernard explains to the Minister the honors available to senior civil servants. Hacker: "Well, what has Sir Arnold to fear, anyway? He's got all the honours he could want, surely?" Bernard: "Well, naturally he has his G." Hacker: "G?" Bernard: "Yes; you get your G after your K." Hacker: "You speak in riddles, Bernard." Bernard: "Well, take the Foreign Office. First you get the CMG, then the KCMG, then the GCMG; the Commander of the Order of St Michael and St George, Knight Commander of St Michael and St George, Knight Grand Cross of St Michael and St George. Of course, in the Service, CMG stands for 'Call Me God' and KCMG for 'Kindly Call Me God.'" Hacker: "What does GCMG stand for?" Bernard: "God Calls Me God."
69. J. M. Hodge (2007) *Triumph of the expert: Agrarian doctrines of development and the legacies of British colonialism* (Athens: Ohio University Press); S. Moon (2007) *Technology and ethical idealism: A history of development in the Netherlands East Indies* (Leiden: CNWS Publications); J. B. Teisch (2011) *Engineering Nature: Water, development, and the global spread of American environmental expertise* (Chapel Hill: University of North Carolina Press); J. Teisch (2005) "Home Is Not so Very far away. Californian Engineers in South Africa, 1868–1915", *Australian Economic History Review*, 45, 139–160. See for an account of experts in Europe M. Kohlrausch and H. Trischler (2014) *Building Europe on expertise: Innovators, organizers, networks* (Basingstoke: Palgrave Macmillan).
70. M. W. Ertsen (2010) *Locales of Happiness: Colonial irrigation in the Netherlands East Indies and its remains, 1830–1980* (Delft: VSSD Press); M. W. Ertsen (2008) "Controlling the Farmer. Irrigation Encounters in Kano, Nigeria," *TD: The Journal for Transdisciplinary Research in Southern Africa*, 4, 209–236; M. W. Ertsen (2007) "The Development of Irrigation Design Schools or How History Structures Human Action," *Irrigation and Drainage*, 56, 1–19.
71. Listed in A. Gaitskell (1964) "Resource development among African countries" in M. Clawson (ed.) 1964 *Natural resources and international development* (Baltimore, MD: Johns Hopkins Press).
72. Chambers (1969; 143).
73. SA, SUD.A.PK1569.2SUD: SPS Report of Directors, dated 17/5/1949.
74. K. M. Barbour (1960) Review of A. Gaitskell (1959) "Gezira. A Story of Development in the Sudan," *Africa: Journal of the International African Institute*, 30, 287–288; W. V. Blewett (1960) Review of A. Gaitskell (1959) "Gezira. A Story of Development in the Sudan," *African Affairs*, 59, 260–261; P. Broadbent (1960) Review of A. Gaitskell (1959) "Gezira. A Story of Development in the Sudan," *International Affairs*, 36, 399–400; W. A. Hance (1961) Review of A. Gaitskell (1959) "Gezira. A Story of Development in the Sudan," *Geographical Review*, 51, 157–158; R. J. Irvine (1960) Review of A. Gaitskell (1959) "Gezira. A Story of Development in the Sudan," *The American Economic Review*, 50, 751–753.
75. Chambers and Moris (1973; 465).
76. A. I. Clarkson (2005) *Courts, councils and citizenship: Political culture in the Gezira scheme in condominium Sudan* (PhD Thesis, Durham University);

S. M. Mollan (2008) *Economic imperialism and the political economy of Sudan: The case of the Sudan Plantations Syndicate, 1899–1956* (PhD Thesis, Durham University).
77. Which is the image that Mollan (2008) suggests.
78. J. Robertson (1974) *Transition in Africa* (London: Hurst & Company). Gaitskell does not mention this direct relationship nor his influence on Robertson in his own 1959 book. A. Gaitskell (1959) *Gezira: A story of development in the Sudan* (London: Faber and Faber).
79. SA, 418/6/1–10; Considerations on trends in Gezira (same document as 418/5/13–27) with accompanying letter from Gaitskell to SPS London Office, dated 25/8/1938. The handwritten comment in the draft note suggested to "omit this para" as it seemed to be clear that the government would not take that option on such short notice—something that Gaitskell seems not to have done, as draft and final texts appear to be the same. "I am taking the liberty of forwarding you another Memorandum on the subject of Devolution which you and General Asquith were kind enough to discuss with me when I was on leave in the early summer. I must apologise for being so verbose and apparently obstinate in again bringing up this subject but I cannot help feeling personally that it is a matter of very great importance and that unless our point of view is represented there is a danger that the action taken on this matter may be in the end a disaster for the Gezira Scheme."
80. SA, 418/5/13–27: A consideration of the general political trend of the Gezira Scheme and its effects on our future interest here. Probably by Gaitskell, includes handwritten comments in margin by (probably) someone else. Undated, although a reference to the document is possibly found in a later (undated) text that refers to a previous memorandum of March 1938 (SA, 418/6/12–35). The same document as 418/5/13–27 is found in SA, 418/6/1–10 with accompanying letter from Gaitskell to SPS London Office, dated 25/8/1938. Texts seem identical, so any suggestions in the other text would not have been incorporated, if indeed the undated text was the draft version for the text sent to London.
81. SA, 418/5/13–27: A consideration
82. SA, 418/3/53: Letter from "J" to "Billy" Accompanying Note on Native Agents to Assist Gezira Administration, undated, probably somewhere during WWII. "I would like to know if you think something like this is wanted for L.O. [London Office] showing the whole subject or whether merely a short statement saying we must have natives to help because of the war & therefore we might as well extend the Hosh Scheme where feasible [...]. My draft won't be wasted as it will [serve] to explain to our people who know nothing of the subject." SA, 418/3/54–62: Handwritten Memorandum on the use of Native agents to assist in the administration of the Gezira Scheme, undated; This last statement is interesting, as Gaitskell suggests in other texts—especially when directing his ideas to British field staff—that the individual efforts and preferences of individual staff members was a key.
83. SA, 418/3/81–91: Note on Village Councils, undated.
84. SA, 418/8/1–15: Letter from Gaitskell to Manager SPS at Barakat, plus Note on Future of the Gezira Scheme, dated 23/4/1943.
85. SA, 693/1/18: Diary F. B. Hunt, entry on 12/11/1938.

NOTES 261

86. SA, 693/1/130–131: Hunt diary, entry on 17/8/1945. We learn from another source that "his wife hates the Sudan and won't come out any more." (SA, 428/3/74: Diary Culwick, dated 28/11/1949).
87. SA, 752/4/3–4: Diary Ms Johnson, January 1949.
88. SA, 752/5/15: Diary Ms Johnson, April 1949.
89. SA, 752/5/13: Diary Ms Johnson, March 1949.
90. Using the term "experiment" itself is not original. Already in 1869, a Sir Samuel Baker traveled along the Nile. A member of the Royal Geographical Society, Baker was working on a "first experiment in the development of Africa" (Times of May 25, 1869, quoted in H. Tilley (2011) *Africa as a living laboratory: Empire, development, and the problem of scientific knowledge, 1870–1950* (Chicago: University of Chicago Press). See also E. J. G. Huxley (1960) *A new earth: An experiment in colonialism* (London: Chatto & Windus) for a late use of the term. In the Netherlands East Indies, experiments were also used when new policies were under consideration (Ertsen 2010).
91. Gaitskell (1959); A. Gaitskell (1955) *What have they to defend?* (The Africa Bureau Annual Anniversary Address); A. Gaitskell (1952) "The Sudan Gezira Scheme," *African Affairs*, 51, 306–313.
92. Gaitskell (1959).
93. Gaitskell (1952; 306–313).
94. W. Hance (1954) "The Gezira. An Example in Development," *Geographical Review*, 44, 253–270.
95. Gaitskell (1952; 306–313).
96. Gaitskell (1959).
97. Hance (1954).
98. C. W. Beer (1955) "Social Development in the Gezira Scheme," *African Affairs*, 54, 42–51.
99. Hance (1954).
100. M. W. Ertsen (2006) "Colonial Irrigation. Myths of Emptiness'," *Landscape Research*, 31, 147–167; Ertsen (2008). See A. Eckert (2008) "'We Are All Planners Now.' Planung und Dekolonisation in Afrika," *Geschichte und Gesellschaft*, 34, 375–397.
101. B. J. Andres (2014) *Power and control in the Imperial Valley: Nature, agribusiness, and workers on the California borderland, 1900–1940* (College Station: Texas A&M University Press); D. Biggs (2010) *Quagmire: Nation-building and nature in the Mekong Delta* (Seattle: University of Washington Press); L. Camprubi Bueno (2011) *Political engineering: Science, technology and the Francoist landscape (1939–1959)* (PhD thesis, University of California, Los Angeles); M. Fiege (1999) *Irrigated Eden: The making of an agricultural landscape in the American West* (Seattle: University of Washington Press); J. Leigh-Smith (2014) *Works in progress: Plans and realities on Soviet farms, 1930–1963* (New Haven, CT: Yale University Press); D. Mosse (2003) *The rule of water: Statecraft, ecology and collective action in South India* (Oxford: Oxford University Press); C. Mukerji (2009) *Impossible engineering: Technology and territoriality on the Canal du Midi* (Princeton, NJ: Princeton University Press); M. Reisner (1993) *Cadillac desert: The American West and its disappearing water* (New York: Penguin); W. D. Swearingen (1987) *Moroccan mirages: agrarian dreams and deceptions, 1912–1986* (Princeton, NJ: Princeton University Press); D. Worster

(1985) *Rivers of empire: Water, aridity, and the growth of the American West* (Panthenon Books: New York).
102. Which could be accompanied very well by (less true) ideas about empty areas, certain activities, etcetera. The use of the Gezira's tenancy model at Mwea served at least short term goals. Veen, Mwea manager from 1962 to 1966, noted how "[d]eveloping the area was facilitated by the almost total lack of population which meant that there were few families to be evicted and they were offered places as tenants." Moris referred to this idea of the empty region as "one of the most enduring management myths of Mwea," as people already lived on the plain. It was certainly cheaper to consider the area as empty, as this saved the colonial government the trouble of compensating the former owners financially. Veen (1973; 116); Moris (1973; 36).
103. V. Bernal (1997) "Colonial Moral Economy and the Discipline of Development. The Gezira Scheme and 'Modern' Sudan," *Cultural Anthropology*, 12, 447–479.
104. Printing Gezira on the United Kingdom obviously confirms the power of the United Kingdom in Sudan to Gezira, but perhaps this was done to allow the British reader to grasp the size of the project. A similar criticism can be made on Foucault's work, as he tends to stress the anonymous structures that overpower the individual. M. Foucault (1995) *Discipline and punish: The birth of the prison* (New York: Vintage Books); M. Foucault (1988) *Madness and civilization: A history of insanity in the age of reason* (New York: Vintage Books). Much of these ideas can be related to images of empires built on irrigation, like in K. Wittfogel (1957) *Oriental despotism: A comparative study of total power* (New Haven, CT: Yale University Press).
105. Bernal (1997).
106. Chambers (1969). Chambers was one of the first researchers whose work was used in the irrigation group at Wageningen University in the early 1980s, in order to move away from the narrow technical focus of irrigation studies at that group. This approach was indeed highly relevant, and I have studied at the group with much pleasure. However, the approach was so successful, that hardly any technical subjects are left in the work of the group.
107. Chambers (1969).
108. See P. Kelemen (2007) "Planning for Africa. The British Labour Party's Colonial Development Policy, 1920–1964," *Journal of Agrarian Change*, 7, 76–98.
109. Chambers (1969).
110. Chambers (1969); Chambers and Moris (1973). Reviews are for example C. E Carr (1973) Review of R. Chambers (1969) "Settlement Schemes in Tropical Africa. A Study of Organizations and Development," *Journal of Asian and African Studies*, 8, 99–101; G. Green (1970) Review of R. Chambers (1969) "Settlement Schemes in Tropical Africa. A Study of Organizations and Development," *African Affairs*, 69, 397; A. H. Hanson (1970) Review of R. Chambers (1969) "Settlement Schemes in Tropical Africa. A Study of Organizations and Development," *The Journal of Modern African Studies*, 9, 143–146; J. A. Hellen (1970) Review of R. Chambers (1969) "Settlement Schemes in Tropical Africa. A Study of Organizations and Development," *International Affairs*, 46, 385–387; D. P. Lumsden (1970) Review of R. Chambers (1969) "Settlement Schemes in Tropical Africa. A Study of Organizations and Development," *Africa: Journal of*

the International African Institute, 40, 290–291; M. P. Miracle (1975) Review of R. Chambers and J. Moris (eds) (1973) "Mwea, an Irrigated Rice Settlement in Kenya," *The International Journal of African Historical Studies*, 8, 319–321; A. F. Williams (1970) Review of R. Chambers (1969) "Settlement Schemes in Tropical Africa. A Study of Organizations and Development," *The Geographical Journal*, 136, 126–127.
111. R. Chambers (2005) *Ideas for development* (London: Earthscan).
112. A selection of this literature is D. Brautigam (2009) *The dragon's gift: The real story of China in Africa* (Oxford: Oxford University Press); Easterly W. (2006) *The white man's burden: Why the West's efforts to aid the rest have done so much ill and so little good* (New York: Penguin); S. Kinsbergen (2014) *Behind the pictures: Understanding private development initiatives* (Phd Thesis, Radboud Universiteit); G. Mohan and M. Power (2009) "Africa, China and the 'New' Economic Geography of Development," *Singapore Journal of Tropical Geography*, 30, 24–28; D. Moyo (2009) *Dead aid: Why aid is not working and how there is a better way for Africa* (New York: Farrar, Straus and Giroux).
113. Chambers (2005).
114. As is done more often by development thinkers, see the excellent recent book by Hodge et al. (2014), especially the introduction. See also E. Akyeampong, R. H. Bates, N. Nunn and J. A. Robinson (2014) *Africa's development in historical perspective* (New York: Cambridge University Press); P. Barron, R. Diprose and M. Woolcock (2011) *Contesting development: Participatory projects and local conflict dynamics in Indonesia* (New Haven, CT: Yale University Press); M. Black (2007) *The No-Nonsense guide to international development* (Oxford: New Internationalist Publications); S. Chari and S. Corbridge (eds) (2008) *The development reader* (London: Routledge); N. Cullather (2000) "Development? It's History," *Diplomatic History*, 24, 641–653; L. Heldring and J. A. Robinson (2012) "Colonialism and Development in Africa," *African Economic History Working Paper Series*, No. 5; P. Hopper (2012) *Understanding development* (Cambridge: Polity Press); U. Kothari (ed.) (2005) *A radical history of development studies: Individuals, institutions and ideologies* (London: Zed Books); M. F. S. Kunkel (2011) "Writing the History of Development. A Review of the Recent Literature," *Contemporary European History*, 20, 215–232.
115. However, as many of those move to the cities, most urban inhabitants would have no say either.
116. Hodge et al. (2014).
117. Quoted in H. L. Wesseling (2003) *Verdeel en heers: De deling van Afrika 1880–1914* (Utrecht: Aula), 17.
118. Wesseling (2003; 77). The Culwick couple, Arthur Theodore and Geraldine Mary (the first female staff member in Gezira), represent two sides of that same argument. After having worked together in Tanganyika and published about that, G. M. Culwick came to represent the side of white superiority helping the backwards to develop. A. T. Culwick went to Kenya, defended the British superiority there, and ended in South Africa. A. T. Culwick (1963) *Britannia waves the rules* (Cape Town: Nasionale Boekhandel); A. T. Culwick (1942) *Good out of Africa* (Livingston: Rhodes-Livingstone Institute); A. T. Culwick and G. M. Culwick (1935) "Culture Contact on the Fringe of Civilization," *Africa: Journal of the International African*

Institute, 8, 163–170; A. T. Culwick and J. C. Oosthuizen (1970) *The inequality principle* (Durban: Dolphin Press); G. M. Culwick (1955) *A study of the human factor in the Gezira Scheme* (Barakat: Sudan Gezira Board). See also V. Berry (1994) *The Culwick Papers 1934–1944: Population, food and health in colonial Tanganyika, by A. T. and G. M. Culwick* (London: Academic Books). In books written by American scholars on Zande and Niger projects, the Africans are represented much more positively and with actual agency and rationality. K. D. S. Baldwin (1957) *The Niger Agricultural Project* (Cambridge, MA: Harvard University Press); C. C. Reining (1966) *The Zande scheme: An anthropological case study of economic development in Africa* (Evanston: Northwestern University Press). See for an intriguing account of racial thinking in the USA P. L. Farber (2011) *Mixing races: From scientific racism to modern evolutionary ideas* (Baltimore, MD: Johns Hopkins University Press).
119. Ertsen (2006; 2008).
120. De Wilde (1967a; 63).
121. De Wilde (1967b; 241). See Ertsen (2008).
122. SA, 415/10/59–62: Ordinary General Meeting SPS, dated 23/11/1921.
123. SA, 851/6/5–6: C. H. Smith 1982 The Gezira Irrigation Scheme. Economic Development. Durham Sudan Historical Records Conference.
124. I assume he is the Charles Smith mentioned by Culwick in her diary entry of 23/4/1949 (SA, 428/3/25).
125. SA, 851/6/25: El Gezira Paper, Issue 1534, 24/5/1984.
126. Chambers (1969; 34). See H. P. Segal (2006) *Technology and Utopia* (Washington: American Historical Association).
127. Quoted in Hodge et al (2014).
128. Morris I. (2013) *The measure of civilisation: How social development decides the fate of nations* (London: Profile Books Ltd). See also Akyeampong et al (2014); Prados de la Escosura l. (2013) "World Human Development: 1870–2007," *EHES Working papers in Economic History*, No. 34.
129. T. Barnett (1977) *The Gezira Scheme: An illusion of development* (London: Frank Cass); T. Barnett and A. Abdelkarim (1991) *Sudan: The Gezira Scheme and agricultural transition* (London: Frank Cass); V. Bernal (1991) *Cultivating workers: Peasants and capitalism in a Sudanese village* (New York: Columbia University Press); G. Hunter (1969) *Modernizing peasant societies. A comparative study in Asia and Africa* (New York: Oxford University Press); J. Matunhu (2011) "A Critique of Modernization and Dependency Theories in Africa: Critical Assessment," *African Journal of History and Culture*, 3, 65–72; W. Rodney (1973) *How Europe underdeveloped Africa* (London: Bogle-L'Ouverture Publications; Dar-Es-Salaam: Tanzanian Publishing House); W. W. Rostov (1960) *The stages of economic growth: A non-communist manifesto* (London: Cambridge at the University Press); O. Rotimi (2013) "A Critique of Walter Rodney's Concept of Development," *International Journal of Education and Research*, 1, 157–166; R. A. Waters (2005) "How Socialism Underdeveloped Africa," *Political Science Reviewer*, 34, 264–320.
130. The preference for settled populations as a measure of civilization is not necessarily a European preference; see work on imperial powers through time. See K. Flannery and J. Marcus (2012) *The creation of inequality: How our prehistoric ancestors set the stage for monarchy, slavery, and empire*

(Cambridge, MA: Harvard University Press); P. Heather (2010) *Empires and barbarians: The fall of Rome and the birth of Europe* (Oxford: Oxford University Press); A. Porter (2012) *Mobile pastoralism and the formation of Near Eastern civilizations: Weaving together society* (New York: Cambridge University Press); S. Stuurman (2009) *De uitvinding van de mensheid: Korte wereldgeschiedenis van het denken over gelijkheid en cultuurverschil* (Amsterdam: Bert Bakker).
131. M. Adas (1989) *Machines as the measure of men: Science, technology, and ideologies of Western dominance* (Ithaca, NY: Cornell University Press). M. Adas (2006) *Dominance by design: Technological imperatives and America's civilizing mission* (Cambridge, MA: Harvard University Press).
132. As claimed in Akyeampong et al., 2014. Still a great read though!
133. B. Latour (1993) *We have never been modern* (Cambridge, MA: Harvard University Press).
134. Hodge et al. (2014).
135. See B. Latour (1996) *Aramis, or the love of technology* (Cambridge, MA: Harvard University Press). In Science in Action Latour referes to the rugby ball metaphor: the object of the game is reshaped in the game itself (B. Latour [1987] *Science in action: How to follow scientists and engineers through society* [Cambridge, MA: Harvard University Press]).
136. B. Latour (2013) *An inquiry into modes of existence: An anthropology of the moderns* (Cambridge. MA: Harvard University Press). B. Latour (2005) *Reassembling the social: An introduction to actor–network theory* (Oxford: Oxford University Press); P. N. Edwards (2010) *A Vast Machine: Computer models, climate data, and the politics of global warming* (Cambridge, MA: MIT Press) is a great example of a (not explicit but still) Latourian story of network development in climate science. See also Barron et al. (2011) and R. Rottenburg (2009) *Far-Fetched Facts: A Parable of Development Aid* (Cambridge, MA: MIT Press) for more development-related scholarship.
137. Latour (2005; 179–180).
138. As seemed to have been the case in Serel (2014). Hopefully, my own details are correct.
139. Compare with Ertsen (2010). See also R. Mrázek (2002) *Engineers of happy land: Technology and nationalism in a colony* (Princeton, NJ: Princeton University Press).
140. One could ask how practices could have changed anyway, as these were executed by many of the people that still had working experience in the colonies, who also trained new development workers. See Ertsen (2010) for a discussion how this worked within education of Dutch irrigation engineers at Delft University of Technology.

References

Abdelkarim A. (1986) "Wage labourers in the fragmented labour market of the Gezira, Sudan," *Africa: Journal of the International African Institute*, 56, 54–70.
Abdelkarim A. (1985) "Why do Gezira tenants withhold their households' labour," *Journal of Asian and African Studies*, 20, 72–88.
Adams D. W. and Coward E. W. (1971) *Small farmer development strategies: A seminar report* (Columbus: Ohio State University).
Adams M. E. (1980) Review of Barnett T. (1977) "The Gezira Scheme: An illusion of development," *Economic Development and Cultural Change*, 28, 633–636.
Adams W. M. (1991) "Large scale irrigation in Northern Nigeria: Performance and ideology," *Transactions of the Institute of British Geographers, New Series*, 16, 287–300.
Adams W. M. and Anderson D. M. (1988) "Irrigation before development: Indigenous and induced change in agricultural water management in East Africa," *African Affairs*, 87, 519–535.
Adas M. (2006) *Dominance by design: Technological imperatives and America's civilizing mission* (Cambridge, MA: Harvard University Press).
Adas M. (1989) *Machines as the measure of men: Science, technology, and ideologies of Western dominance* (Ithaca, NY: Cornell University Press).
Ahmed A. and Tiffen M. (1986) *Water management in the Gezira Scheme, Sudan: A survey of farmers attitudes on two minor canals* (Wallingford: Wallingford Hydraulic Research).
Akyeampong E., Bates R. H., Nunn N. and Robinson J. A. (2014) *Africa's development in historical perspective* (New York: Cambridge University Press).
Ali I. (1988) *The Punjab under imperialism 1885–1947* (Princeton, NJ: Princeton University Press).
Allen R. W. (1926) "The Gezira irrigation Scheme, Sudan," *Journal of the Royal African Society*, 25, 229–236.
Allen R. W. (1924) "Irrigation in the Sudan," *Journal of the Royal African Society*, 23, 257–264.
Al-Rahim M. (1969) *Imperialism and nationalism in the Sudan: A study in constitutional and political development, 1899–1956* (Oxford: Oxford University Press).
Anderson D. (2002) *Eroding the commons: The politics of ecology in Baringo, Kenya, 1890–1963* (Oxford: James Currey).
Andrae G. and Beckman B. (1985) *The wheat trap: Bread and underdevelopment in Nigeria* (London: Zed Books).

Andres B. J. (2014) *Power and control in the Imperial Valley: Nature, agribusiness, and workers on the California borderland, 1900–1940* (College Station: Texas A&M University Press).
Areshian G. E. (ed.) (2013) *Empires and diversity: On the crossroads of archaeology, anthropology, and history* (Los Angeles: Cotson Institute of Archaeology Press).
Arnold E. F. (2013) *Negotiating the landscape: Environment and monastic identity in the medieval Ardennes* (Philadelphia: University of Pennsylvania Press).
Atkinson A. B. (2014) *The colonial legacy: Income inequality in former British African colonies* (Helsinki: UNU-WIDER).
Aubet M. E. (2013) *Commerce and colonization in the ancient Near East* (Cambridge: Cambridge University Press).
Aw D. and Diemer G. (2005) *Making a large irrigation scheme work: A case study from Mali* (Washington: World Bank).
Ax C. F., Brimnes N., Jensen N. T. and Oslund K. (eds) (2011) *Cultivating the colonies: Colonial states and their environmental legacies* (Athens: Ohio University Press).
Bader L. (1927) "British colonial competition for the American cotton belt," *Economic Geography*, 3, 210–231.
Baker K. M. (1989) *Agricultural change in Nigeria* (London: John Murray).
Baldwin K. D. S. (1957) *The Niger agricultural project* (Cambridge: Harvard University Press).
Barbour K. M. (1980) "The Sudan since independence," *The Journal of Modern African Studies*, 18, 73–97.
Barbour K. M. (1960) Review of Gaitskell A. (1959) "Gezira: A story of development in the Sudan," *Africa: Journal of the International African Institute*, 30, 287–288.
Barbour K. M. (1959) "Irrigation in the Sudan: Its growth, distribution and potential expansion," *Transactions and Papers (Institute of British Geographers)*, No. 26, 243–263.
Barnes J. (2014) *Cultivating the Nile: The everyday politics of water in Egypt* (Durham, NC: Duke University Press).
Barnett T. (1979) "Why are bureaucrats slow adopters? The case of water management in the Gezira Scheme," *Sociologia Ruralis*, 19, 60–70.
Barnett T. (1977) *The Gezira Scheme: An illusion of development* (London: Frank Cass).
Barnett T. and Abdelkarim A. (1991) *Sudan: The Gezira Scheme and agricultural transition* (London: Frank Cass).
Barron P., Diprose R. and Woolcock M. (2011) *Contesting development: Participatory projects and local conflict dynamics in Indonesia* (New Haven, CT: Yale University Press).
Bascom J. B. (1990) "Food, wages, and profits: Mechanized schemes and the Sudanese state," *Economic Geography*, 66, 140–155.
Bassett T. J. (1988) "The development of cotton in Northern Ivory Coast, 1910–1965," *The Journal of African History*, 29, 267–284.
Bassett T. J. and Crummey D. (eds) (2003) *African savannas: Global narratives and local knowledge of environmental change* (Oxford: James Curry).
Beattie J., Melillo E. and O'Gorman E. (eds) (2015) *Eco-cultural networks and the British empire: New views on environmental history* (London: Bloomsbury).
Beer C. W. (1955) "Social development in the Gezira Scheme," *African Affairs*, 54, 42–51.

Beinart W. and Hughes L. (2007) *Environment and empire* (Oxford: Oxford University Press).
Bernal V. (1997) "Colonial moral economy and the discipline of development: The Gezira Scheme and 'modern' Sudan," *Cultural Anthropology*, 12, 447–479.
Bernal V. (1991) *Cultivating workers: Peasants and capitalism in a Sudanese village* (New York: Columbia University Press).
Bernal V. (1990) "The politics of research on agricultural development: An instructive example from the Sudan," *American Anthropologist*, 92, 732–739.
Bernal V. (1988) "Coercion and incentives in African agriculture: Insights from the Sudanese experience," *African Studies Review*, 31, 89–108.
Berry V. (1994) *The Culwick Papers 1934–1944: Population, food and health in colonial Tanganyika, by A. T. and G. M. Culwick* (London: Academic Books).
Biggs D. (2010) *Quagmire: Nation-building and nature in the Mekong Delta* (Seattle: University of Washington Press).
Bijker W. E. (1995) *Of bicycles, bakelites, and bulbs: Towards a theory of sociotechnical change* (Cambridge, MA: MIT Press).
Bird A. C. (1983) "The land issue in large scale irrigation projects, some problems from Northern Nigeria" in Adams W. M. and Grove A. T. (eds) *Irrigation in tropical Africa: Problems and problem solving* (Cambridge: Cambridge University Press).
Black M. (2007) *The No-Nonsense guide to international development* (Oxford: New Internationalist Publications).
Blewett W. V. (1960) Review of Gaitskell A. (1959) "Gezira: A story of development in the Sudan," *African Affairs*, 59, 260–261.
Brautigam D. (2009) *The dragon's gift: The real story of China in Africa* (Oxford: Oxford University Press).
Briggs J. A. (1978) "'Farmers' responses to planned agricultural development in the Sudan," *Transactions of the Institute of British Geographers, New Series*, 3, 464–475.
Broadbent P. (1960) Review of Gaitskell A. (1959) "Gezira: A story of development in the Sudan," *International Affairs*, 36, 399–400.
Bush B. (2006) *Imperialism and postcolonialism* (Harlow: Pearson Education).
Butlin R. A. (2009) *Geographies of empire: European empires and colonies, c. 1880–1960* (Cambridge: Cambridge University Press).
Byerley A. (2012) *Africa as laboratory and the complexity of epistemic decolonisation* (Uppsala: Nordic Africa Institute).
Camprubi Bueno L. (2011) *Political engineering: science, technology and the Francoist landscape (1939–1959)* (PhD Thesis, University of California, Los Angeles).
Carr C. E. (1973) Review of Chambers R. (1969) "Settlement schemes in tropical Africa: A study of organizations and development," *Journal of Asian and African Studies*, 8, 99–101.
Casid J. H. (2005) *Sowing empire: Landscape and colonization* (Minneapolis: University of Minnesota Press).
Central Research Farm (1915) *Pump irrigation in the northern Sudan: With special reference to the cotton crop* (Khartoum: Central Research Farm).
Chambers R. (2005) *Ideas for development* (London: Earthscan).
Chambers R. (1969) *Settlement schemes in tropical Africa: A study of organisations and development* (London: Routledge).
Chambers R. and Moris J. (eds) (1973) *Mwea, an irrigated rice settlement in Kenya* (München: Weltforum Verlag).

Chari S. and Corbridge S. (eds) (2008) *The development reader* (London: Routledge).
Cheeseboro A. Q. (1993) *Administration and change in the Gezira scheme and the Sudan, 1938–1970* (PhD Thesis, Michigan State University).
Clarkson A. I. (2005) *Courts, councils and citizenship: Political culture in the Gezira scheme in condominium Sudan* (PhD Thesis, Durham University).
Clawson M (ed.) (1964) *Natural resources and international development* (Baltimore, MD: Johns Hopkins Press).
Cline E. H. and Graham M. W. (2011) *Ancient empires: From Mesopotamia to the rise of Islam* (New York: Cambridge University Press).
Collier P. (2008) *The bottom billion: Why the poorest countries are failing and what can be done about it* (Oxford: Oxford University Press).
Collins R. O. (2008) *A history of modern Sudan* (Cambridge: Cambridge University Press).
Constantini D. (2008) *Mission Civilisatrice: Le role de l'histoire coloniale dans la construction de l'identite politique francaise* (Paris: Editions La Decouverte).
Cookson-Hills C. J. (2013) *Engineering the Nile: Irrigation and the British Empire in Egypt, 1882–1914* (PhD Thesis, Queen's University).
Cross P. (1997) "British attitudes to Sudanese labour: The Foreign Office Records as sources for social history," *British Journal of Middle Eastern Studies*, 24, 217–260.
Crowther E. M. and Crowther F. (1935) "Rainfall and cotton yields in the Sudan Gezira," *Proceedings of the Royal Society of London, Series B, Biological Sciences*, 118, 343–370.
Cullather N. (2000) "Development? It's history," *Diplomatic History*, 24, 641–653.
Culwick A. T. (1963) *Britannia waves the rules* (Cape Town: Nasionale Boekhandel).
Culwick A. T. (1942) *Good out of Africa* (Livingston: Rhodes-Livingstone Institute).
Culwick A. T. and Culwick G. M. (1935) "Culture contact on the fringe of civilization," *Africa: Journal of the International African Institute*, 8, 163–170.
Culwick A. T. and Oosthuizen J. C. (1970) *The inequality principle* (Durban: Dolphin Press).
Culwick G. M. (1955) *A study of the human factor in the Gezira Scheme* (Barakat: Sudan Gezira Board).
Daly M. W. (1991) *Imperial Sudan: The Anglo-Egyptian Condominium, 1934–1956* (Cambridge: Cambridge University Press).
Daly M. W. (1986) *Empire on the Nile: The Anglo-Egyptian Sudan, 1898–1934* (Cambridge: Cambridge University Press).
Daly M. W. (1985) *Modernization in the Sudan: Essays in honor of Richard Hill* (New York: Lillian Barber Press).
Darwin J. (2012) *Unfinished Empire: The global expansion of Britain* (London: Allen Lane/Penguin).
Davis D. K. (2007) *Resurrecting the granary of Rome: Environmental history and French colonial expansion in North Africa* (Athens: Ohio University Press).
Davis D. K. and Burke E. (eds) (2011) *Environmental imaginaries of the Middle East and North Africa* (Athens: Ohio University Press).
De Wilde J. C. (1967a) *Experiences with agricultural development in tropical Africa. Vol I: The synthesis* (Baltimore, MD: Johns Hopkins Press).

De Wilde J. C. (1967b) *Experiences with agricultural development in tropical Africa. Vol II: The case studies* (Baltimore, MD: Johns Hopkins Press).

Dickson H. (2012) *Old reliable in Africa*, Forgotten Books, original publication 1920 (New York: Frederick A. Stokes Company).

Diemer G. (1990) *Irrigatie in Afrika: Boeren en ingenieurs, techniek en kultuur* (Amsterdam: Thesis Publishers).

Diemer G. and Vincent L. (1992) "Irrigation in Africa: The failure of collective memory and collective understanding," *Development Policy Review*, 10, 131–154.

Doran A. (1980) "Agricultural extension and development: The Sudanese experience," *Bulletin (British Society for Middle Eastern Studies)*, 7, 39–48.

Du Bois W. E. B. (1898) "The study of the negro problems," *The Annals of the American Academy of Political and Social Science*, 11, 1–23.

Dufour F. (2010) *De l'idéologie coloniale à celle du développement* (Paris: L'Harmattan).

Dumont R. and Rosier B. (1969) *The hungry future* (London: Andre Deutsch).

Easterly W. (2006) *The white man's burden: Why the West's efforts to aid the rest have done so much ill and so little good* (New York: Penguin).

Eckert A. (2008) "'We are all planners now': Planung und Dekolonisation in Afrika," *Geschichte und Gesellschaft*, 34, 375–397.

Edwards P. N. (2010) *A vast machine: Computer models, climate data, and the politics of global warming* (Cambridge, MA: MIT Press).

Eldaw A. M. (2004) *The Gezira Scheme: Perspectives for sustainable development* (Bonn: German Development Institute).

Ellis G. (1979a) Review of Barnett T. (1977) "The Gezira Scheme: An illusion of development," *ASA Reviews of Books*, 5, 34–36.

Ellis G. (1979b) Review of Barnett T. (1977) "The Gezira Scheme: An illusion of development," *The Journal of Modern African Studies*, 17, 340–342.

Elson R. E. (1984) *Javanese peasants and the colonial sugar industry: Impact and change in an East Java residency, 1830–1940* (Singapore: Oxford University Press).

Ennis P., Healey M. and Purdue F. (2004) *Negotiating development: Rationales and practice for development obligations and planning gain* (London: E&FN Spon).

Ertsen M. W. (2014) "Step after step the ladder is ascended: Human agency in irrigated (anti) landscapes," in Nye D. and Elkind S. (eds) *The Anti-Landscape* (Amsterdam: Rodopi).

Ertsen M. W. (2013) "A poor workman blames his tools or how irrigation systems structure human actions" in Casella E., Evans G., Harvey P., Knox H., Mclean C., Silva E., Thoburn N. and Woodward K. (eds) *Objects and materials: A Routledge companion* (London: Routledge).

Ertsen M. W. (2010) *Locales of Happiness: Colonial irrigation in the Netherlands East Indies and its remains, 1830–1980* (Delft: VSSD Press).

Ertsen M. W. (2008) "Controlling the farmer: Irrigation encounters in Kano, Nigeria," *TD: The Journal for Transdisciplinary Research in Southern Africa*, 4, 209–236.

Ertsen M. W. (2007) "The development of irrigation design schools or how history structures human action," *Irrigation and Drainage*, 56, 1–19.

Ertsen M. W. (2006) "Colonial irrigation: Myths of emptiness," *Landscape Research*, 31, 147–167.

Ertsen M. W., Murphy J. T., Purdue L. E. and Zhu T. (2014) "A journey of a thousand miles begins with one small step: Human agency, hydrological processes and time in socio-hydrology," *Hydrology and Earth Systems Sciences*, 18, 1369–1382.

Etemad B. (2007) *Possessing the world: Taking the measurements of colonisation from the 18th to the 20th century* (New York: Berghahn Books).

Euroconsult, Sir Alexander Gibb and Partners and TCS Sudan (1982) *Sudan Gezira Rehabilitation Project* (6 volumes).

FAO (1987) *Consultation on irrigation in Africa* (Rome: Food and Agricultural Organization).

Farber P. L. (2011) *Mixing races: From scientific racism to modern evolutionary ideas* (Baltimore, MD: Johns Hopkins University Press).

Farbrother H. G. (1979) *Irrigation practices in the Gezira* (London: Cotton Research Corporation).

Farbrother H. G. (1976) *Technical notes on water use* (Agricultural Research Corporation).

Farbrother H. G. (1974) *Irrigation practices in the Gezira* (Khartoum: Cotton Research Report).

Farson N. (1949) *Last chance in Africa* (London: Victor Gollancz).

Ferguson N. (ed.) (1997) *Virtual history: Alternatives and counterfactuals* (New York: Basic Books).

Fiege M. (1999) *Irrigated Eden: The making of an agricultural landscape in the American West* (Seattle: University of Washington Press).

Flannery K. and Marcus J. (2012) *The creation of inequality: How our prehistoric ancestors set the stage for monarchy, slavery, and empire* (Cambridge, MA: Harvard University Press).

Forrest T. (1981) "Agricultural policies in Nigeria, 1900–78" in Heyer J., Roberts P. and Williams G. (eds) *Rural development in tropical Africa* (New York: St. Martin's Press).

Foucault M. (1995) *Discipline and punish: The birth of the prison* (New York: Vintage Books).

Foucault M. (1988) *Madness and civilization: A history of insanity in the age of reason* (New York: Vintage Books).

Francis M. R. and Elawad O. (1986) *A study of the management of minor canals in the Gezira Irrigation Scheme, Sudan* (Wallingford: Wallingford Hydraulics Research).

Francis M. R. and Elaweed O. M. A. (1989) *Diagnostic investigations and rehabilitation of canals in the Gezira Scheme, Sudan* (Manila: Asia Regional Symposium).

Francis M. R. and Hinton R. (1987) *Minor Canal Management in the Gezira Irrigation Scheme, Sudan: An interim report on Diagnostic Investigations during 1986–1987* (Wallingford: Wallingford Hydraulics Research).

Gaitskell A. (1964) "Resource development among African countries," in Clawson M. (ed.) *Natural resources and international development* (Baltimore, MD: Johns Hopkins Press).

Gaitskell A. (1959) *Gezira: A story of development in the Sudan* (London: Faber and Faber).

Gaitskell A. (1955) *What have they to defend?* (The Africa Bureau Annual Anniversary Address).

Gaitskell A. (1952) "The Sudan Gezira Scheme," *African Affairs*, 51, 306–313.

Giddens A. (1990) *The consequences of modernity* (Cambridge: Polity Press).

Giddens A. (1984) *The constitution of society: Outline of the Theory of Structuration* (Oakland: University of California Press).
Giddens A. (1979) *Central problems in social theory: Action, structure and contradiction in social analysis, Contemporary social theory* (London: Macmillan).
Gilman N. (2009) Review of "Modernization as a global project," Diplomatic History, 23 (2009), *H-Diplo Article Reviews* at http://www.h-net.org/~diplo/reviews/, No. 238-B (A two-part Forum Review), published on July 29, 2009.
Goldman M. (2005) *Imperial nature: The World Bank and struggles for social justice in the age of globalization* (New Haven, CT: Yale University Press).
Golkowsky R. (1969) *Bewässerungslandwirtschaft in Kenya: Darstellung grundsätzlicher Zusammenhänge am Beispiel des Mwea Irrigation Settlement* (München: Weltforum Verlag).
Green G. (1970) Review of Chambers R. (1969) "Settlement schemes in tropical Africa: A study of organizations and development," *African Affairs*, 69, 397.
Hance W. A. (1961) Review of Gaitskell A. (1959) "Gezira: A story of development in the Sudan," *Geographical Review*, 51, 157–158.
Hance W. A. (1955) "The Zande Scheme in the Anglo-Egyptian Sudan," *Economic Geography*, 31, 149–156.
Hance W. (1954) "The Gezira: An example in development," *Geographical Review*, 44, 253–270.
Hanson A. H. (1970) Review of Chambers R. (1969) "Settlement schemes in tropical Africa: A study of organizations and development," *The Journal of Modern African Studies*, 9, 143–146.
Hariri J. G. (2012) "The autocratic legacy of early statehood," *American Political Science Review*, 106, 471–494.
Harman G. (2014) *Bruno Latour: Reassembling the political* (London: Pluto Press).
Harman G. (2011) *The prince and the wolf: Latour and Harman at the LSE* (Winchester: Zero Books).
Harman G. (2009) *Prince of networks: Bruno Latour and metaphysics* (Melbourne: re.press).
Headrick D. R. (1988) *The tentacles of progress* (New York: Oxford University Press).
Headrick D. R. (1981) *The tools of empire: Technology and European imperialism in the nineteenth century* (New York: Oxford University Press).
Heather P. (2010) *Empires and barbarians: The fall of Rome and the birth of Europe* (Oxford: Oxford University Press).
Heldring L. and Robinson J. A. (2012) "Colonialism and development in Africa," *African Economic History Working Paper Series*, No. 5.
Hellen J. A. (1970) Review of Chambers R. (1969) "Settlement schemes in tropical Africa: A study of organizations and development," *International Affairs*, 46, 385–387.
Heyer J., Roberts P. and Williams G. (1981) "Rural development" in Heyer J., Roberts P. and Williams G. (eds) *Rural development in tropical Africa* (New York: St. Martin's Press).
Himbury W. H. (1918) "Empire cotton," *Journal of the Royal African Society*, 17, 262–275.
Hoag H. J. (2013) *Developing the rivers of East and West Africa: An environmental history* (London: Bloomsbury).
Hodge J. M. (2007) *Triumph of the expert: Agrarian doctrines of development and the legacies of British colonialism* (Athens: Ohio University Press).

Hodge J. M., Hödl G. and Kopf M. (eds) (2014) *Developing Africa: Concepts and practices in twentieth-century colonialism* (Manchester: Manchester University Press).
Hogendorn J. S. and Scott K. M. (1981) "The East African Groundnut Scheme: Lessons of a large-scale agricultural failure," *African Economic History*, No. 10, 81–115.
Holt P. M. and Daly M. W. (eds) (2011) *A history of the Sudan: From the coming of Islam to the present day*, 6th edn (London: Longman).
Holtham G. and Hazlewood A. (1976) *Aid and inequality in Kenya: British development assistance to Kenya* (London: Croom Helm).
Hopper P. (2012) *Understanding development* (Cambridge: Polity Press).
Hunter G. (1969) *Modernizing peasant societies: A comparative study in Asia and Africa* (New York: Oxford University Press).
Huxley E. J. G. (1960) *A new earth: An experiment in colonialism* (London: Chatto & Windus).
Information Centre of the Ministry of Social Affairs (undated) *The Managil Extension: An achievement of the Republic of Sudan after independence*.
Irvine R. J. (1960) Review of Gaitskell A. (1959) "Gezira: A story of development in the Sudan," *The American Economic Review*, 50, 751–753.
Isaacman A. and Roberts R. (eds) (1995) *Cotton, colonialism and social history in Sub-Saharan Africa* (Portsmouth: Heinemann).
Jennings E. T. (2006) *Curing the colonizers: Hydrotherapy, climatology, and French colonial spas* (Durham, NC: Duke University Press).
Joergensen D., Joergensen F. A. and Pritchard S. B. (eds) (2013) *New Natures: Joining environmental history with science and technology studies* (Pittsburgh: University of Pittsburgh Press).
Kanogo T. (1987) *Squatters and the roots of Mau Mau* (London: James Currey).
Karns Alexander J. (2008) *The mantra of efficiency: From waterwheel to social control* (Baltimore, MD: Johns Hopkins University Press).
Katz C. (1991) "An agricultural project comes to town: Consequences of an encounter in Sudan," *Social Text*, No 28, 31–38.
Kelemen P. (2007) "Planning for Africa: The British Labour Party's colonial development policy, 1920–1964," *Journal of Agrarian Change*, 7, 76–98.
Kenrick R. (ed.) (1987) *Sudan Tales: Recollections of some Sudan Political Service wives 1926–1956* (Cambridge: The Oleander Press).
Kenya Human Rights Commission (2000) *Dying to be free: The struggle for rights in Mwea* (Nairobi).
Kinsbergen S. (2014) *Behind the pictures: Understanding private development initiatives* (Phd Thesis, Radboud Universiteit).
Kirpich P. Z. (1987) "Developing countries: High tech or innovative management?" *Journal of Professional Issues in Engineering*, 113, 150–166; with associated discussion file (1989) by several authors, *Journal of Professional Issues in Engineering*, 115, 67–84 (1989).
Kohlrausch M. and Trischler H. (2014) *Building Europe on expertise: Innovators, organizers, networks* (Basingstoke: Palgrave Macmillan).
Kortenhorst L. F., Van Steekelenburg P. N. G. and Sprey L. H. (1989) "Prospects and problems of irrigation development in Sahelian and sub-Saharan Africa," *Irrigation and Drainage Systems*, 3, 13–45.
Kothari U. (ed.) (2005) *A radical history of development studies: Individuals, insittutions and ideologies* (London: Zed Books).

Kreike E. (2013) *Environmental infrastructure in African history: Examining the myth of natural resource management in Namibia* (New York: Cambridge University Press).

Kunkel M. F. S. (2011) "Writing the history of development: A review of the recent literature," *Contemporary European History*, 20, 215–232.

Lachenmann G. and Dannecker P. (eds) (2008) *Negotiating development in Muslim societies: Gendered spaces and translocal connections* (Plymouth: Lexington Books).

Latour B. (2013) *An inquiry into modes of existence: An anthropology of the moderns* (Cambridge, MA: Harvard University Press).

Latour B. (2005) *Reassembling the social: An introduction to actor–network theory* (Oxford: Oxford University Press).

Latour B. (1996) *Aramis, or the love of technology* (Cambridge, MA: Harvard University Press).

Latour B. (1993) *We have never been modern* (Cambridge, MA: Harvard University Press).

Latour B. (1987) *Science in action: How to follow scientists and engineers through society* (Cambridge, MA: Harvard University Press).

Lees S. H. (1986) "Coping with bureaucracy: Survival strategies in irrigated agriculture," *American Anthropologist*, 88, 610–622.

Leigh-Smith J. (2014) *Works in progress: Plans and realities on Soviet farms, 1930–1963* (New Haven, CT: Yale University Press).

Lilienthal D. E. (1944) *Tennessee Valley Authority (TVA): Democracy on the March* (New York: Penguin).

Lumsden D. P. (1970) Review of Chambers R. (1969) "Settlement schemes in tropical Africa: A study of organizations and development," *Africa: Journal of the International African Institute*, 40, 290–291.

Lund C. (2013) "The past and space: On arguments in African land control," *Africa*, 83, 14–35.

Maathai W. (2009) *The challenge for Africa* (London: Arrow Books).

MacLeod R. (ed.) (2000) "Nature and empire: Science and the colonial enterprise," *Osiris*, 15.

Manig G. G. (1973) "Engineering aspects and water control" in Chambers R. and Moris J. (eds) *Mwea, an irrigated rice settlement in Kenya* (München: Weltforum Verlag).

Matunhu J. (2011) "A critique of modernization and dependency theories in Africa: Critical assessment," *African Journal of History and Culture*, 3, 65–72.

McKnight G. H. (2013) "Harry Johnston's new boot: The Uganda Agreement and ideas of development," *Journal of Historical Sociology*, 26, 234–263.

McLoughlin P. F. M. (1962) "Economic development and the heritage of slavery in the Sudan Republic," *Africa: Journal of the International African Institute*, 32, 355–391.

Memmi A. (1991) *The colonizer and the colonized* (Boston, MA: Beacon Press).

Meredith M. (2005) *The state of Africa: A history of the continent since independence* (London: Simon and Schuster).

Mikhail A. (ed.) (2012) *Water on sand: Environmental histories of the Middle East and North Africa* (Oxford: Oxford University Press).

Mikhail A. (2011) *Nature and empire in Ottoman Egypt: An environmental history* (Cambridge: Cambridge University Press).

Miracle M. P. (1975) Review of Chambers R. and Moris J. (eds) (1973) "Mwea, an irrigated rice settlement in Kenya," *The International Journal of African Historical Studies*, 8, 319–321.

Misa T. J., Brey P. and Feenberg A. (eds) (2003) *Modernity and technology* (Cambridge, MA: MIT Press).

Mishra P. (2013) *From the ruins of empire: The revolt against the West and the remaking of Asia* (London: Penguin).

Mitchell (2002) *Rule of experts: Egypt, techno-politics, modernity* (Oakland: University of California Press).

Mohan G. and Power M. (2009) "Africa, China and the 'new' economic geography of development," *Singapore Journal of Tropical Geography*, 30, 24–28.

Mollan S. M. (2008) *Economic imperialism and the political economy of Sudan: The case of the Sudan Plantations Syndicate, 1899–1956* (PhD Thesis, Durham University).

Moon S. (2007) *Technology and ethical idealism: A history of development in the Netherlands East Indies* (Leiden: CNWS Publications).

Moris J. R. and Thom D. J. (1990) *Irrigation development in Africa: Lessons of experience* (Boulder, CO: Westview Press).

Moris J. (1973) "The Mwea environment" in Chambers R. and Moris J. (eds) *Mwea, an irrigated rice settlement in Kenya* (München: Weltforum Verlag).

Morris I. (2013) *The measure of civilisation: How social development decides the fate of nations* (London: Profile Books Ltd).

Morris I. and Scheidel W. (eds) (2009) *The dynamics of ancient empires: State power from Assyria to Byzantium* (Oxford: Oxford University Press).

Moseley W. and Gray L. C. (eds) (2008) *Hanging by a thread: Cotton, globalization, and poverty in Africa* (Athens: Ohio University Press).

Mosse D. (2003) *The rule of water: Statecraft, ecology and collective action in South India* (Oxford: Oxford University Press).

Moyo D. (2009) *Dead aid: Why aid is not working and how there is a better way for Africa* (New York: Farrar, Straus and Giroux).

Mrázek R. (2002) *Engineers of happy land: Technology and nationalism in a colony* (Princeton, NJ: Princeton University Press).

Mukerji C. (2009) *Impossible engineering: Technology and territoriality on the Canal du Midi* (Princeton, NJ: Princeton University Press).

Ndlovu-Gatsheni S. J. (2013) *Empire, global coloniality and African subjectivity* (New York: Berghahn Books).

NEDECO (1974) *Some considerations and recommendations on the management of the Kano River Project* (Amsterdam: NEDECO).

NEDECO (1970) *Kano River Project feasibility study* (Amsterdam: NEDECO).

Niblock T. (1987) *Class and power in Sudan: The dynamics of Sudanese politics, 1898–1985* (Albany: State University of New York Press).

Noah S. E. L. (1965) "Agricultural extension—its role and importance in agricultural development in the Sudan" in *Agricultural development in the Sudan* (Khartoum: Philosophical Society of the Sudan, Sudan Agricultural Society).

O'Brien J. (1981) "Sudan: An Arab breadbasket?," *MERIP Reports*, No. 99, 20–26.

Oesterdiekhoff P. and Wohlmuth K. (eds) (1983) *The development perspectives of the Democratic Republic of Sudan: The limits of the Breadbasket Strategy* (München: Weltforum Verlag).

Oldenziel R. and Hard M. (2013) *Consumers, tinkerers, rebels: The people who shaped Europe* (Basingstoke: Palgrave Macmillan).
Onyeiwu S. (2000) "Deceived by African cotton: The British Cotton Growing Association and the demise of the Lancashire textile industry," *African Economic History*, No. 28, 89–121.
Oorthuizen J. (2003) *Water works and wages: The everyday politics of irrigation management reform in the Philippines* (Hyderabad: Orient Longman).
Palmer-Jones R. W. (1981) "How not to learn from pilot irrigation projects: The Nigerian experience," *Water Supply and Management*, 5, 81–105.
Pearce R. D. (1982) *The turning point in Africa: British colonial policy 1938–48* (London: Frank Cass).
Pfaffenberger B. (1990) "The harsh facts of hydraulics: Technology and society in Sri Lanka's colonization schemes," *Technology and Culture*, 31, 361–397.
Picon A. (2004) "Engineers and engineering history: Problems and perspectives," *History and Technology*, 20, 421–436.
Picon A. (2000) "Technological traditions and national identities: A comparison between France and Great Britain during the XIXth century" in Nicolaidis E. and Chatzis K. (eds) *Science, technology and the 19th century state* (Athens: Institut de Recherches Neohelleniques).
Pierce S. (2005) *Farmers and the state in colonial Kano: Land Tenure and the Legal Imagination* (Bloomington: Indiana University Press).
Pluesquellec H. (1990) *The Gezira Scheme in Sudan: Objectives, design and performance* (Washington: World Bank).
Pollard N. (1984) "The Sudan's Gezira Scheme: A study in failure" in Goldsmith E. and Hildyard N. (eds) *The social and environmental effects of large dams. Volume 2: Case studies* (Camelford: Wadebridge Ecological Center).
Poncet J. (1961) *La colonization et l'agriculture européennes en Tunisie depuis 1881* (Paris: Mouton).
Porter A. (2012) *Mobile pastoralism and the formation of Near Eastern civilizations: Weaving together society* (New York: Cambridge University Press).
Prados de la Escosura l. (2013) "World human development: 1870–2007," *EHES Working papers in Economic History*, No. 34.
Préfol P. (1986) *Prodige de l'irrigation au Maroc: Le développement exemplaire du TADLA 1936–1985* (Paris: Nouvelle Editions Latines).
Press & Information Officer (ed.) (1959) *The Gezira scheme from within: A collection of articles by heads of departments* (Khartoum: The Middle East Press).
Pritchard S. B. (2012) "From hydroimperialism to hydrocapitalism: 'French' Hydraulics in France, North Africa, and beyond," *Social Studies of Science*, 42, 591–615.
Pritchard S. B. (2011) *Confluence: The nature of technology and the remaking of the Rhône* (Cambridge, MA: Harvard University Press).
Rand L. B. (1989) *High Stakes: The life and times of Leigh S. J. Hunt* (New York: Peter Lang).
Ratcliffe B. M. (1982) "Cotton imperialism: Manchester merchants and cotton cultivation in West Africa in the mid-nineteenth century," *African Economic History*, No 11, 87–113.
Reader J. (1998) *Africa: A biography of a continent* (London: Penguin).
Reij C. and Waters-Bayer A. (2001) *Farmer innovation in Africa: A source of inspiration for agricultural development* (London: Earthscan).

Reining C. C. (1966) *The Zande scheme: An anthropological case study of economic development in Africa* (Evanston: Northwestern University Press).

Reisner M. (1993) *Cadillac desert: The American West and its disappearing water* (New York: Penguin).

Report of the East Africa Commission presented to the British Parliament by the Secretary of State for the Colonies (1925) (London: His Majesty's Stationary Office).

Reuss M. and Cutcliffe S. H. (eds) (2010) *The illusory boundary: Environment and technology in history* (Charlottesville: University of Virginia Press).

Roberts R. L. (1996) *Two worlds of cotton: Colonialism and the regional economy in the French Soudan, 1800–1946* (Stanford, CA: Stanford University Press).

Robertson J. (1974) *Transition in Africa* (London: Hurst & Company).

Robins J. (2013) "Coercion and resistance in the colonial market: Cotton in Britain's African Empire" in Curry-Machado J. (ed.) *Global Histories, Imperial Commodities, Local Interactions* (London: Palgrave Macmillan).

Rodney W. (1973) *How Europe underdeveloped Africa* (London: Bogle-L'Ouverture Publications; Dar-Es-Salaam: Tanzanian Publishing House).

Rostov W. W. (1960) *The stages of economic growth: A non-communist manifesto* (London: Cambridge at the University Press).

Rotimi O. (2013) "A critique of Walter Rodney's concept of development," *International Journal of Education and Research*, 1, 157–166.

Rottenburg R. (2009) *Far-Fetched Facts: A Parable of Development Aid* (Cambridge, MA: MIT Press).

Russell E. (2011) *Evolutionary history: Uniting history and biology to understand life on earth* (New York: Cambridge University Press).

Ryle J., Willis J., Baldo S. and Jok J. M. (2011) *The Sudan Handbook* (Woodbridge: James Curry).

Sagardoy J. A., Bottrall A. and Uittenbogaard G. O. (1986) *Organization, operation and maintenance of irrigation schemes* (Rome: Food and Agriculture Organization).

Sahlins M. (2004) *Apologies to Thucydides: Understanding history as culture and vice versa* (Chicago: University of Chicago Press).

Salzman P. C. (1978) Review of Barnett T. (1977) "The Gezira Scheme: An illusion of development," *Annals of the American Academy of Political and Social Science*, 440, 174–175.

Samoff J. (1978) Review of Barnett T. (1977) "The Gezira Scheme: An illusion of development," *The American Historical Review*, 83, 1068–1069.

Sauder R. A. (2009) *The Yuma reclamation project: Irrigation, Indian allotment, and settlement along the lower Colorado river* (Reno: University of Nevada Press).

Schreyger E. (1984) *L'Office du Niger au Mali 1931 à 1982: La problématique d'une grande enterprise agricole dans la zone du Sahel* (Stuttgart: Beitrage zur Kolonial- und Überseegeschichte 27).

Scott J. M. (2009) *The art of not being governed: An anarchist history of upland Southeast Asia* (New Haven, CT: Yale University Press).

Scott J. C. (1999) *Seeing like a state: How certain schemes to improve the human condition have failed* (New Haven, CT: Yale University Press).

Scott J. M. (1985) *Weapons of the weak: Everyday forms of peasant resistance* (New Haven, CT: Yale University Press).

Scott J. M. (1976) *The moral economy of the peasant: Rebellion and subsistence in Southeast Asia* (New Haven, CT: Yale University Press).

Seddon D. (1978) Review of Barnett T. (1977) "The Gezira Scheme: An illusion of development," *International Affairs*, 54, 161–163.

Segal H. P. (2006) *Technology and Utopia* (Washington: American Historical Association).

Serels S. (2013) *Starvation and the State: Famine, slavery, and power in Sudan, 1883–1956* (New York: Palgrave Macmillan).

Serels S. (2007) "Political landscaping: Land registration, the definition of ownership and the evolution of colonial objectives in the Anglo-Egyptian Sudan, 1899–1924," *African Economic History*, No. 35, 59–75.

Sewell W. H. (2005) *Logics of History: Social theory and social transformation* (Chicago: University of Chicago Press).

Sharkey H. J. (20030 *Living with colonialism: Nationalism and culture in the Anglo-Egyptian Sudan* (Berkeley: University of California Press).

Shaw D. J. (1965a) "The development and contribution of irrigated agriculture in the Sudan," in *Agricultural Development in the Sudan* (Khartoum: Philosophical Society of the Sudan, Sudan Agricultural Society).

Shaw D. J. (1965b) "The Managil South-Western extension: An extension to the Gezira scheme: An example of an irrigation development in the Republic of the Sudan," *Koninklijk Nederlands Aardrijkskundig Genootschap*, 82.

Siann J. M. (1983) "Labour constraints in the implementation of irrigation projects," in Adams W. M. and Grove A. T. (eds) *Irrigation in tropical Africa: Problems and problem solving* (Cambridge: African Studies Center).

Simpson I. G. (1970) "New approaches to irrigated farming in the Sudan: Organization and management," *Land Economics*, 46, 287–296.

Simpson I. G. (1965) "Economic aspects of diversification in the Gezira," in *Agricultural Development in the Sudan* (Khartoum: Philosophical Society of the Sudan, Sudan Agricultural Society).

Sluyter A. (2002) *Colonialism and landscape: Postcolonial theory and applications* (Lanham: Rowman and Littlefield).

Smyth R. (2004) "The roots of community development in Colonial Office: Policy and Practice in Africa," *Social Policy and Administration*, 38, 418–436.

Sörlin S. and Warde P. (2011) *Nature's end: History and the environment* (Basingstoke: Palgrave).

Spaulding J. (1988) "The business of slavery in the central Anglo-Egyptian Sudan, 1910–1930," *African Economic History*, No. 17, 23–44.

Spitz G. (1949) *Sansanding: Les irrigations du Niger* (Paris: Société d'Etitions Géographiques, Maritimes & Coloniales).

Stone I. (1984) *Canal Irrigation in British India: Perspectives on technological change in a peasant society* (Cambridge: Cambridge University Press).

Strong T. H. and Paton P. (1968) *Economic appraisal and analysis of requirements for development of the Talata Mafara area, Sokoto Rima Drainage Basin* (Rome: United Nations Development Program and Food and Agriculture Organization).

Stunden Bower S. (2011) *Wet prairie: People, land, and water in agricultural Manitoba* (Vancouver: University of British Colombia Press).

Stuurman S. (2009) *De uitvinding van de mensheid: Korte wereldgeschiedenis van het denken over gelijkheid en cultuurverschil* (Amsterdam: Bert Bakker).

Sudan Gezira Board (1967) *The Sudan Gezira Board: What it is and how it works*.

Sunseri T. (2001) "The Baumwollfrage: Cotton colonialism in German East Africa," *Central European History*, 34, 31–51.

Swearingen W. D. (1987) *Moroccan mirages: Agrarian dreams and deceptions, 1912–1986* (Princeton, NJ: Princeton University Press).

Swearingen W. D. (1984) *In search of the granary of Rome: Irrigation and agricultural development in Morocco, 1912–1982* (PhD Thesis, University of Texas).

Taha T. J. (1965) "Optimisation of production in the Gezira Scheme," in *Agricultural Development in the Sudan* (Khartoum: Philosophical Society of the Sudan, Sudan Agricultural Society).

Tait J. (1979) "Interner Kolonialismus und ethnisch-soziale Segregation im Sudan: Nigerianisch-Westafrikanische Arbeitsmigranten und das Arbeitsmarktsystem in der Gezira," *Africa Spectrum*, 14, 361–382.

Tait J. (1978) "Diversifizierung, Mechanisierung und Kapitalisierung der Produktion im Sudan Gezira Scheme: Auf dem Weg zur Überwindung kolonial deformierter Agrarstrukturen?," *Africa Spectrum*, 13, 165–178.

Teisch J. B. (2011) *Engineering Nature: Water, development, and the global spread of American environmental expertise* (Chapel Hill: University of North Carolina Press).

Teisch J. (2005) "'Home is not so very far away': Californian engineers in South Africa, 1868–1915," *Australian Economic History Review*, 45, 139–160.

Thornton D. S. (1966) "Contrasting policies in irrigation development, Sudan and India," *Development Studies*, No. 1, University of Reading.

Tignore R. L. (1966) *Modernization and British colonial rule in Egypt, 1882–1914* (Princeton, NJ: Princeton University Press).

Tilley H. (2011) *Africa as a living laboratory: Empire, development, and the problem of scientific knowledge, 1870–1950* (Chicago: University of Chicago Press).

Tilley H. and Gordon R. J. (eds) (2007) *Ordering Africa: Anthropology, European imperialism, and the politics of knowledge* (Manchester: Manchester University Press).

Tropp J. A. (2006) *Natures of colonial change: Environmental relations in the making of the Transkei* (Athens: Ohio University Press).

Tvedt T. (2004) *The river Nile in the age of the British: Political ecology and the quest for economic power* (London: I. B. Taurus).

Ubels J. and Horst L. (eds) (1993) *Irrigation design in Africa: Towards an interactive method* (Ede: Technical Center for Agricultural and Rural Cooperation).

UNESCO (1964) "Social aspects of African resource development," *International Social Science Journal*, 16.

Van Beusekom M. M. (2002) *Negotiating development: African farmers and colonial experts at the Office du Niger, 1920–1960* (Portsmouth: Heinemann).

Van Beusekom M. M. (1989) *Colonial rural development: French policy and African response at the Office du Niger, Soudan Français (Mali), 1920–1960* (PhD Thesis, Johns Hopkins University).

Van de Poel I. (2003) "The transformation of technological regimes," *Research Policy*, 32, 49–68.

Van Reybrouck D. G. (2010) *Congo: Een geschiedenis* (Amsterdam: De Bezige Bij) (or (2012) *Congo: Une histoire* (Arles: Editions Actes Sud) or (2014) *Congo: The epic history of a people* (New York, Harper Collins).

Veen J. J. (1973) "The production system" in Chambers R. and Moris J. (eds) *Mwea, an irrigated rice settlement in Kenya* (München: Weltforum Verlag).

Versluys J. D. N. (1954a) "The Gezira Scheme in the Sudan and the Russian Kolkhoz: A comparison of two experiments, Part II," *Economic Development and Cultural Change*, 2, 120–135.

Versluys J. D. N. (1954b) "The Gezira Scheme in the Sudan and the Russian Kolkhoz: A comparison of two experiments, Part III," *Economic Development and Cultural Change*, 2, 216–235.
Versluys J. D. N. (1953) "The Gezira Scheme in the Sudan and the Russian Kolkhoz: A comparison of two experiments," *Economic Development and Cultural Change*, 2, 32–59.
Von S. J. (1978) "Cotton in Kassala: The other scheme," *Journal of African Studies*, 5, 205–243.
Wallace T. (1981) "The Kano River Project, Nigeria: The impact of an irrigation scheme on productivity and welfare" in Heyer J., Roberts P. and Williams G. (eds) *Rural development in tropical Africa* (New York: St. Martin's Press).
Wallace T. (1979a) "The impact of a large-scale irrigation scheme on two wards in a small town in Kano state, Nigeria: Its implications for rural development" in Southall A. (ed.) *Small urban centers in rural development in Africa* (Madison: University of Wisconsin).
Wallace T. (1979b) *Rural development through irrigation: Studies in a town on the Kano River Project*, Research Report 3, Ahmadu Bello University.
Wallach B. (1988) "Irrigation in the Sudan since independence," *Geographical Review*, 78, 417–434.
Waters R. A. (2005) "How socialism underdeveloped Africa," *Political Science Reviewer*, 34, 264–320.
Wells J. C. (1974) *Agriculture policy and economic growth in Nigeria, 1962–8* (Ibadan: NISER).
Wesseling H. L. (2003) *Verdeel en heers: De deling van Afrika 1880–1914* (Utrecht: Aula).
Widgren M. and Sutton J. E. G. (eds) (2004) *Islands of intensive agriculture in Eastern Africa: Past & present* (London: British Institute in Eastern Africa; Athens: Ohio University Press).
Williams A. F. (1970) Review of Chambers R. (1969) "Settlement schemes in tropical Africa: A study of organizations and development," *The Geographical Journal*, 136, 126–127.
Willis J. (2011) "Tribal gatherings: Colonial spectacle, native administration and local government in Condominium Sudan," *Past & present*, 211, 243–268.
Wittfogel K. (1957) *Oriental despotism: A comparative study of total power* (New Haven, CT: Yale University Press).
Wood A. (1950) *The groundnut affair* (London: The Bodley Head).
World Bank (1983a) *Sudan Gezira Rehabilitation Project*, Staff Appraisal Report No. 4218.SU.
World Bank (1983b) *Sudan Gezira Rehabilitation Project: Implementation*, Volumes I and II.
Worster D. (1985) *Rivers of empire: Water, aridity, and the growth of the American West* (Panthenon Books: New York).
Wright C. S. (ed.) (1962) *Africa and irrigation* (Salisbury: Proceedings of an International Symposium).
Yousif G. M. (1997) *The Gezira Scheme: The greatest on earth under one management* (Khartoum: Africa University House for Printing).
Zeisler-Vralsted D. (2015) *Rivers, memory and nation-building: A history of the Volga and Mississippi Rivers* (New York: Berghahn Books).

Index

Abdel Magid, 41–3, 94–6, 102, 224n99, 249n51
Abu Ishreens, 51, 53, 70, 71, 75, 78, 242n62
Abu Usher, 64, 170, 221n41, 223n71
Advisory Council for the Northern Sudan, 60, 136, 137, 138, 147, 224n104, 240n31, n34, 241n37, 242n57, 245n100
Aglen, Ms., 125, 193, 235n33, 237n84, n90
alcohol, 8, 107, 114, 122, 123, 175, 237n77
Allan, William Nimmo, 151, 169, 193, 195n4, 204n2, 205n10, n17, 216n49, 224n94, n96
Allen, 151, 226n138, 246n114, 247n11
Allenby, Edmund, 66, 220n9, 228n32–3
Alternative Livelihood Schemes. *See* White Nile Schemes
Archdale, William Porter Palgrave, 93, 94, 120, 134, 175, 231n102
Asquith, Arthur, 43, 102, 175, 231n103, 233n147, 260n79
Aswan, 29, 32, 209n127

Barakat, 25, 27, 28, 93, 113, 117, 118, 119, 120, 145, 147, 175, 182, 237n86, 245n104, 260n84
Barnett, Tony, 7, 10, 173, 198n44, 200n60, 201n68, 227n6, 233n143
BCGA. *See* British Cotton Growing Association
Bernal, Victoria, 7, 10, 186, 198n44
Bernard, 28, 29, 209n104, n109, n110, n126, 212n182–9, n205, 213n206, 219n130

Black Arm. *See* crop disease
Block Inspectors. *See* inspectors
Blue Nile, 4, 10, 15, 18, 27, 29, 66, 68, 88, 89, 103, 151, 192, 210n141–3, 214n13, 232n111, 252n112
Blue Nile (political), 91, 96, 100, 137–41, 144, 200n58, 228n25–6, n29, 229n54, 231n89, 238n118, 241n43
Bonham Carter, Edgar, 19, 28, 29, 205n15, 208n96, 209n119, 211n157, n160–1, n170, n172, n174
Bredin, George Richard Frederick, 96, 138, 228n46, 229n54, 241n43
British Cotton Growing Association, 16, 18, 21, 31, 175, 206n38, 207n45, 212n196, 227n15, 240n17, 243n63, 255n10
British Guiana, 167, 177, 178
Brown, F. A., 176–8, 256n22, n25–6
Butcher, Arthur Douglas Deane, 69–72, 74, 76, 183, 221n30–6

Cairo, 1, 7, 20, 23, 25, 26, 30, 31, 32, 35, 205n23, 206n26, n40, 207n60–1, 214n3, 228n31
Carmichael, John, 145, 146, 149, 151, 152, 157, 168, 169, 193, 224n93, 226n143, 243n66, n68–70, 244n88–9, 245n92–3, n96, 246n115, n2, 247n6–7, n10, n12, n15, n17, 248n43, n45, 249n46, n48, n55, n59, 251n97, 252n113
Central Office of Information, 165, 227n17, 242n54
Chambers, Robert, 186–8, 262n106

Chick, A. L., 141, 144, 223*n*82, *n*84, 226*n*138, 242*n*53–5, *n*57, 244*n*81–2
Civil Secretary, 133, 135, 137, 139, 229*n*64–5, 234*n*18, 237*n*86, 239*n*4, 240*n*10
civilization, 2, 93, 104, 129, 155, 185, 190, 226*n*138, 229*n*72, 264*n*130
Clayton, Gilbert Falkington, 25, 26, 193, 205*n*7, *n*9, *n*11–12, *n*16, 206*n*29, *n*30, *n*36, *n*39, 207*n*52, *n*56, 208*n*92–4, *n*97–8, 209*n*107, *n*111–14, *n*117, *n*120, *n*122–4, *n*127, 210*n*128–9, *n*132–3, *n*137–8, 211*n*151, *n*153, 217*n*70, *n*74, 220*n*5
Club Messellemia. *See* clubs
clubs, 8, 53, 104, 110, 121, 122, 123, 126, 130, 132, 232*n*115, 236*n*57, 237*n*75, 239*n*5
collective arrangements, 36, 59, 60, 97, 98, 105, 136, 160, 161, 167, 168, 184, 185
Colonial Office, 167, 168, 256*n*31
commonwealth, 152, 181, 247*n*22
concession, 4, 6, 11, 15, 20–2, 24, 28, 31, 32, 36–40, 43, 47, 58, 59, 60, 66, 81, 82, 96, 97, 99, 124–6, 130–4, 167, 175, 182, 183, 189, 197*n*40, 208*n*78, 212*n*204, 215*n*31, 216*n*60, 219*n*147, 230*n*75
cotton, 1, 2, 4–10, 12, 15–20, 22, 24, 25, 27–33, 35–8, 40, 41, 43–50, 51–6, 57–61, 63, 68, 73, 74, 78, 82, 86–8, 90–6, 98, 99, 101, 103–7, 109, 110, 115, 117, 119, 122, 126, 129, 131–3, 135–9, 143–7, 153–66, 168, 169, 174, 175, 177, 178, 180, 184, 185, 191–3, 195*n*7–9, 196*n*11, 197*n*26, 197*n*40, 205*n*14, *n*24, 206*n*41, 207*n*45, 210*n*143, 211*n*180, 212*n*181–2, *n*196, 215*n*31, 216*n*65, 217*n*81, 218*n*94, 219*n*147, 226*n*138, 227*n*8, *n*10, *n*15–16, 234*n*9, 236*n*63, 239*n*130, 242*n*53, *n*63, 243*n*68, 244*n*81, *n*90, 249*n*61–2, 252*n*112, 255*n*10
cotton picking, 50, 53, 55, 78, 92, 93, 100, 105, 106, 111, 119, 157

cotton sowing, 50, 53–6, 60
Crompton, Charles William Lee, 1–2, 5, 114–15, 193, 195*n*1–3, *n*6, 234*n*20
crop disease, 9, 26, 43, 47, 49, 54, 161, 215*n*37
crop rotation. *See* rotation
Culwick, Geraldine Mary, 10, 104, 105, 126, 148, 174, 193, 231*n*105–6, 232*n*115, *n*123–9, 233*n*130–3, *n*138–9, *n*143, 234*n*24, *n*26–7, *n*29, 235*n*30, 236*n*65, *n*68, *n*71–4, 237*n*75, *n*87, *n*92–5, 246*n*108–9, 254*n*6, 255*n*10, 261*n*86, 263*n*118, 264*n*124

Design Sheet, 75, 222*n*53–5, *n*57–60
development, 2–13, 16, 20, 21, 27, 37, 44, 45, 60, 63, 64, 79–81, 83–6, 89, 92–5, 99–104, 106, 107, 122, 125, 128–34, 137, 139, 146, 148, 149, 151–3, 155–66, 170, 171, 173–8, 180–2, 184, 185, 187–92, 195*n*4, 196*n*14, 197*n*26, 200*n*60, 204*n*2, 205*n*10, *n*17, 210*n*143, 211*n*177, 213*n*208, *n*221, 216*n*60, 223*n*82–3, 224*n*95–7, *n*101, *n*109, 225*n*113–15, *n*121–2, *n*133, 226*n*138, 227*n*6, 229*n*54, *n*63, 230*n*82, *n*86, 231*n*102, 232*n*109, 238*n*114, 241*n*44, 242*n*62, 247*n*6, *n*17–19, *n*21–2, 248*n*24, *n*26, *n*29, *n*38–9, 249*n*46–9, *n*56, 250*n*67, *n*82, 252*n*107–8, 255*n*14, 256*n*26, *n*30–1, *n*34–5, 257*n*36, *n*43, *n*50, 261*n*90, 263*n*114, 264*n*123, 265*n*136, *n*140
devolution, 94, 97, 99, 101, 102, 126, 128, 133, 147, 175, 182, 229*n*57–8, *n*61, 231*n*89, 260*n*79
Dew, Roddie, 126, 238*n*107, 252*n*110–11
Dongola, 15, 44, 65
drinking. *See* alcohol
drive, 79, 87, 88, 118, 119, 138, 148, 180
Dupuis, Charles Edward, 15, 16, 65, 66, 204*n*2, 205*n*3, *n*14
durra, 18, 45, 46, 49, 50, 53, 56, 58, 59, 72, 73, 86–8, 95, 103, 151,

155, 157, 160, 162, 163, 166, 167, 176, 255*n*17

ECC. *See* Empire Cotton Committee
Eckstein, Frederick, 6, 7, 18, 19, 21, 22, 23, 28, 33, 38, 43, 59, 174, 184, 185, 198*n*40, 207*n*48, 208*n*78, 212*n*183, *n*186, 215*n*21, *n*28, 216*n*54, 219*n*130, 227*n*21
Egypt, 2, 5, 9, 15, 16, 17, 18, 19, 26, 27, 29, 30, 32, 35, 36, 38, 44, 50, 60, 64, 65–6, 67, 69, 73, 74, 81, 84, 85, 95, 103, 113, 114, 152, 156, 159, 160*n*182, 189, 195*n*3, *n*9, 205*n*9, *n*14, *n*16, 206*n*24, 207*n*45, 209*n*127, 210*n*143, 211*n*169, *n*175, 212*n*181–2, 213*n*217, 214*n*4, 215*n*20, 218*n*94, 220*n*4, 221*n*13, 222*n*56, 227*n*15, 237*n*77, *n*83
Egypt Department of Public Works, 2, 15, 21, 25, 26, 65–6, 69
Egypt Farmers, 9, 26, 44, 67, 103, 189
Empire Cotton Committee, 29, 30, 212*n*181
England, 17, 97, 109, 127, 169, 195*n*3, 222*n*46
Ethiopia, 169, 206*n*41, 252*n*121
Executive Council, 142, 143, 223*n*82
expatriate, 84, 85, 113, 149, 178
experiment, 2, 6, 16, 17–19, 38, 47, 73, 82, 94, 95, 96, 99, 102, 131, 136, 138, 159, 160, 161, 168, 170, 177, 183, 184, 187, 206*n*33, 213*n*218, 218*n*86, 228*n*49, 229*n*54, *n*62, 230*n*78, *n*87, 249*n*56, 251*n*102, 261*n*90
extensions
  as in enlarging Gezira, 5, 11, 12, 15, 35–41, 43, 47, 49, 62–4, 66, 67, 69, 73, 74, 78, 80–3, 85, 86, 90, 92, 102, 113, 127–9, 132, 146, 148, 153–8, 197*n*40, 212*n*182, 214*n*2, 215*n*19, *n*37, 216*n*53, 222*n*52, 224*n*95, *n*98–102, *n*105–7, 225*n*116, 226*n*137, 238*n*128, 242*n*62, 248*n*23–4, *n*27, *n*31, *n*37, *n*41–2, 249*n*46
  as in training, 9, 161, 163, 164, 165, 171, 180

fallow, 18, 43, 45–50, 52, 82, 159, 160, 161, 210*n*143
FAO. *See* Food and Agricultural Organization
Ferguson, Hugh, 162, 164, 168, 169, 193, 249*n*62, 252*n*114–25
field canals. *See* Abu Ishreens
Field Outlet Pipes, 40, 51, 52, 53, 70, 71, 72, 77, 78, 80, 81, 223*n*73
Financial Secretary, 31, 37, 40, 66, 82, 127, 132, 133, 137–45, 209*n*126, 215*n*28, 216*n*46, 219*n*140, *n*146, 224*n*104, 225*n*113, *n*121, 226*n*140, 227*n*20, 232*n*109, 240*n*34, 241*n*35–7, 242*n*53, 244*n*81, 245*n*99
Food and Agricultural Organization, 169, 179, 180, 258*n*67
FOP. *See* Field Outlet Pipes
fringe crops, 47, 86, 141

Gaitskell, Arthur, 7, 8, 10, 60, 92–4, 96, 97, 98, 101, 102, 104, 107, 111, 118, 119, 121, 123, 127, 128, 130, 132, 133–41, 143–8, 168, 169, 171, 173, 181–5, 186, 187, 193, 197*n*31, 198*n*44, 200*n*56, 205*n*18, 216*n*58, 228*n*43, 229*n*54, *n*70, 230*n*78, *n*86, 231*n*97–9, *n*102, 235*n*39, *n*48, 236*n*60, *n*65–6, 238*n*103, *n*120, 240*n*25, 241*n*37, *n*40, *n*43, *n*51, 243*n*65, *n*70, 244*n*76, *n*81, 245*n*96, 252*n*108, *n*119, *n*121, *n*123, 253*n*138, 255*n*12, *n*17, 260*n*78–81, *n*84
Gaitskell, Hugh, 136
Garstin, William, 1, 15, 17, 29, 65, 88, 205*n*14, *n*20, 209*n*115, 211*n*167, *n*175, *n*178–9
Gash Delta, 63, 64, 166, 251*n*93
Gezira Advisory Board, 138–41, 224*n*105, 241*n*48–51, 242*n*52
Gezira Local Committee, 140, 141, 142
Gezira Scheme Bill, 135, 143, 242*n*61–3
Gezira Tenants Representative Body, 142, 242*n*57
ghaffirs, 51, 63, 72, 78, 80

Gibbons, Edward Philip, 110, 111, 113–15, 119, 121, 122, 125, 127–9, 155, 168, 193, 234*n*12–13, *n*23–4, 235*n*36, 236*n*62, *n*67, *n*70, 237*n*83, *n*86, 238*n*103–6, *n*109, *n*112–21, 252*n*109–10
gin. *See* alcohol
Governor of Blue Nile Province. *See* Blue Nile (political)
Groundnut Scheme, 144, 176, 255*n*17
groundnuts, 160, 161
Group Inspectors. *See* inspectors

habub, 1, 114, 123, 236*n*73
Hag Abdulla, 32, 73
Hawkesworth, D., 137, 138, 241*n*39
hedgehogs, 19, 23
Hosh, 46, 96, 102, 184, 229*n*54, *n*62, 260*n*82
howasha (field), 45, 46, 55
Huddleston, Arthur, 40, 68, 215*n*39, 219*n*139, 221*n*20–1, *n*27
Huddleston, Hubert Jervoise, 131, 239*n*2, 240*n*13
Hunt, F. Bertram, 54, 55, 75, 99, 109, 110, 115–17, 119, 121–3, 127, 136, 183, 193, 230*n*83–5, 231*n*102, 232*n*118–20, *n*122, 233*n*1–5, 234*n*8, *n*14, *n*21, *n*25, 235*n*31–2, *n*34–5, *n*37–8, *n*40–7, 236*n*57, *n*65, *n*69, 237*n*76–80, *n*82, 238*n*102, 240*n*30, 260*n*85, 261*n*86
Hunt, Leigh, 1, 5–7, 15, 193, 196*n*20, 196–7, *n*23–6, 197*n*29, *n*34–5
Hutton, J. Arthur, 16, 21, 29, 31, 206*n*38, 208*n*72, 211*n*164–5, 212*n*195

IBRD. *See* International Bank of Reconstruction and Development
indenting, 43, 51–3, 99, 173
India, 9, 67, 85, 91, 169, 170, 178, 200*n*57, 233*n*5, 252*n*115, *n*126, 253*n*131, *n*133, *n*135
Inspector General of Irrigation, 15, 65, 205*n*3, *n*5, 206*n*41
inspectors, 5, 8, 9, 10, 11, 12, 32, 36, 44, 46, 51, 52, 53, 55, 56, 58, 62, 68, 78, 80, 81, 87, 90, 91, 92, 93, 94, 96, 99–107, 109–11, 113–30, 135, 136, 141, 142, 147–9, 159, 161, 165, 168, 169, 174, 175, 178, 181, 182, 183, 186, 189, 192, 200*n*56, 213*n*218, 226*n*1, 228*n*37–40, 230*n*75, 231*n*102, 233*n*5, 234*n*16, 235*n*34, *n*37, *n*48–9, *n*53, 237*n*83, 239*n*130, 243*n*70, 245*n*106, 255*n*10
International Bank of Reconstruction and Development, 128, 153, 156–9, 164, 165, 167, 248*n*27, *n*38–40, *n*42, *n*44, 249*n*49–53, *n*63, 250*n*82–4, 251*n*98, 252*n*124, 257*n*43

Jebel Aulia Dam, 94, 95
Johnson, Ms., 126, 127, 144, 148, 149, 183, 193, 225*n*125, 233*n*5, 236*n*54–6, *n*65, 237*n*54–6, *n*65, 237*n*81, *n*88–9, *n*91, *n*96–9, 238*n*100–1, *n*129, 243*n*70, 244*n*71–4, *n*77–81, *n*83–7, 245*n*104–6, 246*n*107, *n*110–13, 261*n*87–9
Johnson, Thewlis Clarkson, 110, 126, 127, 149
Johnstone, Harmood Victor Carruthers, 39, 66
junior inspectors. *See* inspectors

K-88 Club. *See* clubs
Kano, Nigeria, 176, 179, 180, 185, 189
Kano River Project. *See* Kano, Nigeria
Kassala, 25, 119, 167, 197*n*40, 208*n*71, 212*n*199, 251*n*102
Kassala Cotton Company, 40, 47, 49, 132, 166, 193, 197*n*40, 215*n*31, 234*n*9, 255*n*10
KCC. *See* Kassala Cotton Company
Kenana, 83, 155, 157, 158, 159
Kennedy, Macdougall Ralston, 17, 18, 27, 32, 33, 45, 205*n*24, 209*n*126, 210*n*145, 213*n*209, *n*217
Kenya, 5, 168, 176, 177, 178, 186, 187, 256*n*24, *n*26, *n*31, *n*33–7, 257*n*47, 258*n*67, 263*n*118
Khartoum, 1, 4, 18, 22, 25, 26, 28, 30, 64, 80, 88, 103, 113, 116, 117, 122, 144, 148, 151, 161, 170, 192, 196*n*10, 205*n*24, 207*n*60, 216*n*60, 217*n*79, 236*n*57, 242*n*53,

n58, 244n1, 247n7, 249n54, n60,
n65, 250n90, 254n4
Kitchener, Lord, 19–21, 23–8, 30, 184,
206n43–4, 207n46–7, n55, n60,
208n76, n87, n90–1, n95, 209n102,
n105, n115–17, n121, n126
KRP. *See* Kano, Nigeria

Lancashire, 29, 36, 153, 175, 243n63
Latour, Bruno, 13, 191, 192,
265n135–6
leaf crinkle. *See* crop disease
leaf curl. *See* crop disease
leave (for inspectors and SID), 77, 109,
111, 118, 125, 127, 128, 169,
235n53, 260n79
Lloyd George, David, 27, 210n136,
211n156
London, 7, 15, 16, 17, 18, 19, 21, 23,
25, 26, 28–31, 32, 35, 37, 38, 118,
134, 151, 152, 167, 174, 182, 194,
195n3, 200n64, 208n74, 211n165,
214n3, 215n21, 227n21, 239n130,
240n21, 242n54, 260n79–80,
n82
Lovat, Lord, 19, 21–3, 25, 27, 33,
197n38, 207n48, 208n80–3, n89,
n95, n100, 209n101, 211n150
lubia, 15, 18, 44, 45, 46, 47, 48, 50,
56, 72, 95, 136, 160, 249n62

MacDonald, Murdoch, 25–31, 32, 37,
65, 209n115, n126–7, 210n141–3,
n146–7, 211n157, n167–9,
212n190–1, 213n209, n216,
220n4, 232n111
MacGillivray, D. P., 6, 7, 16, 21,
22, 24, 25, 27, 30, 31, 44, 45,
58, 184, 197n34–5, 209n101,
210n143, n145–6, 211n152, n154,
n160, 216n67–8, 217n69, n76–7,
219n127
MacGregor, R. M., 66, 68, 69, 71, 72,
86, 214n14, 215n20, n24, 216n48,
221n17, n22–4, n26, n28–9,
n37–40, 228n31
MacIntyre, Alexander, 6, 16, 31, 39,
40, 41, 47, 49, 58, 68, 102, 107,
119, 132–4, 175, 185, 212n205,
213n206, 215n33, 216n68,
219n139, 221n20–1, n27, 223n73,

229n68, n70, 231n102–3,
233n147, 239n2, 240n12–13,
255n12
Managil, 5, 41, 47, 80, 81, 83, 86,
127–9, 146, 151, 153–9, 161, 171,
173, 222n52, 224n97, 225n116,
238n128, 248n23–7, n31, n37,
n41–2, 249n46–7
Managing Director of SPS/SGB, 6, 7,
127, 132, 134, 140, 141, 143–6,
181, 211n148, 235n50, 243n70,
253n138
maroor, 46, 53, 55, 102, 117, 119
Mather, William, 15, 16, 18, 206n27,
n37
mechanized durra schemes, 151, 167,
176, 255n17
Mekki Abbas, 146, 168, 245n90,
249n59
minor canals, 50, 51, 53, 70, 71, 72,
78, 81, 83
mud, 2, 9, 10, 51, 122, 123, 148, 174,
236n74
Mwea-Tebere, 5, 176–80, 185–7,
189, 256n32, 257n47, 258n67,
262n102

NAP. *See* Niger Agricultural Project
Niger Agricultural Project, 5, 176,
264n118
Nigeria, 5, 105, 176, 179, 180, 189,
257n50
night storage, 64, 67–75, 79, 81–3, 86,
99, 158, 170, 173, 183
Nile, 5, 6, 16, 29, 32, 35, 37, 41, 50,
65, 66, 68, 73, 82, 113, 153,
155, 156, 159, 165, 175, 206n24,
n41, 209n115, 210n131, n141–3,
261n90, 214n4, 217n79, 221n13
Nile Commission, 38, 215n20
Nile Control, 29, 37, 211n167
Nile Project Commission, 32–3,
213n214–15

Pakistan, 85, 128, 168, 196n10
partnership in Gezira, 4, 6, 17, 19, 21,
22, 24, 28, 30, 31, 32, 36, 40, 41,
56, 58, 59, 60, 92, 134, 142, 158,
159, 160, 163, 166, 167, 183, 184,
186, 206n26, 208n78, 217n83,
243n68, 253n102

pipe outlets. *See* Field Outlet Pipes
polo, 111, 118, 121, 123, 237*n*84
pumping stations and schemes, 4, 6, 17, 18, 27, 32, 35, 37, 43, 64, 67, 68, 95, 165, 205*n*24, 213*n*218
Punjab, 66, 91, 92, 170, 228*n*31

Raby, George, 130, 144–6, 239*n*130–1, 244*n*74, *n*76, *n*83, *n*88–90, 245*n*96–7, 255*n*17
rain, 19, 20, 36, 47, 49, 51–5, 74, 88, 89, 97, 104, 122, 123, 142, 161, 206*n*29, 215*n*37
Roberts, W. D., 65, 68, 69, 86, 221*n*22–4, *n*26, *n*28–9, *n*37–40
Robertson, James Wilson, 114, 135, 137, 138, 140, 182, 193, 228*n*47–8, *n*51, 229*n*64–5, 231*n*100, 234*n*18, 236*n*63, 239*n*4, 241*n*37, *n*39–41, 243*n*65, *n*67, 260*n*78
Roseires, 27, 151, 153–7, 159, 160, 162, 164, 171, 225*n*127, 248*n*45, 249*n*46–7, 253*n*135
rotation, 12, 36, 41, 43, 45–7, 49, 50, 64, 73–5, 82, 83, 95, 136, 155, 156, 158–61, 163, 168, 184, 189, 218*n*86, 230*n*87, 248*n*30

sand storm. *See* habub
Schedule X, 95, 96, 140, 160, 229*n*53
Schuster, George Ernest, 37, 38, 66, 71, 214*n*10–12, *n*14–15, 215*n*16, *n*24, 221*n*37–40, 232*n*109, 241*n*35
sediment, 12, 53, 65, 67, 68, 69, 73, 78, 79, 86, 102, 158, 170, 173
Select Committee of the Legislative Assembly, 142, 143, 242*n*59, 245*n*101
senior inspectors. *See* inspectors
Sennar, 4, 5, 12, 15, 16, 18, 25, 26, 27, 28, 29, 30, 33, 35, 36, 37, 41, 43, 47, 50–1, 63–8, 73, 74, 77, 81, 83, 84, 94, 153, 154, 158, 166, 169, 184, 187, 196*n*11, 214*n*13, 218*n*108, 222*n*56, 253*n*133
SEPS. *See* Sudan Plantations Syndicate
SGB. *See* Sudan Gezira Board
SGB inspectors. *See* inspectors

sheikhs, 89, 91, 94, 95, 96
SID. *See* Sudan Irrigation Department
silt. *See* sediment
Simpson, Stanhope Rowton, 116, 122, 178, 193, 234*n*28, 236*n*58, 237*n*77
Smith, C. H., 168, 189, 193, 252*n*111, 255*n*10, *n*15, 264*n*123–4
Smith, R. J., 12, 63, 64, 66, 67, 71, 73, 76, 82, 85, 86, 193, 223*n*62, *n*68, *n*79, *n*89, 224*n*98, *n*105–6, 225*n*113–15, *n*121–2, *n*126, *n*128–9, *n*131, 226*n*140–1, 232*n*108, 251*n*97
social development, 10, 37, 93–5, 99, 100, 102–4, 107, 128, 131, 132, 139, 141–4, 148, 157, 165, 181, 192, 230*n*82, 231*n*102, 232*n*109, 242*n*62, 249*n*56
Soda & Mezza Clubs. *See* clubs
South Sudan, 5, 146, 149, 151, 153, 166, 189, 246*n*116
Sparrow, Jack, 2, 13, 192
Special Committee. *See* Advisory Council for the Northern Sudan
spoon feeding, 37, 103, 107, 232*n*109
SPS. *See* Sudan Plantations Syndicate
SPS inspectors. *See* inspectors
Stack, Lee, 20, 37, 66, 91, 206*n*29–30, 207*n*50, *n*54, *n*58, *n*62–6, *n*68–9, 208*n*71, *n*73, 209*n*126, 210*n*130, 212*n*199, 213*n*211, 214*n*6, *n*8–9, 220*n*9, 228*n*32–3
Sudan Agriculture and Forestries Department, 20, 21, 27, 37, 80, 95, 136, 137, 138, 139, 197*n*29, 217*n*72–3
Sudan Civil Service, 10, 22, 36, 40, 110, 114, 116, 122, 125, 133, 135, 137, 139, 152, 168, 178, 183, 235*n*33, 237*n*77, 259*n*68
Sudan Experimental Plantations Syndicate, 6, 15, 197*n*26
Sudan Gezira Board, 5, 7, 11, 41, 82, 102, 103, 104, 107, 110–13, 126–30, 131, 132, 135, 143–9, 152, 154–66, 168, 171, 176, 181, 182, 185, 193, 200*n*63, 201*n*69, 217*n*83, 218*n*83, *n*89, 219*n*132,

230n81, 231n104, 234n9–10,
n26, 238n109, 239n130, 243n68,
244n90, 245n99, 247n6, 248n31,
249n56, 251n93, 253n138,
255n17, 264n118
Sudan Gezira Board Inspectors. *See* inspectors
Sudan Irrigation Adviser (Advisor), 26, 63, 64, 66, 67, 155, 169, 228n31
Sudan Irrigation Department, 12, 15, 51, 63–8, 72–86, 99, 102, 132, 147, 148, 151, 152, 155, 158, 165, 169, 171, 183, 205n5, 215n37, 218n108, n110, 220n1, n6, n8, 221n13–16, n18, 222n46, 223n78, 225n117, n123, n128–9, n131, 254n7
Sudan Irrigation Service. *See* Sudan Irrigation Department
Sudan Plantations Syndicate, 4–12, 16–33, 35–41, 43–5, 46, 47, 49, 51, 53, 54, 56, 58–62, 64–9, 71–5, 78, 81, 82, 86–8, 90, 97–105, 107–30, 131–6, 138, 140, 141, 143, 147, 148, 152, 165–8, 174–6, 181–6, 189, 191, 193, 198n40, 206n33, n38, 208n76, 213n218, 216n45, 217n69, 219n147, 227n24, 229n67, 230n75, n78, 231n89, n102, 232n109–10, n113, 233n5, 234n6, n9, n12, n14, n16, n22, 235n34, n37, n48–53, 237n77, n83, n86, n88, 239n3, n6–9, 240n11, n14–16, n18, n24–5, n27, 241n37, 255n8–11, 259n73, 260n79–80, n84, 264n122
Sudanization (Sudanisation), 83–5, 125, 127–9, 143, 146–9, 154, 225n117, n123, n127, 239n130, 246n115–16
syndicate. *See* Sudan Plantations Syndicate
Syndicate inspectors. *See* inspectors

Tayiba, 16, 18–23, 25, 27, 28, 44–5, 59, 93, 94, 104, 175, 184, 191, 205n24, 207n51, n53, 213n218, 216n68–9, n71

tenant account, 56, 57, 58, 103, 160, 219n129
tenant debts, 8, 58, 59, 60, 208n78
Tenants Collective Funds, 59, 60, 97, 98, 142, 184
Tenants Equalization Fund. *See* Tenants Collective Funds
Tenants Reserve Fund. *See* Tenants Collective Funds
Tenants Welfare Fund. *See* Tenants Collective Funds
Tennessee Valley Authority, 135, 137, 138, 139, 141, 171, 241n47
tennis, 53, 121, 122, 123
Tokar, 25, 32, 208n71, 212n181, n199, 213n208
Tottenham, Percy Marmaduke, 18, 20, 65, 193, 204n1, 205n8, 206n42, 209n15, 220n3
TVA. *See* Tennessee Valley Authority

village council, 96, 97, 99–102, 106, 126, 163, 229n53, n72, 230n70, 231n91–2, 260n83

Wad el Nau, 32, 46, 51, 223n66–7
Wad Medani, 15, 51, 64, 66, 67, 77, 81, 88, 89, 121, 122, 128, 223n70–1, 238n119
WB. *See* World Bank
Webb, Arthus Lewis, 25, 29, 209n115
weed, 49, 53–6, 65, 78, 79, 86, 95, 102, 125, 159, 161, 173
West Africa. *See* Westerners
Westerners, 88, 105, 106, 228n38
wheat, 5, 15, 16, 27, 43, 44, 47, 48, 50, 95, 160, 161, 163, 179, 180, 205n14
White Fly. *See* crop disease
White Nile, 4, 15, 20, 26, 29, 65, 66, 94, 95, 192, 206n41, 209n115, 210n143
White Nile Board. *See* White Nile Schemes
White Nile Schemes, 94, 95, 154, 228n46, n50
Willcocks, William, 32, 213n212, n215, n217

Wingate, Reginald, 5, 15, 16, 18–28, 31, 32, 44, 45, 65, 184, 193, 205n7, n9, n11–12, n15–16, 206n27–8, n36, n39, 207n47, n58–60, n65, n70, 208n72, n73, n80–4, n87, n89–100, 209n101–2, n104–7, n111–14, n117, n120–7, 210n128–9, n132–3, n136–8, n143, n146, 211n150–4, n156–7, n161, n164, n167–8, n170, n179, 212n182, n187–90, n194–5, n199, n201, 213n210–12, n216, 216n68, 217n69–70, n74–8, 220n5, 232n112

Working Party, 162–4, 169, 250n67, n69–78, n80

World Bank, 73, 156, 173, 248n38

World War I, 11, 16, 27

World War II, 5, 9–12, 41, 43, 47, 60, 63, 79, 80–6, 88, 95–7, 100, 106, 109, 111, 120, 124, 125, 130–3, 135, 149, 151, 152, 165, 166, 170, 174–6, 182, 184, 187, 192, 200n58, 220n12, 229n53, 237n83, 248n38, 250n87

WP. *See* Working Party

yeoman (yeomen), 88, 92, 97, 101, 157, 159

Zande Scheme, 5, 151, 166, 251n93, 264n118

Zeidab, 1, 6, 16, 17, 19, 21, 26, 27, 36, 43–5, 47, 93, 118, 134–5, 184, 212n181, 216n60, n65, 235n39

The manufacturer's authorised representative in the EU is Springer Nature Customer Service Centre GmbH, Europaplatz 3, 69115 Heidelberg, Germany. If you have any concerns regarding our products, please contact ProductSafety@springernature.com

Printed and bound by CPI Group (UK) Ltd, Croydon, CR0 4YY
23/03/2026
02076449-0020